Nano–Microsystems

Science and applications

Online at: https://doi.org/10.1088/978-0-7503-3111-1

Nano–Microsystems

Science and applications

Akinobu Yamaguchi

Faculty of Science and Engineering, Toyo University, Kawagoe, Saitama, Japan

IOP Publishing, Bristol, UK

ISBN   978-0-7503-3111-1 (ebook)
ISBN   978-0-7503-3109-8 (print)
ISBN   978-0-7503-3112-8 (myPrint)
ISBN   978-0-7503-3110-4 (mobi)

DOI   10.1088/978-0-7503-3111-1

Version: 20250801

IOP ebooks

British Library Cataloguing-in-Publication Data: A catalogue record for this book is available from the British Library.

Published by IOP Publishing, wholly owned by The Institute of Physics, London

IOP Publishing, No.2 The Distillery, Glassfields, Avon Street, Bristol, BS2 0GR, UK

US Office: IOP Publishing, Inc., 190 North Independence Mall West, Suite 601, Philadelphia, PA 19106, USA

Contents

Preface

The French bacteriologist Louis Pasteur said 'Le hasard ne favorise que les esprits préparés'. In English, it presents 'Chance favours the prepared mind'. The inventor Thomas Edison, on the other hand, said 'I find out what the world needs. Then I go ahead and invent it.' The former are unexpected discoveries made by chance and ingenuity and are called serendipity. The latter is the act of a curious mind that enables it to see beyond what it knows and is called creation or dreaming. They are two sides of the same coin and both are called originality. Both are indispensable foundations of learning and the basis of technical applications. For example, the discovery of the point-contact transistor was a discovery. Later, circuitry was developed into the central processing unit (CPU) and the graphics processing unit (GPU), which were developed into network systems and blockchains. That is an invention. Inventions and discoveries have evolved in a complementary way, leading to innovations in each generation. What we are describing in this book are Nano–Microsystems, being systems that realize the desire to precisely control, reproducibly measure and understand what we want to know about physical and chemical phenomena. On the one hand, the realization of such requirements not only enables scientific developments, but can also be applied to medical applications and chemical materials exploration platforms. On the other hand, if a device designed and built for medical applications does not work as expected, close examination reveals a strong dependence on the surface treatment of the device, which can lead to the discovery of new electronic states. Invention and discovery are one and the same, and basic and applied research are also inextricably linked, and vice versa. I have written this book to help the reader understand these things as much as possible.

Nano–Microsystems cover a very wide range of research objectives, from solid-state devices and microchemical systems to medical devices, cell culture and drug screening. Rather than researching Nano–Microsystems themselves, they are better built and used by researchers and engineers who are interested in precisely controlling the physical and chemical phenomena they want to understand, and who want to carry out experimental research quantitatively and with as little effort as possible. On the other hand, company engineers who want to fabricate medical devices or chips for environmental analysis may want to build a Nano–Microsystem to carry out their operations under automatic control. In this way, as there are a thousand different research objects and purposes, Nano–Microsystems will be constructed to suit each purpose and application. Thus, to summarize them in a book in this way, the content of each field tends to be shallow because of the wide range of fields to be dealt with. On the other hand, it would be impossible to explain Nano–Microsystems if they were limited to specific areas.

For these reasons, the aim of this book is to provide the reader with guidelines on how to use Nano–Microsystems for their own research interests. The intended readers are of the following three types: (1) researchers who have their own area of expertize and wish to systematize experiment and theory by reviewing their existing research methods or creating controlled experimental systems as a new endeavour;

(2) engineers in companies who wish to carry out feasibility studies or screening prior to plant design; (3) university graduates and postdoctoral students who wish to use Nano–Microsystems in their own research and postdoctoral programmes. For these readers, the aim is to provide guidance on (1) what fields and research are using Nano–Microsystems, (2) what can be achieved by using Nano–Microsystems, and (3) how to design, fabricate, analyze and evaluate Nano–Microsystems to use them in research topics of interest. Therefore, it may not be sufficient for specialists looking for the latest research cases or in-depth research content. However, for the above-mentioned beginners, if the readers themselves have a research or development objective they wish to pursue and can use Nano–Microsystems as an approach to achieve it, they will be able to pursue their own research content in a deeper and more systematic way. Therefore, this book explains the concept of Nano–Microsystems and how to actually perform and construct experimental systems. It also shows how they can be used and how they are developing as a new research trend. The author has tried to write the book in such a way that it is informative for the reader. Readers wishing to learn more about more specialized or cutting-edge research should refer to the references given.

I would like to thank the many people involved in the writing of this book. First, I would like to thank IOP Publishing for giving me the opportunity to write this book. I would particularly like to thank Caroline Mitchell, who first approached me about writing the book, and Mia Foulkes, for her unstinting support and encouragement, despite the slow progress and considerable inconvenience. I am very grateful to Chris Benson for his help with proofreading. I would also like to express my deepest gratitude to all the staff who are helping me to complete the book and get it out into the world.

In writing this book, I would like to express my deepest gratitude to all those involved in the Nano–Microsystems Group at the University of Hyogo. I would also like to express my deepest gratitude to Professor Yuichi Utsumi and the master's course and PhD course students. I would like to thank the graduated students of the University of Hyogo, including Hideki Kido, Yasuto Arisue and Dr Masaya Takeuchi (currently Assistant Professor at the Faculty of Engineering, the University of Hyogo), for their research results on the design, fabrication and evaluation of Nano–Microsystems, which were introduced as research examples to the beginners in writing this book. Some of the fabrication methods were taken from their master's and PhD theses. For surface acoustic wave (SAW)-based devices and systems, Professor Tsunemasa Saiki of the University of Fukuchiyama, Research Fellow Dr Satoshi Amaya of the University of Toyo, and Masatoshi Takahashi of the University of Hyogo have made significant contributions. For the surface enhanced Raman scattering (SERS) systems, I have introduced the construction of the system and examples of its use, based on some of the results of joint research with Ryo Takahashi, Ryo Hara and Dr Shunya Saegusa of the University of Hyogo, as well as with Takao Fukuoka of Kyoto University and Dr Masayuki Naya of Keio University. I was able to carry out operando synchrotron radiation experiments in collaboration with Dr Masaki Oura, team leader of the Soft X-ray Spectroscopy Instrumentation team at RIKEN SPring-8, Professor Takuo

Ohkochi and Professor Satoru Suzuki of the University of Hyogo, Associate Professor Ikuya Sakurai of Nagoya University and Research Fellow Ikuo Okada of Nagoya University, and others. I would like to thank Kumi Kusumoto of University of Hyogo for her administrative support with our research. The cooperation and support of many others has made it possible for me to continue our research. Once again, I would like to express my deepest gratitude.

My own expertize was originally in condensed matter physics, nanomagnetism and spintronics. In these fields, I had reached the end of my research into bulk physical properties and moved on to research into the creation of new materials by using semiconductor microfabrication to create an ideal experimental environment to create the physical phenomena I wanted to clarify, or by creating artificial metallic lattices or multilayers. With this background, I myself entered the field of Nano–Microsystems and was able to shift to a style of precise control, clarification and application of phenomena that interested me. I am very grateful to Toyo University for accepting me into the field of Nano–Microsystems and providing me with an environment where I can carry out semiconductor microfabrication and Bio-Nano Electronics research. Readers are in different positions and situations, but when they get stuck in their area of expertize, new ways of thinking and different experimental approaches can often solve problems. I sincerely hope that the Nano–Microsystems presented in this book will help.

Finally, with regard to writing, I am deeply grateful to my wife and family for giving me time to write the book.

We publish this book in the hope that readers will build their own Nano–Microsystems to study phenomena of interest to them, leading to new discoveries and inventions. We hope that the readers will obtain the serendipity.

2025 March
Akinobu Yamaguchi
Faculty of Science and Engineering, Toyo University

Author biography

Akinobu Yamaguchi

Akinobu Yamaguchi completed a doctoral course at the Graduate School of Engineering Science, Osaka University in March 2005 after obtaining a bachelor's degree in physics from Himeji Institute of Technology and a master's degree in physics from Osaka University Graduate School of Science. In April 2006, he was a teaching assistant and New energy and Industrial Technology Development Organization (NEDO) Project Research associate at the Institute of Chemistry, Kyoto University. April 2006 Assistant Professor, Department of Physics, Keio University, Faculty of Science and Technology. From October 2007 to March 2010, he was additionally JST PRESTO researcher. From April 2011, He was a Researcher, National Institute of Advanced Industrial Science and Technology (AIST). Since October 2012, he has been an associate professor at the Laboratory of Advanced Science and Technology for Industry, University of Hyogo. He has been a visiting researcher at RIKEN since 2018, Institute for Advanced Medical Engineering, University of Hyogo from April 2022 to 2024. Since April 2024, he has been a professor at the Faculty of Science and Engineering, Toyo University. His awards include the Young Scientist Encouragement Award from the Physical Society of Japan and the Magnetics Society of Japan's Best Paper Award and Encouragement Award. He is currently engaged in research on Nano–Microsystems, spintronics, synchrotron radiation optical reactions, and molecular sensing. He is a member of societies such as the Physical Society of Japan, the Institute of Electrical Engineering of Japan and the IEEE, He is currently an IEEE Senior Member.

IOP Publishing

Nano–Microsystems
Science and applications
Akinobu Yamaguchi

Chapter 1

Introduction of Nano–Microsystems

This section outlines the elements of conceiving, designing, building and evaluating systems that lead to social and ideational originality using the platform of Nano–Microsystems, which are dealt with in each chapter. It is noted that the chapters are structured with the aim of equipping the reader with philosophical guidelines and specific methods for platform formation and post-creation evaluation. Some chapters also include a few examples of recent research to give the readers an idea of how their own systems might develop.

1.1 Introduction

In 2022, the human genome, a person's genetic information, was fully sequenced by an international team of researchers including the US National Institutes of Health, announced in the journal *Science* [1]. The decoding of the human genome had been declared complete by an international project in 2003, but about 8% of the genome remained difficult to decode with the technology available at the time. The human genome consists of about 3 billion pairs of nucleotide sequences, which are contained in 46 chromosomes in the cell. Sequences at the ends and in the middle of the chromosomes are difficult to decipher due to their many repetitions, and the international team of researchers developed a new method that allowed the complete sequence to be deciphered. The results have the potential to facilitate the discovery of diseases and health conditions linked to the human genome. In particular, the combination with artificial intelligence (AI)-based machine learning is particularly adept at revealing correlations between proteins and genome sequences, and is expected to lead to new discoveries in a wide range of fields, from genetic information to the history of human evolution and the mechanisms of cell division and their relationship to disease. Indeed, the 2024 Nobel Prize in Physics and the Nobel Prize in Chemistry were related to AI-based machine learning [2].

In the preceding years, however, a pandemic struck the world and economic and social activities were severely restricted. Even under such restrictions, research and

doi:10.1088/978-0-7503-3111-1ch1

development of internet-based web conferencing systems and AI progressed, leading to the aforementioned Nobel Prize; real-time polymerase chain reaction (PCR) for testing for COVID-19 became widespread, and antigen test kits using immunochromatography were also rapidly introduced to the market. In particular, the revolutionary method of actively using messenger ribonucleic acid (mRNA) in vaccines for rapid and efficient immunity against infectious diseases can now be addressed. This approach has shown that vaccines can be developed and produced in a matter of months rather than years using conventional methods. Coronaviruses are particularly mutable viruses, with mutant strains emerging all the time, and it is worth remembering that it is now possible to provide vaccines against them. Eventually, when an infectious disease caused by an unknown virus occurs, it will be possible to provide a vaccine by artificially synthesizing mRNA corresponding to the spike protein of the virus to produce antibodies against the virus, once the genetic information of the virus has been analyzed. However, a number of technical issues and an understanding of the biological response characteristics are required to achieve this, so it is not an immediate possibility at present. However, as science and technology evolve and develop, it is expected that this will become possible.

Mathematics, physics and chemistry theoretically systematize the principles underlying science and technology and explain various phenomena around us. Conversely, it can be said that the natural sciences developed to understand the various phenomena that occur around us. The natural science knowledge gained in this way accumulated and formed the soil for engineering applications. If we think along the process of the formation of human civilization, we can say that we have evolved from a hunting society (Society 1.0), an agricultural society (Society 2.0), an industrial society (Society 3.0) and an information society (Society 4.0) to the present day, a world in which the formation of a social system that is highly integrated with cyberspace and physical space is taking place [3]. This can be predicted from the fact that information in physical space, such as the aforementioned protein and genome sequences, can be analyzed by AI in cyberspace to construct freely designed DNA and RNA sequences and predict the proteins that will be produced when they are returned to the body. Such a system, which highly integrates cyberspace and physical space, is expected to achieve both economic development and solutions to social problems, and is thought to be capable of realizing a human-centred society (Society 5.0) [3]. So, how can we build a system to make Society 5.0 a reality?

In this book, the question is primarily outlined in terms of how to build systems for science and engineering applications. Although science has been systematized and has become a basic discipline taught in primary, secondary, high school and university, it is still evolving and new discoveries are being made all the time. AI is now readily available on smartphones, but energy is used to perform the calculations. Internet traffic has become very high, which also increases energy consumption. Therefore, small nuclear modules will be placed next to data centres, which will soon evolve into fusion reactors and so on. In other words, there are always two sides to a discipline: a systematized discipline and an evolving discipline. Thus, even as this book is being written, science and technology are gradually advancing, new knowledge is being discovered, and applied systems are being brought to market to solve economic activities and social problems. Therefore, theories, methods, systems and empirical results that are considered state-of-the-art at the time of writing may become obsolete after a few years. This tends to be the case with academic articles, as they report on the latest research. However, books, especially textbooks, need to serve the needs of readers over a somewhat longer period of time. This book has therefore been published as an introductory text for undergraduates, rather than as a report on cutting-edge research.

The natural sciences also start by building experimental systems that are fit for purpose to ensure automatic measurement and reproducibility of phenomena. Engineering applications, needless to say, build mechanical and electrical controls that are fully usable by the user. The control and structure fabrication methods of this engineering field have contributed significantly to the development of the natural sciences. Micro-electro-mechanical-systems (MEMS), for example, are microscopic electro-mechanical composite systems consisting of electrical inputs and outputs and mechanical elements. Micro-electro-mechanical complex systems fabricated by semiconductor microfabrication can, for example, perform mechanical actions by applying a voltage, as shown in figure 1.1. For example, the element in figure 1.1 can be incorporated into synchrotron radiation

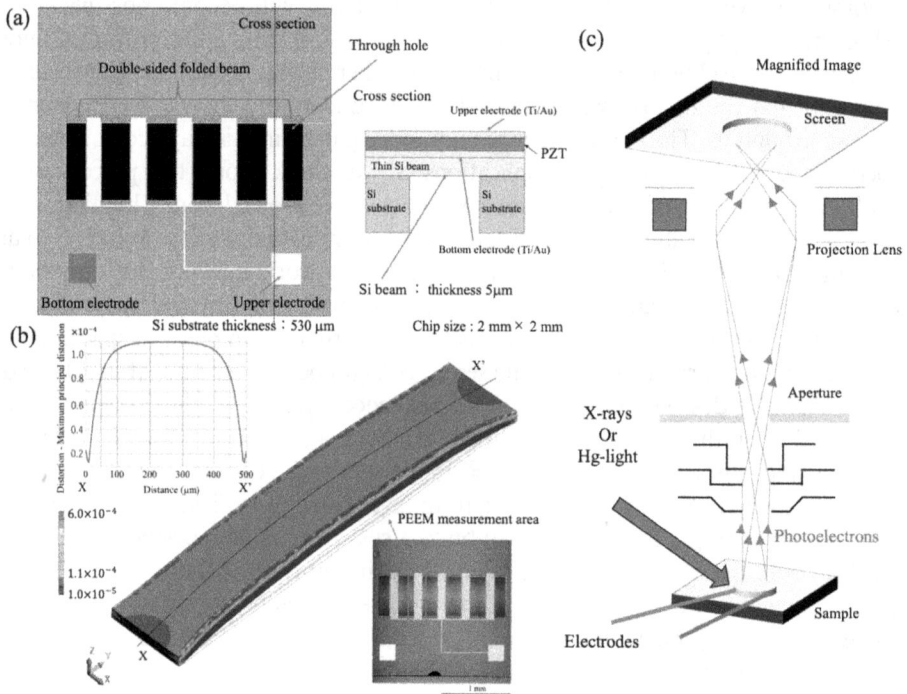

Figure 1.1. (a) Schematic of a MEMS device with a double-sided beam structure. The substrate size is 2×2 mm^2. Multiple double-sided beams are arranged in a row, and when voltage is applied to the top and bottom electrodes, the beams bend to enable bending tests. A schematic of the MEMS cross-sectional shape with respect to the beam structure is also shown. (b) Simulation results of strain evaluation for MEMS and optical photographs of fabricated MEMS devices with double-sided beam structure type. The spatial variation of distortion along the positions $X–X'$ is shown. (c) Schematic of device arrangement and operando x-ray photoelectron emission microscopy (PEEM) observation setup at synchrotron radiation facility SPring-8 observation setup. Relevant content is presented in reference [4].

absorption spectroscopy measurements and micro-Raman spectroscopy systems to apply mechanical strain to materials such as magnetic materials and graphene to deform their crystals and study transitions in their electronic and chemical bonding states [4]. The examples presented here are part of a book on MEMS applications in science.

Research into MEMS itself has been carried out by many scientists and engineers who have entered the field to understand the mechanical and physical phenomena at the micro- and nanoscale that are different from those at the scale of human life. The technology developed there has matured and is now used in accelerometers and microphones in smartphones. What was originally researched as a MEMS system is now being used as a basic element in smartphones and cars. In this way, what was originally built as a system to conduct ideal experiments based on principles and to demonstrate phenomena, once completed, becomes an element in its own right and is used to build new systems. Fundamentals and applications are two sides of the same coin, and it is necessary to think of a scientific system as something that is constantly evolving. On the other hand, there is also a warm and new element when considering fundamentals and applications, or science and engineering. The following is an example from some time ago concerning mechanical watches [5].

Silicon is often used as a structural material in MEMS. This is because silicon has excellent mechanical properties (high specific modulus of elasticity, thermal conductivity and low thermal expansion) that exceed those of steel materials. However, silicon is also a brittle material and has a very high yield stress, but yield = fracture, which is a catastrophic phenomenon. Various studies have shown that the fracture of silicon is in most cases controlled by the surface roughness. Statistical analysis is essential because of the variability of this process roughness. To digress for a moment, let us consider the semiconductors that underpin today's information society in the context of silicon. Computations in computers and smartphones are performed by silicon-based semiconductor devices. The basic principle of arithmetic is based on binary numbers, and computers perform calculations using information in digital form. However, physical unclonable functions (PUFs) use variations in digital information due to variations in the processing accuracy of semiconductor devices [6]. In other words, there is a very analogue element visible and hidden beneath the digital information [7]. Furthermore, if we reduce the scale even further, down to the world of atoms and electrons, we arrive at a world dominated by the uncertainty principle. This is a world digitized by Planck's constant, while in the region stretched by Planck's constant it is an analogue world. Very often, digital or analogue is chaotic, and depending on the observer's point of view, it is a different world depending on which one is chosen or used.

Interestingly, single-crystal silicon components produced by semiconductor micromachining are now used in the gears, escapement wheels and balance springs of wristwatch movements. There are three reasons for this. One is that the machining accuracy is greatly improved by the use of semiconductor microfabrication. The second is that silicon components, due to their light elemental composition, have a lower density, lower mechanical torque and better energy efficiency than metal components. Also, that the above mechanical properties are superior to those of metals and other materials at scales of a few millimetres or less and can be fully exploited: it is unlikely that centimetre and metre scales components will be made from silicon in the foreseeable future, but at scales of mm or less it is quite feasible. This suggests that it is possible to build a new world without preconceptions.

Thus, the world you see changes depending on where you place your point of view and how you think about the whole system. This book has been written with the intention of giving guidance to beginners on what perspective to take, how to build a system, how to use the system they have built, and how to develop their own system into the next system.

The subject matter of this book is limited to nanoscience and its applied research, i.e. the fundamentals and applications of Nano–Microsystems. Nano–Microsystems is a multidisciplinary field that combines various fields such as mechanical, electrical, electronic, fluidic, interfacial, nanostructural, physical, chemical and biological. It is not possible to cover them all, so the scope is really limited, but will cover areas related to microfluidics, sensor applications using photonics, etc. The following sections give an overview of what is covered in each chapter. Readers are encouraged to read the chapters according to their own interests, objectives, previous experience and basic knowledge. Readers can skip chapters if they wish.

1.2 Systems for microfluidics and related fields

Microfluidics is an interdisciplinary field spanning engineering, physics, chemistry, biochemistry, nanotechnology and biotechnology. The field began as a fundamental science of microvolume fluids and has developed mainly around the formation of laminar flows and their applications due to the low Reynolds number. The ability to control and manipulate laminar flow enables chemical unit operations, and the multiplexing and automation of these chemical unit operations has practical applications such as drug discovery and high-throughput screening in infectious disease testing [8, 9]. Microfluidics emerged in the early 1980s and has been used in developments such as inkjet print heads, deoxyribonucleic acid (DNA) chips and lab-on-a-chip technology. The field is concerned with the behaviour and precise control of geometrically confined fluids on small, typically sub-millimetre, scales. Because of the characteristics described here, it is used for the analysis of very small amounts of specimens. For example, shown in figure 1.2 is an example of a pre-treatment chip for analyzing amino acids from meteorites, which is made of acid- and alkali-resistant polytetrafluoroethylene (PTFE) [10].

Microfluidic channels are formed using semiconductor microfabrication technology. For environmental analysis and drug screening, more recently, 3D printing and other technologies have also been used to form custom and application-specific chemical reaction circuits, such as application-specific integrated circuit (ASIC). Here, ASIC is a type of electronic component, a generic term for integrated circuits that combine multiple functions into one circuit for a specific application. Passive fluid control technologies such as capillary action are used in the applications also presented in this book, as are active fluid control technologies such as micropumps and microvalves. Microchemical systems are often constructed from glass substrates or plastic mouldings, but micropumps and similar devices are often used in the laboratory or R&D phase because of the ease and precision with which fluids can be controlled.

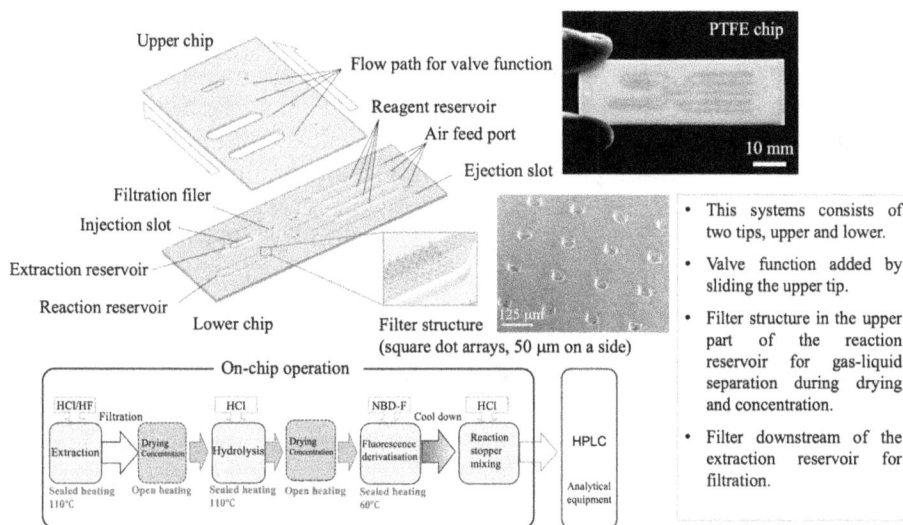

Figure 1.2. Lab-on-a-Chip for the analysis of amino acid extracted from meteorites; the polytetrafuluoroethylene (PTEF) structure was fabricated by anisotropic pyrochemical microetching initiated by synchrotron radiation x-ray-induced scission of molecular bonds. The scanning electron microscope (SEM) image shows the magnified structure of the fine-filter; there are square depressions measuring 50 μm on each side. The chip consists of two parts. Sliding upper chip provides a functional valve, achieving the on-chip operation. Relevant content is presented in reference [10].

However, it is often observed that the use of micropumps and other devices reduces the performance efficiency of the overall system, as some fluid residues of reagents and test results may remain in the tubes connecting the microchemical system to the micropump, or the fluid volume, including the volume of the tubes, is required. Therefore, in some practical applications, external drive means are used to assist in the transport of media. One of the most practical examples is Lab-on-a-Disc (LOD), which uses a centrifugal pump with a rotary drive. However, in the research and development phase, when designing chemical operations, designing system operation, checking operation, etc, it is relatively easy to create systems by combining explicit operation with active (micro) elements such as micropumps and microvalves. With micropumps, fluid can be pumped continuously or injected at will. System design is also easier with microvalves, as the direction of flow and movement of the pumped fluid can be arbitrary. It is therefore often used to miniaturize chemical manipulation processes, such as those normally carried out in laboratories using glass beakers, to be carried out on a single chip to improve efficiency and portability, and to reduce the amount of chemicals used. In addition, LOD has been proposed as a system specifically designed for multi-specimen, rapid and microanalysis, taking into account residual liquid and other factors mentioned above.

As mentioned above, due to the inability to perform residual liquid and powder manipulation in microfluidics and the problems with the pump function, microchemical systems using surface acoustic waves (SAWs) have also been proposed as a new pump.

1.3 Analysis and Nano–Microsystems

In microchemical systems, chemical operations and combinatorial chemical reactions are often performed for analysis. It is then also necessary to analyze the compounds chemically synthesized in the microchemical system and the intermediates obtained in the reaction process. In such cases, the products are evaluated with existing instrumentation, i.e. scanning electron microscopy and x-ray structural analysis, to assess whether the desired reaction is controlled and whether the targeted products are synthesized. Therefore, they are often used in combination with analytical equipment, in particular, spectrometers such as FT-IR, Raman spectroscopy and UV–vis, etc.

On the other hand, Nano–Microsystems including microchemical systems are also built to realize microfluidic chemical reactions and Enzyme-Linked Immunosorbent Assay (ELISA) for medical engineering and point-of-care testing (POCT). These reactions must be monitored and evaluated, usually by optical detection. In ELISA, fluorescent dyes are used, but these dyes have a lifetime and fade. Therefore, a mechanism for reaction monitoring and molecular sensing with high sensitivity was required. Therefore, sensing mechanisms using surface-enhanced Raman scattering (SERS) are being implemented and integrated into systems. SERS is a spectrometric method with very high detection sensitivity and can be combined with fluorescent labelling, which is commonly used in biological experiments. However, nanostructures are required to express SERS. The aim of this textbook is to convey guidelines that will enable readers to design, fabricate, evaluate and use Nano–Microsystems in this book. We want to enable readers to build not only microchemical systems as Nano–Microsystems for analysis, but also to combine them with analytical instruments or to make them analytical systems in themselves. Although the SERS-active structure is one example, it is possible to pre-build SERS-active structures in Nano–Microsystems. It is also possible to dynamically create SERS-active structures while the system is running. Thus, this book also provides the readers with the clue to the construction of systems that enable *in situ* observation of chemical reactions.

1.4 About the structure of the book and how it is read

The structure of the book and how it is read is outlined and explained for each chapter.

Chapter 2: This chapter outlines the basic design guidelines and elementary technologies for building microchemical systems and provides guidance on systems thinking. The overview is easy to understand for beginners learning about microchemical systems. Readers with basic knowledge will already be familiar with much of the content, but we hope that the book will serve as a guide to systems thinking.

Chapter 3: This chapter provides an explanation of the general microfabrication techniques required to build systems, including both solid-state devices and microchemical systems. Although it may not be sufficient for readers familiar with microfabrication techniques, the principles and methods associated with each method and technique are explained for beginners just starting out in microfabrication. Methods for constructing simple flow paths and making moulds are also described.

Chapter 4: Fluid mechanics provides a basic overview of fluid mechanics. Readers wishing to learn more are advised to read a specialist book on fluid mechanics. Since the number of jet nozzles in microfluidic channels is small and laminar flow is the main concern, this chapter describes the minimum necessary basics for the beginner entering microchemical systems.

Chapter 5: The basic principles and fundamentals of LOD as a microfluidic system are described. More recently, some systems have become multi-layered, incorporating very complex chemical operations, and are being integrated with WiFi connectivity and smartphones. The complete system, including connections to electronic devices, has only been introduced; the connection of devices such as Arduino and Raspberry Pi will be described at another time.

Chapter 6: Fluid propulsion and powder transport by surface acoustic waves is described as a means of solving the problems associated with micropumps in microfluidic and Nano–Microsystems. Surface acoustic waves are solid-state devices created using piezoelectric elements, while microchemical systems are passive systems consisting of microfluidic channels, but when the two are integrated they become new active systems. Surface acoustic waves have a piezoelectric function and can be used for sensing applications, etc, but the pumping function is described here.

Chapter 7: Microchemical systems are capable of automatically repeating a fixed operation with a small amount of fluid in a precise and reproducible manner. In addition, the small amount of fluid has a low heat capacity and chemical reactions etc can be easily excited in a controlled and uniform manner. Therefore, it is expected to be used for combinatorial chemical synthesis. After introducing examples of applications such as drug screening, examples of fusion with reaction acceleration technology using microwave irradiation as a special system will be presented.

Chapter 8: Methods for introducing SERS sensing mechanisms into microfluidic systems are described. In particular, the system configuration for inducing SERS by actively constructing nanopores, e.g. by applying a voltage, is presented.

Chapter 9: The chapters have been outlined as discrete systems, but in fact provide guidelines for rethinking them as elementary technologies and integrating them as a new system. It reiterates that the development of science and technology does not stand still, but evolves to meet needs, and provides guidance for the first-time learner.

1.5 Summary

This book is written on the fundamentals and applications of Nano–Microsystems for first-time students, especially those in their third year of university up to postgraduate level, who are about to start their professional studies. It is also aimed at new entrants to industry. This book has been written so that it can be used as a textbook for first-time students in university seminars and lectures, explaining the basics as simply as possible, while at the same time providing a model case for

Nano–Microsystems. Progress in the field of Nano–Microsystems is so rapid that if I write about something that is too cutting-edge, it will be out of date within a few years. Therefore, I have tried to make sure that the reader understands and uses the basic scientific principles and thinking guidelines that have remained unchanged over time. If they understand the principles, can design the structures and systems they want to build, and can evaluate them after building them, they will be able to build new systems. Therefore, in this book we have tried to write in such a way that, once the basic guidelines are understood, they can be applied. We have also included some examples of recent studies that report on the practical application of the principles in this book. It is hoped that readers will be able to relate this book to their own research and development.

References

[1] Nurk S *et al* 2022 The complete sequence of a human genome *Science* **376** 6588
[2] https://nobelprize.org/all-nobel-prizes-2024/
[3] https://www8.cao.go.jp/cstp/english/society5_0/index.html
[4] Yamaguchi A, Ohkochi T, Oura M, Yokomatsu T and Kanda K 2024 Consideration of experiment to introduce MEMS devices into spectroscopic systems for bending and three-point tension tests *Sensors Mater.* **36** 3465–77
[5] https://orient-watch.jp/orientstar/brand-value_manufacture.php
[6] Maes R 2013 Physically unclonable functions: properties *Physically Unclonable Functions* (Berlin: Springer) pp 49–80
[7] Fukuoka T, Yasunaga T, Namura K, Suzuki M and Yamaguchi A 2023 Plasmonic nanotags for on-dose authentication of medical tablets *Adv. Mater. Interfaces* **10** 2300157
[8] Yager P, Edwards T, Fu E, Helton K, Nelson K, Tam M R and Weigl B H 2006 Microfluidic diagnostic technologies for global public health *Nature* **442** 412–8
[9] Shembekar N, Chaipan C, Utharala R and Merten C A 2016 Droplet-based microfluidics in drug discovery, transcriptomics and high-throughput molecular genetics *Lab Chip* **16** 1314
[10] Yamaguchi A, Kido H, Ukita Y, Kishihara M and Utsumi Y 2016 Anisotropic pyrochemical microetching poly(tetrafluoroethylene) initiated by synchrotron radiation-induced scission of molecule bonds *Appl. Phys. Lett.* **108** 051610

IOP Publishing

Nano–Microsystems
Science and applications
Akinobu Yamaguchi

Chapter 2

Design and concept for system thinking

Microfluidic chips are often used primarily for analytical purposes. The feature of the microfluidic chip is consisting of microchannels (micro-scale flow path) created by nano/microfabrication. On the chip, by consolidating and integrating laminar flow and unit chemical operations using laminar flow realized by a small Reynolds number, chemical experiments can be precisely performed. From the beginning of research and development of microfluidic chip or Lab-on-a-chip (LOC), many combinations with electrochemical and electrical control mechanisms have been proposed. Considering the electrical control system as well, it is possible to expand the concept as an ideal experimental system including not only microfluidic chips but also solid-state devices and micro electro mechanical systems (MEMS) elements. That is, it will be treated as a system called a Nano–Microsystem. Here, we will outline the basic ideas and examples of creating an ideal experimental system and its application and utilization by combining various measurement systems including microfluidic elements and solid devices. In addition, we outline technologies and trends for Nano–Microsystems mainly microfluidics including Organ-on-a-Chip. Design and concept approaches for realization and assessment of these systems are also described. We shed light on a discussion on system thinking for achieving the current significance/ relevance, trends, limitations, challenges and future prospects in terms of revolutionary impact on biomedical research, preclinical models and drug development.

2.1 Introduction

Nano–Microsystems have advanced beyond scientific and engineering applications such as life science and environmental analysis [1–21]. A growing trend of these systems has emerged thanks to the nano/microfabrication precision, time-effectiveness, and cost-effectiveness of advanced nano/micro systems. The state-of-the-art nano/microsystems can become significant integral platforms further investigated not just for fundamental science and mechanism for the engineering applications. This chapter outlines Nano–Microsystems or LOCs for creating ideal experimental systems and conducting scientific experiments. On the other hand, the system for conducting scientific experiments can be a tool for conducting chemical/environmental analysis or a medical test kit. In other words, it can be said that LOC is an experimental device that has both basic science and applied use. LOC is often used to refer to microfluidic chips that mainly handle fluids. However, ideal unit chemistry operations are not the only experiments. In other words, it can be said that the development of solid-state devices and gas sensors is also in the category of LOC. The chip is both an integration of designed functions and a pre-integrated element with functionality. Therefore, the word 'chip' corresponds to the concept corresponding to the central operator of a computer or a chip or chip set such as a north bridge or a south bridge. A computer is formed by adding various functions such as a storage device, an input device, a display mechanism, a communication mechanism, etc, as well as a central operator and various chipsets. That is, it can be said that a computer is a collection of chips that integrate various functions, and is a system that integrates and aggregates such chips. The system is not limited to computers, but humans themselves are systems in which cells and tissues are integrated and aggregated. When individual humans gather to form a society, it becomes a social system. Thinking in this way, it can be said that the Earth and the Universe are also systems. Here, we focus on phenomena such as physics, chemistry, and biology on the nanoscale or micro-scale, as well as scientific verification and empirical research, and create and use a mechanism, that is, a system, that realizes basic research and applied use. This book introduces the philosophy, policies, methods and examples of creating and verifying simple, small-scale processes for scientific phenomena and engineering applications that require clarification. Therefore, we will introduce various LOCs as a Nano–Microsystem from a bird's-eye view. In order to create a Nano–Microsystem, the reader considers what he or she wants to achieve, and at what level the function or role is viewed, the system components are identified, and the design, placement, aggregation, and integration are performed. The systems that will be created will vary widely. Combining functional three-dimensional structures and elements to build an experimental system that you want to realize will be a very fulfilling practice. And when it opens the door to environmental analysis, medical tests, new substance creation, etc, it becomes a creator and an evaluator. In the following, we will outline the examples of LOC reported so far, and introduce various application developments and their uses for basic research.

2.2 Lab-on-a-chip

An LOC is a device performing on a miniaturized scale one or several analyses commonly carried out in a laboratory. The microchannel circuits are integrated and automated multiple high-resolution laboratory process such as synthesis, analysis of chemicals and fluid testing into a system that fits on a chip. The combination of various unit chemical operations is integrated into a chip as schematically illustrated in figure 2.1 [1], resulting in being many advantages to operating as designed in advance. LOC is provided not only as a microchannel circuit, but also as a chip structure equipped with various technologies such as MEMS, dielectrophoresis, and AC electroosmotic flow control. It is often used not only for applied research but

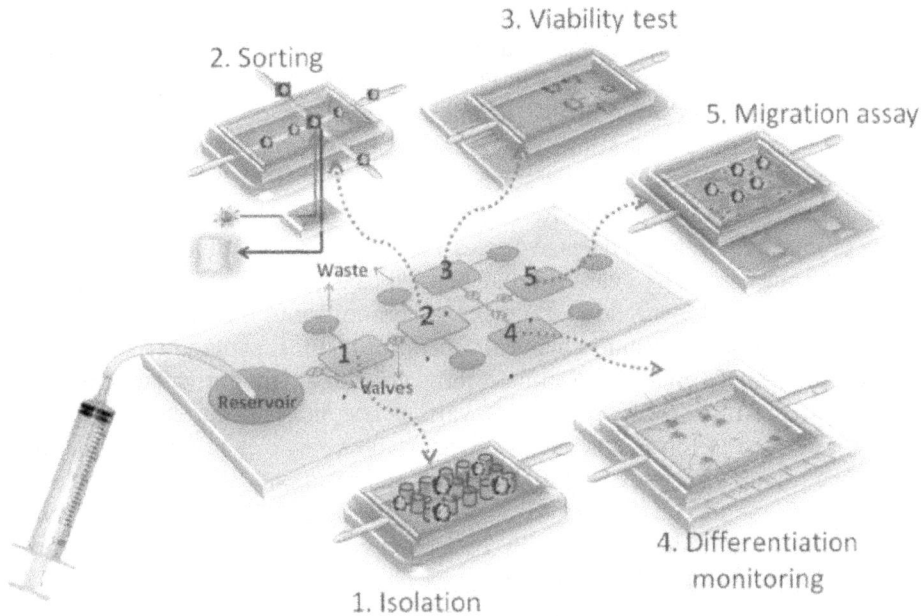

Figure 2.1. Example of a modular LOC for stem cell studies. Several microfluidic components and sensing modules are integrated together for cell isolation, detection and counting, viability or migration assays and differentiation studies. (Figure reprinted from reference [1] with permission from The Royal Society of Chemistry.)

also for basic scientific research because it can be an ideal chemical experiment device as well as an analytical application [2]. It is well established as a broad general concept. One of the candidates is a circulatory system constructed from a network of blood vessels. The blood vessels mediate a wide range of functions including cellular and biochemical transport, nutrient and oxygen exchange, and temperature regulation while maintaining a high degree of plasticity throughout the life of an organism. Recently, foundation of these blood vessels has been performed by a lot of research groups. The details will be presented later. Another state-of-the-art development of LOC is digital fluidics [3, 4, 20, 22–26]. The droplet microfluidics enables precise, high-throughput and highly controlled fluid handling. The digital fluidics based on the droplet microfluidics will be outlined. Based on the concept of LOC, it is possible to form a structure that imitates the human body or a system like a chemical plant. Here, we notice the most important key concepts for designing, fabrication, evaluation, and application of the desired LOC. This is the understanding of the behaviour of fluids within microfluidic channels, which may be broadly described by ratios of specific physical properties or forces that define a set of dimensionless numbers. Table 4.1 summarizes dimensionless numbers that describe the behaviour of fluid transport and droplet formation in microfluidic devices in chapter 4. The basic fluidic behaviour and equations are also described in chapter 4. The behaviour of fluids within microfluidic channels may be broadly described by ratios of specific physical properties or forces that defined a set of dimensionless numbers described in table 4.1. Interfacial forces which act as fluid–fluid, fluid–solid, and fluid–gas interfaces are of significant importance in LOC for multiphase flow. Details are given in chapter 4, where the Reynolds number, Re, which is often used as an indicator in microchannels, is discussed. Re, a ratio of inertial to viscus force, quantifies the relative contribution of each force for given flow conditions. In general, two flow states are described: one is laminar flow in $Re < 1$, and the other is turbulent flow in $Re > 1000$. Turbulent flow is dominated by inertial forces, producing random eddies, vortices and other chaotic fluctuations, and the resulting turbulent mixing enables rapid high-efficiency mixing. However, there are weakness in the turbulent flow. It is the precision control of fluids required for quantitatively controlled unit chemical operations. In microfluidics, limiting the characteristic length creates a laminar flow condition. In particular, micro-scale processing makes it possible to combine various chemical operations, such as semi-conductor devices, to form a chemical laboratory on a single chip. On the other hand, mixing operations, which are essential for chemical reactions, often use diffusion and other methods, as the usual method of creating turbulence in microfluidic channels requires a certain amount of ingenuity. In such cases, the Reynolds number must be taken into account in the design of the channel structure.

In such microfluidic systems, the bulk flow is usually laminar. On the other hand, digital fluidic systems, based on generation and utilization of droplets, have been proposed in recent years. When using such droplets, the hydrodynamics are more complex compared to bulk fluids in the microchannels. To generate a droplet, two immiscible fluid interfaces have to be formed. Reynolds numbers characterize the dynamics of single-phase flows in microfluidic channels well, but are not often used to describe droplets. When dealing with droplets, it should be taken into account that the velocity of the multiphase flow through the microchannel determines the mechanism of droplet formation. The Weber number, We, and capillary number, Ca, are important in this context. The We is the ratio of the inertia to the interfacial tension. $We < 1$, as fluid inertia is negligible in bulk fluid flows in microfluidic channels. In the cases of high velocity jets or in the vicinity of droplet formation, as they pinch off from the continuous phase fluid. Then, the most critical parameter for applications discussed is the capillary number, Ca, representing the relative effect of viscous forces over surface tension across an interface. We must consider that the rate of multiphase flow through the microchannels will dictate the mechanism of droplet formation. Thus, the parameters governing the system depend on the fluid and multiphase flow conditions used and what functionality is given to the flow path. Therefore, it indicates that the physical parameters need to be evaluated and the channel structure designed according to the channel structure to achieve the functionality to be fabricated.

Precise and effective control of fluidics is critical for various applications of life science, environmental analysis, food analysis, and medical diagnosis. A micro-fluidic chip is a pattern of microchannels, moulded or engraved. There are two mainstream methods for transporting liquids. One is a transport systems using nano/micro-scale channels. The other is a liquid feeding systems using droplets. In either case, a pump mechanism for transportation of liquid is essential. The transport pump mechanism includes an active control mechanism and a passive control mechanism. The passive control mechanism is a mechanism that utilizes capillarity, pressure difference, etc, and is determined at the stage of designing and manufacturing the flow path structure. The active control mechanism is a pump mechanism operated by an external input. For example, there are diaphragm pumps and syringe pumps in the active control mechanism. Centrifugal fluid, which is often used in Lab-on-a-disc (LOD) [5, 14], uses capillary valves and siphon valves, and its performance and characteristics are determined at the stage of designing and manufacturing the flow path structure, but it rotates to execute the operation. Power for the rotation is needed in LOD. In this case, it can be said that it is a pump mechanism that uses the characteristics of both the passive control mechanism and the active control mechanism.

These microfluidic devices were often fabricated out of polydimethylsiloxane (PDMS) using soft lithography and replica moulding. In chapter 3, the details for the processes are described. While PDMS is easy to use, it is porous and requires careful handling. Details will be mentioned in each section and chapter with respect to the actual microfluidic devices. Basically, PDMS is not suitable for mass production, and is often used at the stage of research at universities and basic research at companies. Here, we will outline examples of various microfluidic systems and Nano–Microsystems without going into detail. When the readers fabricate the systems and devices, please refer to chapter 3 'Nano/Microfabrication for creating nano/micro-systems'. Some of the recent developments and examples of implementation are presented with a view to helping systems thinking.

2.3 Digital fluidics

Droplet-based microfluidic systems have emerged as a versatile tool for widespread applications because of the following characteristic advantages such as a small volume of reagents, monodisperse droplets, and high surface-area-to-volume ratio that can facilitate fast reaction. In addition to it, each droplet can be independently controlled. The digital fluidics is new and old. As the reader can notice from the word digital, it becomes a fluid chip that utilizes discrete states. Why are fluids discrete even though they are continuous? Many readers may be wondering. The reason is that it is a fluidic chip using droplets. Droplets are a technique that has existed before the study of microfluidic systems. However, by combining it with microfluidics, it becomes more controllable and various things can be done. For example, the droplet platform can accelerate antibiotic susceptibility test (AST). Microfluidic droplets in particular have reduced AST time down to the timescale of bacterial cell replication via single-cell measurements, because they can develop a multiplexed droplet platform for rapid single-cell antibiogram by testing pathogens against multiple antibiotic conditions within a single device. Zhang *et al* demonstrated a multiplex droplet platform [6]. The details will be described later. The droplet platform is very useful and is developed to the digital microfluidic systems. Below, we will outline the generation of droplet, designs to produce the droplet, and how to use it.

Recently, Zhu and Wang have published an excellent review article on droplet generation and flow path systems [3]. Those who are interested or who want to learn more should read the review article. Here is a brief overview with reference to their reviews. According to their review, to generate the droplets, some devices take advantages of shear forces, of which the most-frequently-used structure types are (a) cross-flow, (b) co-flow, (c) flow-focusing geometries, (d) step emulsification, (e) microchannel emulsification, and (f) membrane emulsification, as shown in figure 2.2.

Droplet formation in microfluidic devices involves the deformation and breaking of liquid–liquid interface, as schematically shown in figure 2.2. Analyses on droplet formation have been calculated in several ways and reported to be able to explain the phenomena. One is the numerical solution of the basic equations, which are described below. First, the momentum equation is described by the well-known Navier–Stokes equation of incompressible Newtonian fluids [3, 22, 27]:

$$\rho_s \frac{\partial u_s}{\partial t} + \rho_s (u_s \cdot \nabla) u_s = - \nabla P_s + \eta_s \nabla^2 u_s + f_s. \tag{2.1}$$

Here, ρ, u, P, η, and f are the density of the fluid, pressure, dynamic share viscosity, and body force per unit volume, respectively. The subscript 's' denotes either 'd' or 'c', corresponding to dispersed or continuous phase fluid, respectively. As the details are explained, the left side of equation (2.1) described inertial acceleration composed of time-dependent acceleration $\rho_s \frac{\partial u_s}{\partial t}$ and convective acceleration $\rho_s (u_s \cdot \nabla) u_s$. is the velocity of the fluid. The right side of equation (2.1) describes forces densities composed of the force resulting from pressure difference ∇P_s, diffusion term $\eta_s \nabla^2 u_s$, and body forces f_s from various effects from those like gravity, electrical, magnetic, and centrifugal forces, and so on. Fluid behaviour can be reproducible by solving this Navier–Stokes equation. Liquid–liquid two-phase microflows are well described with continuous hypothesis. Here, the problem described is straightforward to set up with the laminar two-phase flow, which is described in chapter 4. Liquid–liquid two-phase microflows are well described with continuum hypothesis. The incompressible continuity equation is given by

$$\nabla \cdot u_s = 0, \tag{2.2}$$

where u_s is the velocity of the fluid. The following items need to be considered for the boundary conditions at the interface of the droplet:

$$u_d \cdot n = u_c \cdot n. \tag{2.3}$$

This equation (2.3) is required for the continuity of normal velocity at the immiscible interface. n is the unit vector outward to the interface. Secondly, following two equations are necessary in order to satisfy with the conditions that the tangential viscous stress should be continuous and the normal stress difference between the dispersed and continuous phases is balanced by capillary pressure:

$$\sigma_d \cdot t = \sigma_c \cdot t \tag{2.4}$$

and

$$T_d \cdot n - T_c \cdot n = -\gamma \kappa n. \tag{2.5}$$

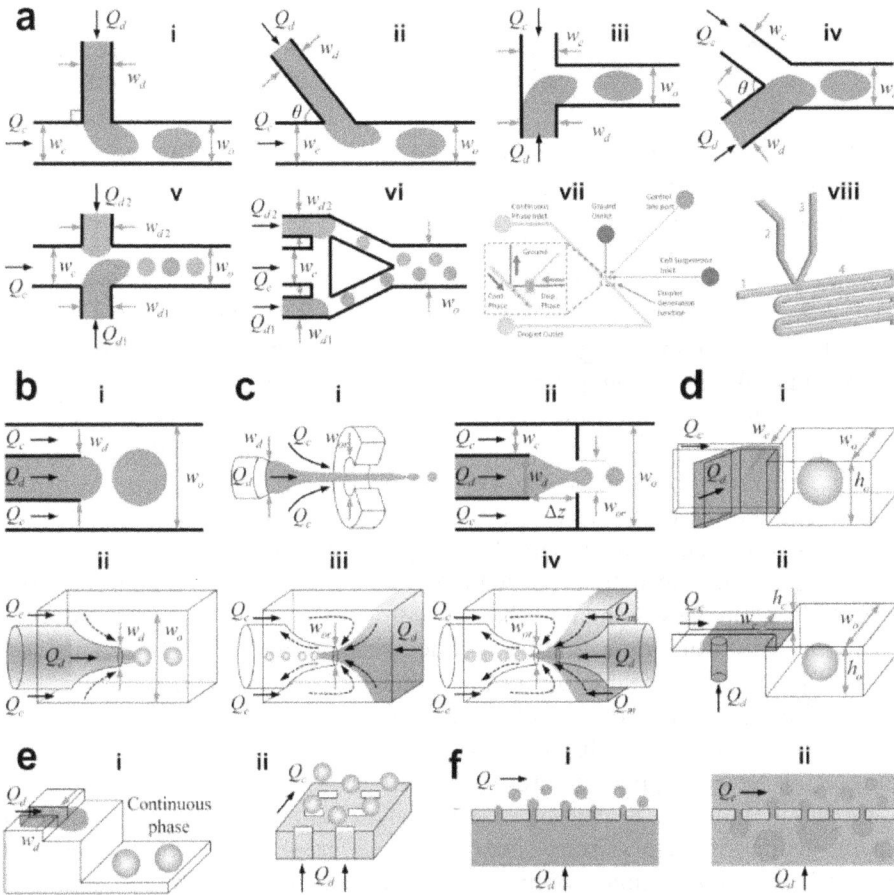

Figure 2.2. Schematic of various microfluidic device geometries (not to scale). (a) Cross-flow. (i) 'T-junction' where the continuous and dispersed phase fluids meet perpendicularly ($\theta = 90°$). (ii) 'T-junction' in which the two fluids intersect at an angle θ ($0° < \theta < 90°$). (iii) 'Head-on' geometry ($\theta = 180°$). (iv) Y-shaped junction with intersection angle of θ ($0° < \theta < 180°$). (v) Double T-junction with droplet pairs generated at the same location. (vi) Double T-junction that produces droplet pairs at separated parallel T-junctions. (vii) 'K-junction'. (viii) 'V-junction'. (b) Co-flow. (i) Quasi-2D planner co-flow. (ii) 3D co-flow. (c) Flow-focusing. (i) Axisymmetric flow-focusing geometry. (ii) Planner flow-focusing geometry. (iii) Microcapillary flow-focusing device. (iv) Microcapillary device combining co-flow and flow-focusing geometries. (d) Step emulsification. (i) Horizontal step. (ii) Vertical step. The inlet channel has a high aspect ratio, and the reservoir is wider and deeper, with an abrupt topographic step in between the inlet channel and the reservoir. (e) Microchannel emulsification. (i) Grooved-type microchannel. (ii) Straight-through microchannel. (f) Membrane emulsification. (i) Direct membrane emulsification. (ii) Premix membrane emulsification. Q, w, h, and Δz denote the volumetric flow rate, channel width, channel height, and horizontal distance from the end of the dispersed microchannel to the orifice entrance, respectively. For planar devices, the channel height h is uniform. In the case of the geometry with a circular cross-section, w represents the channel diameter. The subscripts 'c', 'd', 'o', 'or' stand for the continuous phase, dispersed phase, outlet channel, and orifice, respectively. (This figure is reprinted from [3] with permission from The Royal Society of Chemistry.)

Here t is the unit tangential vector at the interface. $T_s = -P_s I + \sigma_s$ denotes the stress tensor that contains pressure P_s and viscous stress σ_s, where I is the indemnity matrix. γ and κ are the interfacial tension and twice of the mean curvature of the interface, that is $\kappa = R_1^{-1} + R_2^{-1}$, with R_1 and R_2 being the principle radii of the curvature. Recently, computers with a combination of graphical processing unit and central processing unit performance and a large amount of memory have become available at low prices. The generation and behaviour of droplets can also be reproduced sufficiently, and it is often used for device design and analysis. The simulation of droplet generation and transportation can be easily performed by using some simulation packages. For example, using COMSOL multiphysics [22], the problem described is straightforward to set up with the laminar two-phase flow, level-set multiphysics coupling feature. The level-set method is a general-purpose contour extraction algorithm that can be applied to any dimension. In general, this method detects the boundary surface of an object in n-dimensional space by time-evolving a surface (hypersurface) in $(n + 1)$-dimensional space. In a droplet, the main solvent and the dispersed phase are created in the channel, and the boundary between them is the edge of the droplet. By giving a function describing the closed surface and calculating it together with Navier–Stokes and continuity equations, COMOSL multiphyics simulates the formation and dynamic behaviour of the droplet. Droplet formation was also simulated with the lattice Boltzmann method, a method suitable for modelling on the mesoscale by van der Graff *et al* [23].

During the process of the droplet generation, three major steps are normally involved. Comparisons of different droplet generators are well summarized in reference [3]. Several typical microfluidic geometries used in generating droplets are shown in figure 2.2. Among these devices, cross-flow, co-flow and flow-focusing geometries use viscous shear forces to break off droplets. The others employ variations of channel confinement to facilitate or drive droplet generation. The cross-flow geometry is commonly used for droplet generation and is the first proposed flow path geometry. There are various geometries for the cross-flow geometry, and these geometries are schematically shown in (i)–(vii) of figure 2.2(a). For example, T-junction, where the dispersed and continuous fluids are fed orthogonally (figure 2.2(a,i)), was firstly reported by Thorsen *et al* to produce monodisperse water droplet in oil surroundings for generating ordered dynamic patterns in pressure-controlled laminar flows. For its simplicity and ability to produce monodisperse droplets, this geometry have been widely used. The coefficient of variation (CV), which is defined as the ratio of standard deviation to the mean of the droplet radius is usually less than 2% in T-junctions. Even when the junction is not orthogonally but at an arbitrary angle θ, it is possible to generate a droplet as similar to the orthogonal T-junction geometry. In the case shown in figure 2.2(a,ii), the shear stress at the time of droplet generation is dependent on the angle θ, so that the droplet generation changes. Another variation from the typical T-junction is the 'head-on device' as shown in figure 2.2 (a,iii). Droplet formation in these head-on devices has similar features to that in typical T-junctions. The extended head-on device reduces into a Y-shaped junction as shown in figure 2.2(a,iv). To perform chemical reactions via merging two droplet microreactors that contain different reagents or indexing the targeted droplet by the addition of droplet marker, it is desired to generate droplet alternative multiples in microchannels. To achieve the desired result, the fabrication of a double T-junction is suggested and demonstrated as shown in figure 2.2(a,v). Next, in co-flow geometry, the dispersed and continuous fluids meet in parallel streams, as shown in figure 2.2(b). Firstly, Umbranhowar *et al* demonstrated the co-flow microfluidic device [24]. There are some co-flow geometries such as quasi-two-dimensional (2D) planar (figure 2.2(b,i)), or three-dimensional (3) coaxial (figure 2.2(b,ii)).

Droplet formation can also be achieved through a variety of emulsification methods. As described the above, the precision and control of microfluidic fabrication and fluid flowing within microchannels enables precise control of droplet size, allowing droplets to either be monodisperse or of a desired polydispersity. The droplet formation can be actively or passively performed on the microsystesm. Vesicles formed by droplet microfluidics have tended to be 1–100 μm in diameter. Giant unilamellar vesicles (GUVs) can be formed by using the droplet formation technique based on the microfluidic system. Recently, there have been increasing efforts in forming sub-micron vesicles, which are needed for many applications such as drug delivery, although nano/micro-scale precise control for the fabrication have proved to be challenging.

Pautot *et al* firstly demonstrated an on-chip translation of the emulsion phase transfer method [25]. Their method relies on the transfer of lipid monolayer coated with and without droplets across a second monolayer lying at the interface of a water–oil column. This method also enables formation of bilayer droplet. Microfluidic versions of this method rely on microfabricated features such as triangular posts and step junctions to transfer droplets across the water/oil interface [26]. The reverse process can also be controlled and used for formation of a stationary droplet trapped in a hydrodynamic trap. The multiple emulsion can be formed as droplet precursors by using this method. Funakoshi *et al* firstly demonstrated the assembly of a bilayer between two aqueous streams inside a microfluidic flow-focusing junction containing lipid-in-oil, as shown in figure 2.3 [27]. The most novel property of droplet interface bilayers is the ability to connect droplets together to form a bilayer network. The assembly of bilayer networks form several different droplet types is particularly interesting if the bilayers can be functionalized with membrane proteins to yield network architectures.

Droplet size can be controlled by controlling growth time, altering channel dimension or detachment time, and by altering the ratio of the flow rates. The viscosity ratio of the phases can also affect the growth time and thereby result in larger droplets. These methods are often used to generate higher-order architectures. For example, Li and Barrow demonstrated formation of double emulsion droplets, which can encapsulate inner cores of carious size and reagent composition using a droplet-forming fluidic junction named 'bat-wing junction', as shown in figure 2.4 [28].

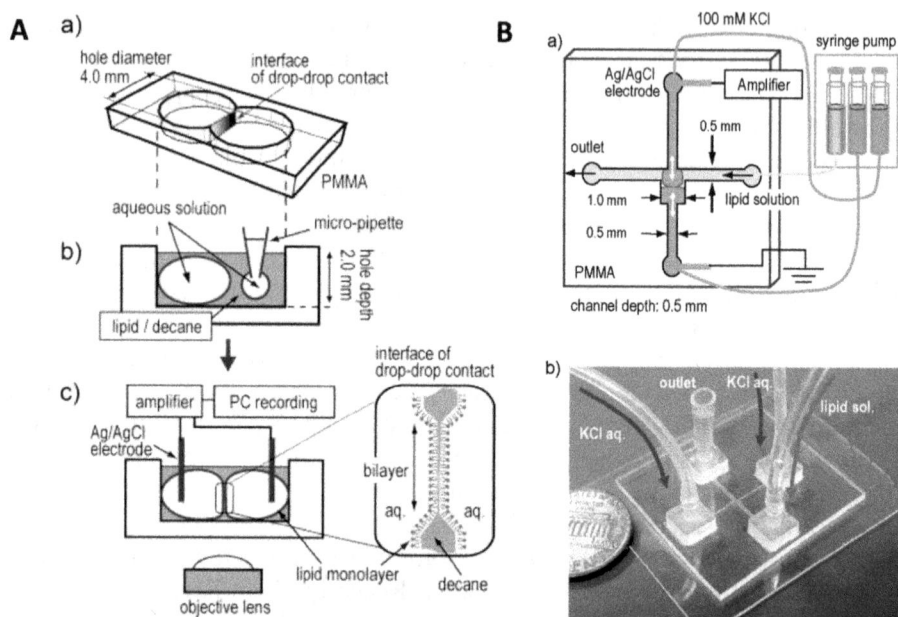

Figure 2.3. (A) (a) Schematic of the double well chip, connected two wells. (b) Procedure of the bilayer formation. Two droplets are injected in each well filled with lipid solution, forming a water/lipid solution/water interface at the section. (c) Schematic diagram of the experimental setup. (B) (a) Schematic of the microfluidic chip with cross channel. (b) Photograph of the chip. Reprinted with permission from [27], copyright 2006 the American Chemical Society.

Its bifurcation structures oscillate the side flows during droplet formation, and the droplet break-up point and droplet size were regulated by adjusting only the continuous phase inflow rate. These methods provide a suitable process for the modification to generate higher-order structures with a fine degree of a control. There have been many demonstrations to generate vesicles of any size from small unilamellar vesicles (SUVs) to GUVs.

Incorporating proteins in model membranes is a challenge that must be addressed in order to realize the artificial cell or controlled model. While there have been excellent proof-of-concept studies using water-soluble peptides and proteins in compartmentalized systems including alpha-hemolysim, OmpG, and gramisidin, etc, there is increasing demand for membranes to be functionalized with more complex channels like mammalian cells. One major problem is that the proteins are notoriously difficult to express in sufficient yields to support reconstitution into model membranes. To date, several groups have reported protein incorporation in membranes by using these microfluidic systems. In the cases, the *in vitro* channels have low probabilities and relatively short opening times. It seems highly likely that many of the current challenges facing protein expression and bilayer incorporation will be addressed using *in vitro* protein expression systems, given that stabilizing agents or molecular chaperons can be freely added.

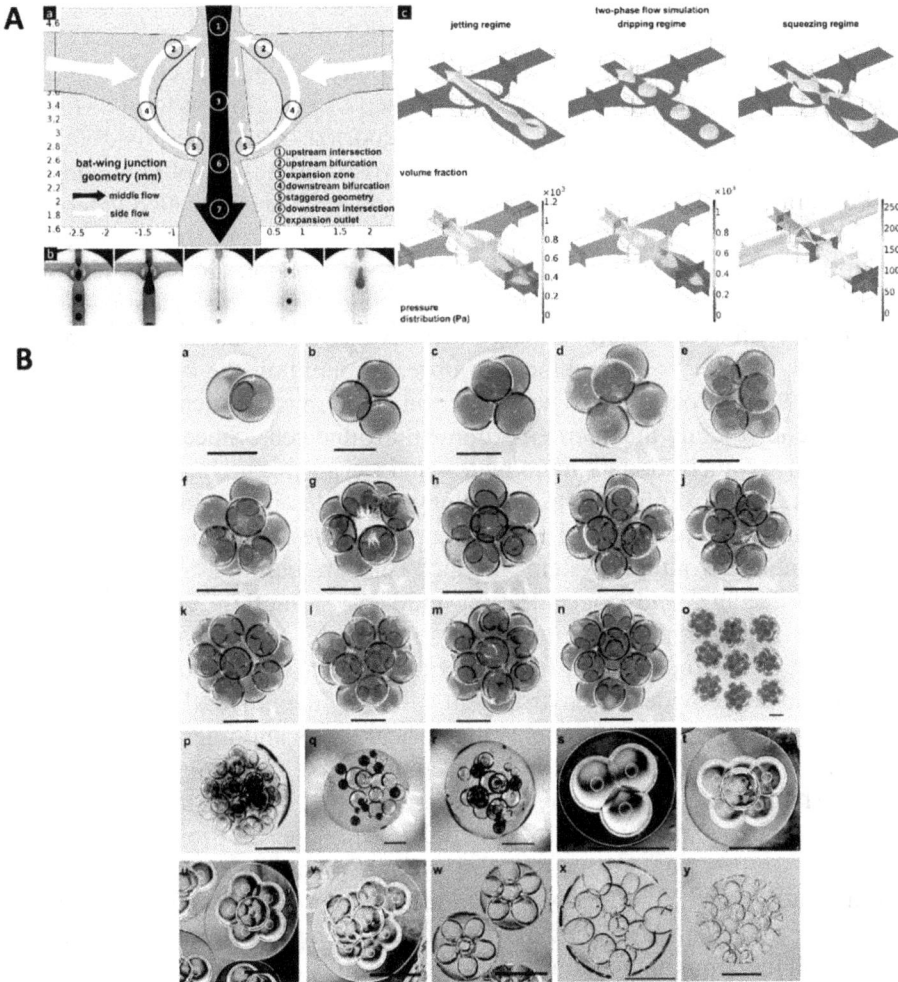

Figure 2.4. (A) (a) Schematic of geometry of the bat-wing junction. The numerical scale of the images is in mm. The depth of the microfluidic channels is 600 μm. Trimethylolpropane triacrylate (TMPTA), photo-initiator 2-hydroxy-2-methylpropiophenone, glycerol (CAS Number: 8042-47-5), oil red O (CAS Number: 1320-06-5), thymol blue (CAS Number: 76-61-9), potassium carbonate (CAS Number: 584-08-7, 3% w/w aqueous solution), and pure water was used for this demonstration. (b) Images of different droplet-forming regimes of the bat-wing junction. The two left images show the water droplet (blue) formation in mineral oil (red), in dripping and squeezing regimes. The three right images show the TMPTA (red) droplet formation in mineral oil (transparent) in jetting, dripping and squeezing regimes. (c) Simulation results for various conditions. The top row images show the droplet interface coloured by yellow, illustrating the boundary between the water phase and the oil phase. The blue vacancies in the yellow films illustrate the water phase touching the top channel wall. The bottom row images show the pressure distribution. (B) (a)–(o) Photographs of microcapsules fabricated by the microfluidic device. All the scale bars indicate 500 μm. (p)–(r) Photographs of capsules produced by a two-step method. (s)–(v) Solid TMPTA capsules. (w)–(y) Encapsulated droplet interface bilayers in water/squalene/TMPTA double emulsion droplets. The squalene phase contains 2%, 1,2-diphotanoyl-sn-glycero-3-phosphocholine. Reprinted from [28] CC BY 3.0.

2.4 Organ-on-a-chip

In medical researches such as drug discovery, toxicity testing, medical and biological research, *in vitro* evaluation as a preliminary step before moving to *in vivo* clinical trials is very important. However, it is often reported that the results obtained by *in vitro* tests differ from the cellular response *in vivo*, and it is essential to construct a highly accurate model. In the usual culture method, cells are cultivated in a plane on a plastic culture dish. This environment is very different from the *in vivo* environment in which cells normally survive. Therefore, the functions and responses of cells are significantly different from those in the living body. The function of cells is closely controlled by their surrounding microenvironments, and it is very important to reproduce the *in vivo* environment in order to maintain the cell function. LOC environmental control technologies based on nano/microfabrication are effective tools for reconstructing the microenvironment around cells. Since the size of cells is on the order of micrometres, nano/micro-engineering technology based on semi-conductor nano/microfabrication technology facilitates micro-environmental control and enables reproduction of an environment close to that of a living body. In particular, in a microfluidic system in which cells are cultured in a microchannel, local liquid feeding and concentration gradient formation can be performed precisely by unit chemical manipulation, so that cell–cell interactions caused by cell secretion can be reproduced. In addition, by controlling the flow of the solution, it is possible to provide an environment close to the internal environment. In particular, by using a microchannel system, shear stress and interstitial flow can be reproduced. In fact, as described in the section on soft moulding in this book, by using PDMS, which is an elastomer, it is possible to easily form microchannels, and since it can expand and contract as a material, it stimulates the growth of cells. These simulation controls enable one to give cells a dynamic environment such as breathing and sedition. In other words, by using a microfluidic system, it is possible to construct a microenvironment that imitates a living body, and it is possible to maintain cell functions and conduct experiments at the level of cell tissue rather than a single cell. Therefore, it is expected to be applied and developed for disease modelling and drug discovery. The development of these LOCs in cell tissues is called organ-on-a-chip, and research and development of various microfluidic systems are currently underway.

Organ-on-a-chip can imitate the *in vivo* microenvironment [6–16]. In particular, by imitating the tissue structure, chemical state, and mechanical environmental factors in the living body, it is possible to reproduce the cell function at the tissue level. Therefore, organ-on-a-chip technology has great potential to facilitate drug discovery and development, in addition, to model pharmacokinetics and pharmacodynamics in cells and organs. Here, we describe the research contents and their important results on the recently highly developed organ-on-a-chip. To date, most tissues and/or organ types have been successfully modelled to reproduced corresponding functional subunits, such as the brain, heart, lung, liver, vasculature, and kidney etc, as shown in figures 2.5 and 2.6. In particular, figure 2.6 summarizes typical examples of organ-on-a-chip reported until now. These chips mimic the

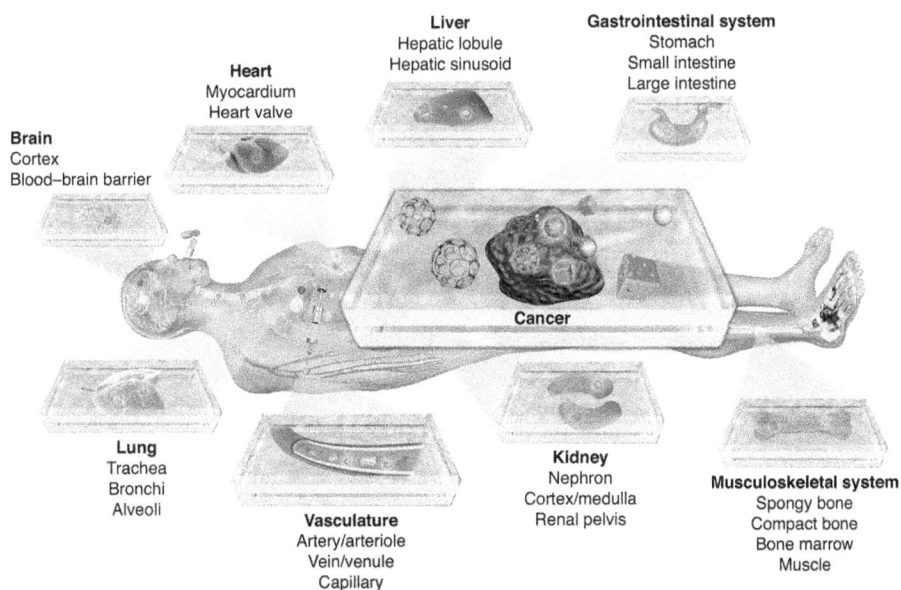

Figure 2.5. Organ-on-a-chip platform for drug screening and their potential applications. Major organs are described. Reprinted from [6], copyright 2017 with permission from Elsevier.

cellular environment in the human body on a chip and operate more like the situation inside the human body, making them successful not only for basic research but also for practical applications such as drug efficacy and cellular response control. These individual devices enable to investigate and understand pharmacokinetic and pharmacodynamics by linking such that the integrated platforms further mimic the physiology and compartmentalization of the human system. As described later, one of the most distinct advantages of using organ-on-a-chip for nanomedicine or pharmacodynamics evaluations is precise assessment of transport and translocation of medicine across tissue–tissue interfaces under *in vivo*-relevant shear flow, which makes a considerable difference compared to the conventional method based on 2D LOC. Organ-on-a-chip has the great potential to meet the demand of creating a robust preclinical screening *in vitro* model for the evaluations of state-of-the-art medicine. Various companies have pursued the translation and commercialization of organ-on-a-chip technology. Below, some investigation examples are outlined.

2.4.1 Three-dimensional formation of blood vascular network

A method has been developed in which angiogenesis is induced by mimicking the environment of microangiogenesis in the body and vascular endothelial cells spontaneously form a three-dimensional vascular network [9, 10]. This method solves the problems and provides a platform that enables biomimetics in microfluidic systems.

The circulatory system is constructed from a network of blood vessels which mediate a wide range of functions including cellular and biochemical transport,

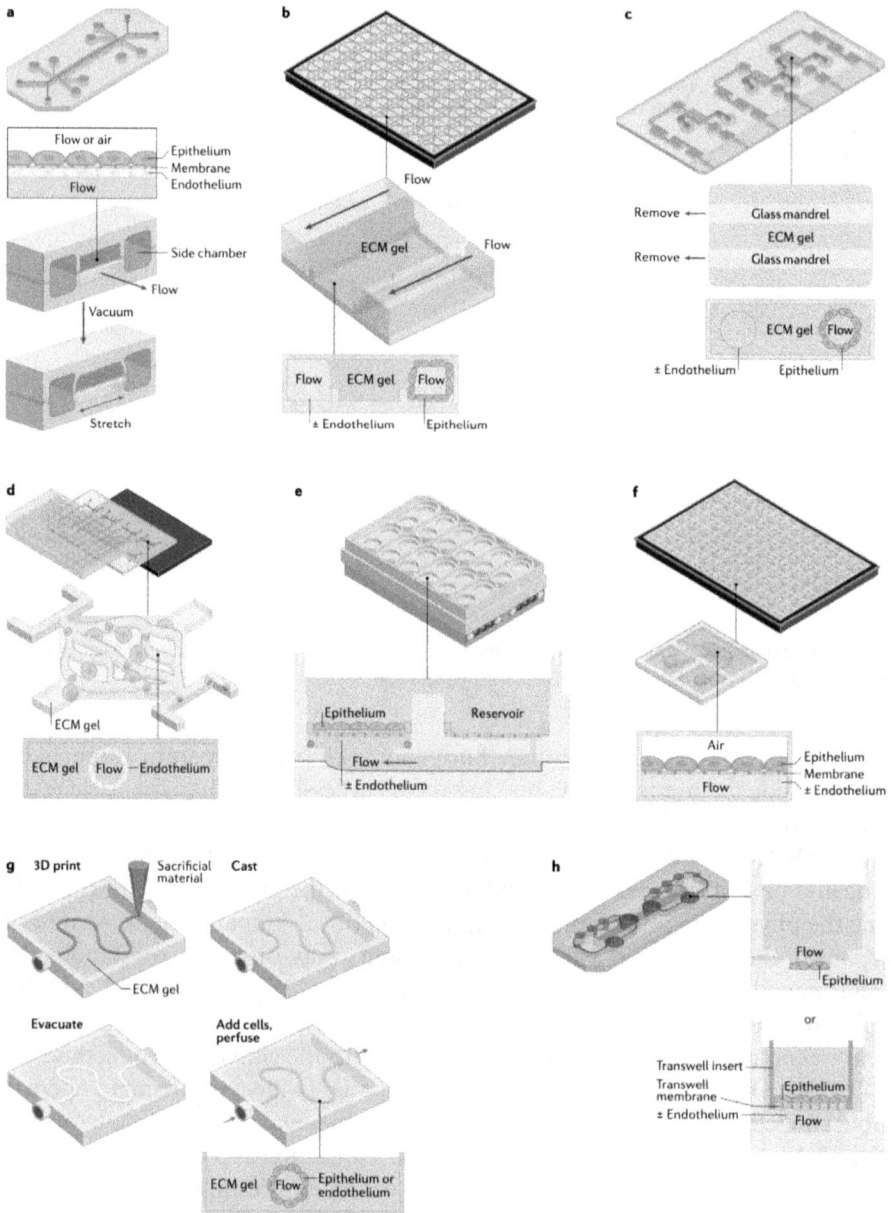

Figure 2.6. Summary of microfluidic organ chip designs. (a) Mechanically actuatable organs-on-chip fabricated from PDMS using soft lithography. Periodic suction applied to the hollow lumen causes the epithelium to distort rhythmically. (b) Multiple array of three-channel plastic (polystyrene) chips. The central channel contains a thick extracellular matrix (ECM) gel. Cells can be cultured in one or both of the channels as well as in the ECM gel. (c) After gelation, the cylindrical mandrel is removed to create one or more hollow channels within the thick 3D ECM gel material and cells are cultured on the inner surface. (d) Multiple PDMS microfluidic devices, including two endothelium-lined PDMS microfluidic devices, containing two endothelium-lined channels separated by a third rhombic chamber filled with ECM gel. Can be used for capillary

neovascularisation and formation of 3D microvascular networks. (e) A plastic multi-well format organotip system incorporating multiple bioreactor chambers. (f) A plastic multi-chamber organ-on-chip in which multiple mini-biroreactor chambers positioned on a flat pate can be cultured for high-throughput screening. (g) Organ-on-chip fabricated using 3D printing. Cylindrical sacrificial material is deposited in an arbitrary pattern in an ECM gel with or without embedded cells. When gelation is complete, the material is removed and epithelial or endothelial cells are cultured on the inner surface of the channel. (h) Plastic multi-chamber organ-on-chip system. Multiple mini-bioreactor chambers arranged on a flat plate can be cultured individually or fluidically linked together. Reprinted from [7] with permission from Sringer Nature.

nutrient and oxygen exchange, and temperature regulation. The life of an organism is maintained due to the circulatory system. The formation of vascular networks is essential to mimic the biological environment. There are three types of devices that can form a vascular network. One is a 2D culture method in which a thin film or a two-dimensional sheet of vascular endothelial cells is formed on a substrate in a flow path. This method is simple and excellent in quantification. The other two are 3D culture methods. Both 3D culture methods can form a three-dimensional tubular blood vessel that resembles the structure in the living body. One of the 3D culture methods is a method of forming a vascular network by culturing the movement of vascular endothelium so as to cover the inside of the scaffold prepared in the flow path. Another 3D culture method is a method in which angiogenesis is induced so that vascular endothelial cells spontaneously form a tube structure. Since this method reproduces the environment for forming minute defects in the living body, it is possible to form a vascular network structure similar to that in the living body. Each of these three methods has different characteristics, and angiogenesis is selected and used according to the purpose of evaluation.

Generating 3D microvessels *in vitro* is a very important goal for tissue engineering because of reliable modelling of blood vessel function. Kim *et al* demonstrated a microfluidic-based platform whereby they model natural cellular programs found during normal development and angiogenesis to form perfusable networks to intact 3D microvessels as well as tumour vasculatures based on the spatially controlled co-culture or endothelial cells with stromal fibroblasts, as shown in figure 2.7 [9, 10].

2.4.2 Gut-on-a-chip and intestinal imitation device

Gut-on-a-chip is a device that reproduces the function of the intestine by imitating the tissue structure and incitement of the intestine using a diaphragm-type micro-fluidic device [10]. By culturing intestinal epithelial-like cancer cells (Caco-2 cells) in the microfluidic device, the dynamic environment of the load of extension stimulus and the load of shear stress due to the flow of solution is mimicked to reproduce the villous structure of the intestine and with bacteria. The construction of a co-culture system has been achieved. The intestine is an organ responsible for absorption and metabolism of nutrients, and is indispensable for drug discovery and toxicity evaluation. In particular, intestinal bacteria are associated with metabolic function, immune response, and various diseases, and it is important to construct an intestinal environment imitation system by a co-culture system. However, in most cases, the intestinal environment cannot be reproduced by a general culture method, and

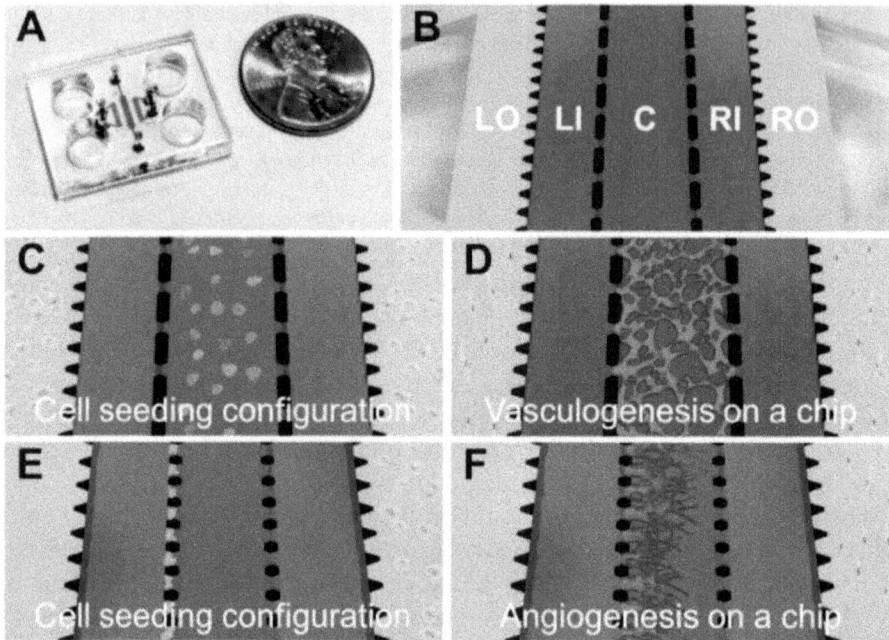

Figure 2.7. Microfluidic chip design and cell-seeding configurations for microvascular network and angiogenic sprout formation. (a) Photograph of the microfluidic chip, filled with coloured fibrin matrix. (b) Schematic of the microfluidic channels partitioned by microposts. The central channel (C, blue) is flanked by two fluidic channels (LI and RI) and two outside stromal cell culture channels (LO and RO). (c, d) Cell-seeding configuration for the vasculogenesis experiment. Human umbilical vein endothelial cells (HUVECs) are embedded in a 3D fibrin matrix placed in the central channel, and LFs with fibrin matrices are placed in the LO and RO channels. (e, f) Cell-seeding configuration for the angiogenesis experiment. HUVECs are coated on the side of central channel that is filled with acellular fibrin matrix. Reprinted from [9] with permission from the Royal Society of Chemistry.

co-culture cannot be performed. Here, by introducing the vascular structure by organ-on-a-chip and performing mechanical environmental control, Caco-2 cells form a structure similar to three-dimensional intestinal villi, and accordingly, intestinal bacteria co-culture with and became feasible. By using this device, it has been difficult to evaluate the drug evaluation model for pathogenic bacteria, the bacterial overgrowth model caused by the load of mechanical stimuli, and the functional evaluation of bacteria in the immune response. Various experiments have become feasible in an *in vitro* environment. In other words, it can be said that it has become possible to construct ideal model experiments, and drug discovery and drug evaluation can be performed more quickly and accurately. Furthermore, it is a very useful model experimental system because it can evaluate safety more accurately before clinical trials of drug administration to the human body or living organisms. It has been shown that in the functional reproduction of the intestine, imitation of the three-dimensional structure of the tissue is important for maintaining the function, and it is very important to introduce the vascular structure into the microchannel system.

The utilization of the gut-on-a-chip described above is performed to obtained deep understanding of contributions of microbiome and mechanical deformation to intestinal bacterial overgrowth and inflammation, as shown in figure 2.8 [12]. Although higher levels of intestinal differentiation can be obtained using recently developed 3D organoid cultures, it is not possible to expose these cells to physiological peristalsis-like motions or living microbiome in long-term culture. There are some big problems to solve. The major limitation is that bacterial overgrowth occurs rapidly compromising the epithelium. To develop an experimental model that overcomes the limitation, they prepared a human gut-on-a-chip microfluidic device that enables human intestinal epithelial cells (Caco-2) to be cultured in the presence of physiologically relevant luminal flow and peristalsis-like mechanical deformations. Readers who are interested in their studies should read some of the references for more details. Here, only the outline is shown below. They analyzed how probiotic and pathogenic bacteria, lipopolysaccharide, immune cells, inflammatory cytokines, vascular endothelial cells and mechanical forces contribute individually, and in combination, to intestinal inflammation, villus injury, and compromise of epithelial barrier function.

Figure 2.8(a) shows an intestine-on-a-chip (Intestine Chip) composed of optically clear, flexible PDMS polymer with three parallel hollow microchannels. The central channel is split into an upper (lumen) and lower (capillary) channel by a flexible, ECM-coated PDMS membrane containing an array of pores (7 μm diameter, 40 μm spacing) lined by human Caco-2 intestinal epithelial cells. Here, ECM is composed of

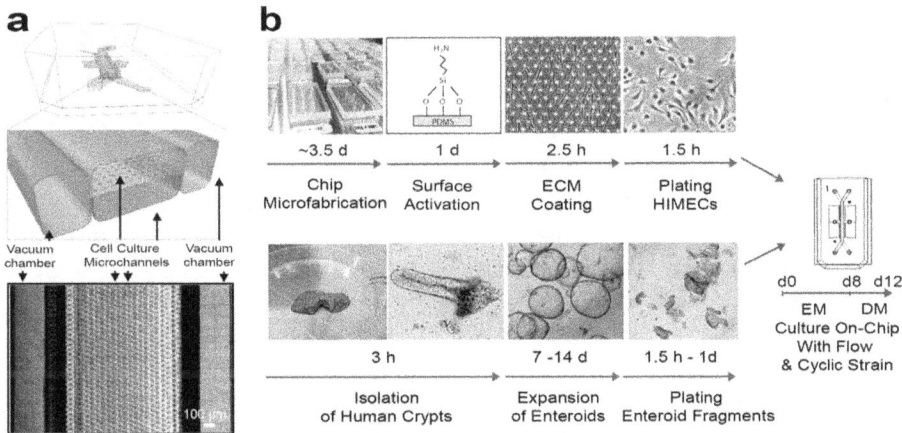

Figure 2.8. Fabrication of the primary human Intestine Chip. (a) A schematic cross-sectional view (top) and a phase contrast micrograph of the chip viewed from above (bottom) showing the upper (epithelial; blue) and lower (microvascular; pink) cell culture microchannels separated by a porous, ECM-coated, PDMS membrane sandwiched in between. The membrane is elastic and can be extended and retracted by the application of cyclic vacuum to the hollow side chambers. This actuation causes outward deflection of the vertical side walls and lateral extension of the attached horizontal, porous elastic membrane, which induces mechanical deformation of the adherent tissue layers cultured in the central channels. (b) Schematic representation of the step-by-step procedure involved in the establishment of microfluidic co-cultures of primary human intestinal epithelium and intestinal microvascular endothelium in the Intestine Chip. (Reprinted from [12] CC BY 4.0.)

type I collagen and Matrigel. Each microchannel has a dedicated inlet and outlet for the inoculation of human cells, molecules or microbes as well as for the precise control of physicochemical parameters through the perfusion of laminar flow of appropriate culture medium. In addition, to analyse human intestinal physiology and pathophysiology in a more *in vivo*-relevant culture microenvironment, the biopsy-derived epithelium, intestinal endothelial cells, physiological fluid flow and peristalsis-like mechanical motions are integrated in the chip. They discovered that the loss of physiological mechanical deformation is itself sufficient to induce small intestinal bacterial overgrowth similar to that observed in some patients with ileus and inflammatory bowel disease (IBD) by leveraging the modular capability of gut-on-a-chip to vary mechanical parameters independently. Figure 2.8(b) shows bacterial overgrowth induced on-chip by cessation of peristalsis-like mechanical deformations. According to references [11, 12], the 10% in cell strain with 0.15 Hz in frequency can cause drastic decline in the bacterial density. The strain-induced mechanical deformation plays a significant role in preventing the overgrowth of bacteria. It was well understood by making the device that the incitement movement in the actual intestine contributed greatly to the balance of the intestinal bacteria [13, 14].

In addition, for mimicking devices such as the liver and kidney, the introduction of vascular structure is indispensable for growing into a three-dimensional tissue and closer to the actual situation inside the living body. The following is an example of kidney-on-a-chip, a kidney mimicking device [15]. In the renal tubules, epithelial cells are exposed to shear stress and osmotic gradient due to solution flow. This environment has become a reproducible environment with the recent development of organ-on-a-chip. As introduced in gut-on-a-chip, by culturing proximal tubular epithelial cells on a multi-layered thin film in the device, the cell side of the thin film is on the luminal side and the lower side is on the stromal side. By applying a shear stress similar to that in the living body only on the epithelial cell side, the proximal tubule functions. Using this device promotes cell polarity and primary pili formation, increases albumin transport and glucose reabsorption, and is toxic to similar drugs *in vivo*, compared to conventional culture techniques. The effects can be reproduced. By further applying these devices, it can be developed into a glomerular model. Glomerular clearance function by co-culturing glomerular epithelial cells and endothelial cells induced by iPS cells via a porous thin film and loading similar shear stress growth trees and tension stimuli *in vivo* [16]. Moreover, drug toxicity can be also evaluated by using these methods based on the microfluidic systems. By administering a drug to the flow path on the blood vessel side, it becomes possible to evaluate the functional change of the glomerular epithelial cells, and it is possible to evaluate the glomerular function similar to that in the living body and to model the disease. The glomerulus is a tissue responsible for blood filtration, and the introduction of vascular structure is indispensable. In recent years, kidney mimicry devices using three-dimensional culture have also been developed. A proximal tubular model incorporating a vascular network has been developed by using a 3D structure construction technology using a 3D printer, as shown in figure 2.9 [8]. As shown in figure 2.8, two tube mechanisms are created in the gel, and epithelial cells and intravascular

Step 1. Bioprinting of microfibrous scaffold encapsulating endothelial cells.

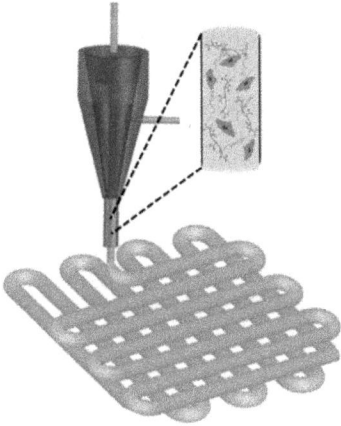

Step 2. Formation of the endothelialized structure and the vascular bed.

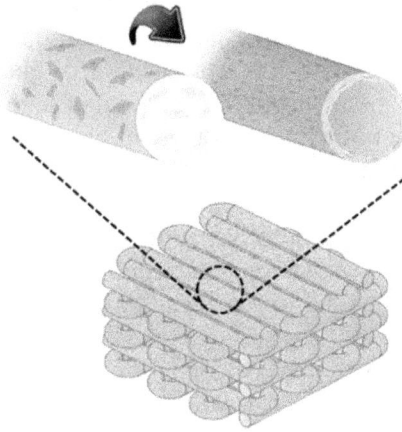

Step 3. Seeding with cardiomyocytes.

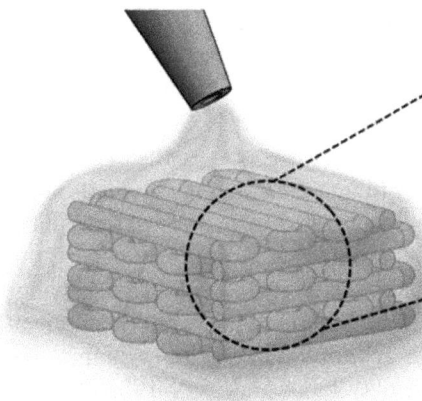

Step 4. Formation of endothelialized myocardium.

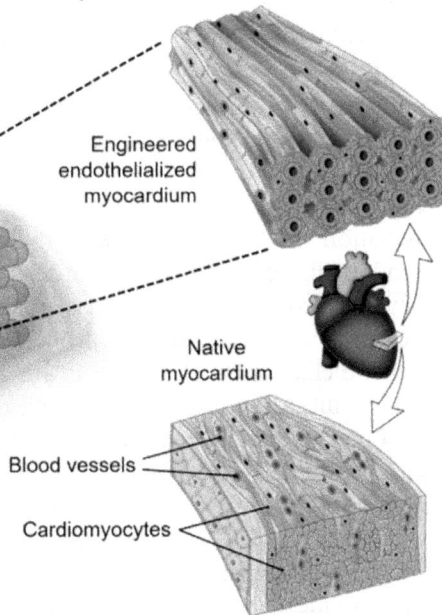

Engineered endothelialized myocardium

Native myocardium

Blood vessels

Cardiomyocytes

Figure 2.9. Schematics showing the procedure of fabricating endothelialized myocardium using the 3D bioprinting strategy. Step 1: bioprinting of a microfibrous scaffold using a composite bioink encapsulating endothelial cells. Step 2: formation of the vascular bed through migration of HUVECs to the peripheries of the microfibers. Step 3: seeding of cardiomyocytes into the interstitial space of the endothelialized scaffold. Step 4: formation of engineered endothelialized myocardium structurally resembling the native myocardium. Reprinted from [8], copyright 2016 with permission from Elsevier.

cells are cultured so as to cover the surface of each tube structure, and the two network structures of epithelial cells and endothelial cells are formed. The distance and shape between each network can be controlled arbitrarily, and the structure in the living body can be imitated. With the development of this system, the absorption of albumin and the reproduction of glucose reabsorption function can be reproduced. It was shown that a model of dysfunction of vascular endothelial cells due to the effect of hyperglycemia can be constructed.

Even in the kidney model chip [16], when iPS cell-derived kidney organoids are cultured in an environment loaded with mechanical stimulation by flow, the number of vascular endothelial progenitor cells increases and the formation of vascular network depends on the magnitude of shear stress. It turns out that promotion occurs. In other words, it can be seen that the mechanical environment is a factor for inducing the original functions of cells. Therefore, it is indispensable to reproduce the *in vivo* environment by organ-on-a-chip, and in order to evaluate a drug that reflects the human body, an organ that can maintain cell morphology and cell function similar to that in the living body. It can be said that the creation of organ-on-a-chip is indispensable. In addition, many organ-on-a-chips have been produced for the purpose of constructing a cancer model incorporating a device that mimics the blood–brain barrier and a vascular network. The migration of cancer cells from the vascular network to the outside is closely related to the metastasis of cancer, and its mechanism investigation and control are attracting attention. By inducing angiogenesis in the gel in the microfluidic device to construct a vascular network that can be perfused, and by introducing cancer cells into the blood vessels and culturing them, an extravasation model of cancer is also constructed. In this device, the extravasation of cancer cells can be quantitatively evaluated by imaging, so that the effect of promoting the extravasation of cancer cells due to the deterioration of the barrier function of vascular endothelial cells by inflammatory cytokines can be evaluated. Using this device, they revealed that integrin β1 has an important role in extravascular migration and is an essential factor for passing through the basement membrane underneath the vascular endothelial cells. Here, there are research reports that cannot be fully introduced. As introduced here, the readers may have understood that the artificial reconstruction of the microenvironment around the cell is essential for the evaluation of the cell and the induction of its inherent function. In future research including medical applications and science, it will be very important to construct a co-culture system with immune cells such as cancer-related fibroblasts and macrophages in addition to the vascular network. The development of such a reconstructed model is expected to provide a more reliable *in vitro* evaluation system.

The development of new drugs has been a long and tedious process of chemical synthesis of potential drugs and a steady and diligent study of their efficacy. In other words, the process of drug discovery and development is a very inefficient, long and costly process. Indeed, establishing a single new drug usually requires several years of research and hundreds of millions of dollars/euros in funding. New drug development involves specialized personnel, a strictly regulated clinical phase and a validation procedure characterized by high risk in front of such a huge investment [17]. Success rates for new compounds are as low as 5%, and significant resources are lost every year. Furthermore, another key issue in the search for novel drugs to improve therapeutic efficacy relates to the availability of effective routes of drug administration to patients, such as oral/rectal, inhalation, transdermal and intravenous administration. Organ-on-a-chip devices can provide useful support in both of these directions. It is now reported that organ-on-a-chip can be created for most organs, tissues, etc. This is because organ-on-a-chip can reproduce the intrinsic and extrinsic features of an organ, tissue or structure, its microenvironment and biological barriers, and can test drug efficacy, solubility, permeability, targeted delivery and toxicity in a more appropriate and reliable manner.

Today, pharmaceutical research relies on conventional *in vitro* 2D cultures and animal experiments. Cell culture and animal experiments are inadequate to reproduce human physiology and therefore cannot adequately predict the clinical efficacy, toxicity and side effects of therapeutic agents in humans. For these reasons, organ-on-chip technology has recently attracted attention as an alternative platform for drug development and research is growing rapidly worldwide. This interest is expected to lead to a paradigm shift, as summarized in figure 2.10 [18], with the

Model properties/ applications	Animals	2D cell cultures	3D cell cultures	Spheroids/ Organoids	Single OOC	Multiple OOC
Biomimetics / Recapitulation	(unclear)	✗	(unclear)	✓	✓	+
Complexity	✓	✗	(unclear)	✓	✓	+
Disease models	(unclear)	(unclear)	(unclear)	✓	✓	✓
Drug research	(unclear)	(unclear)	(unclear)	✓	✓	+
Cell-cell interactions	✓	✗	✓	✓	✓	✓
Organ-organ interactions	✓	✗	✗	✗	✗	✓
Vascularization	✓	✗	✗	(unclear)	(unclear)	✓
Integration of biosensors for real time	(unclear)	✓	✓	✓	✓	✓
Throughput	✗	✓	✓	✓	✓	✓
Costs	✗	✓	✓	(unclear)	(unclear)	(unclear)
Ethics	✗	✓	✓	✓	✓	✓

Figure 2.10. Advantages of single and multiple organ-on-a-chip in comparison to other methodologies. Reprinted from [18] CC BY 4.0.

potential to be used for high-throughput, high-content and resource-efficient screening [19, 20]. It is important to review recent innovations and advances in this area in order to capture the opportunities presented by the most relevant disease models that are becoming accepted by pharmaceutical companies. As presented here, organ-on-a-chip is gaining acceptance by pharmaceutical companies as a new tool that can accelerate the drug development process to more accurate and ultimately personalized standards.

2.5 Fundamentals of microfluidic channel: surface tension and surface energy

As described above, LOCs are well used in chemical analysis as well as in bioscience and biotechnology. They consist of microfluidic channels, whereas the basic structure of a microfluidic channel is a capillary tube. In constructing the LOC, microfluidic channels are formed and combined. So, the first step is to develop a basic understanding of capillaries. In this section, basic knowledge on the handling of capillaries forming microfluidic channels should be studied. The basic principles and mathematical treatment of fluid mechanics are also summarized in chapter 4. In the following, the basic principles and handling of surface tension and surface energy in relation to capillaries are discussed [29–31].

When liquid is added to a capillary or test tube, a meniscus is formed. A meniscus is a liquid bridge that forms in a very small gap between objects. For example, if you pour water on a table, the water will form round droplets due to surface tension. If you put your finger close to this drop of water, when your finger and the drop get close to a certain distance, the drop of water jumps to your finger and forms a bridge.

Surface tension is the property of a liquid or solid to make its surface as small as possible, and is a type of interfacial tension. Quantitatively, it expresses the surface free energy per unit area, with units mJ m^{-2}, dyn cm^{-1} or mN m^{-1}. The symbols γ and σ are often used. The interface is the boundary where one liquid or solid phase is in contact with another. If one of these phases is a liquid or solid and the other a gas, the interface is called the surface. Historically, the 1805 report by Thomas Young, 'An Essay on the Cohesion of Fluids', marked the beginning of the study [31].

Attractive forces, known as intermolecular forces, act between molecules. Molecules in a liquid have low free energy due to the action of intermolecular forces from other molecules from all directions. Molecules on the surface, on the other hand, are acted upon by the molecules inside, but hardly by the molecules in the gas. This means that the molecules on the surface have a larger free energy than those inside and are more unstable. As a result, a tendency to make the surface as small as possible emerges. Surface tension increases the more unstable the interface is, and therefore varies depending on the influence of surfactants and other agents. There are various theoretical methods for determining surface tension, but here it is defined from a thermodynamic and statistical mechanics perspective. Considering the interface as shown in figure 2.11(a), microscopically it is formed by a group of molecules. To stabilize the interface, we consider the molecular population energy,

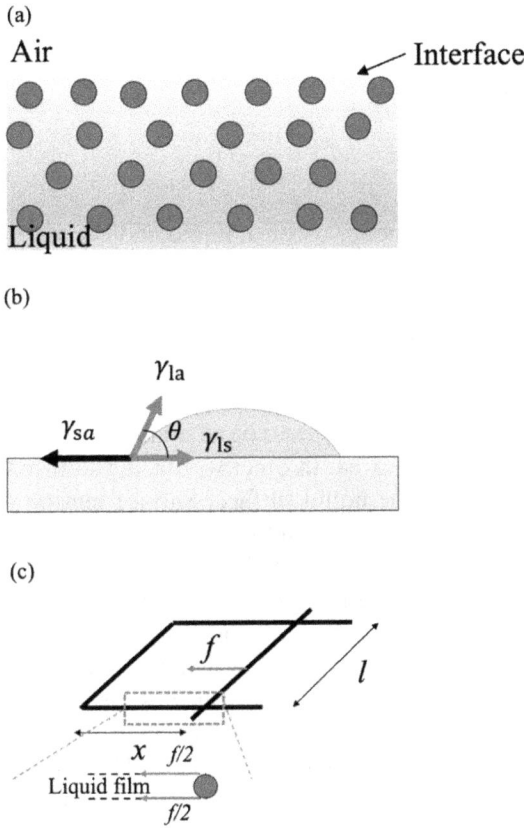

Figure 2.11. (a) Schematic diagram of the structure and arrangement of molecules and atoms near the interface between air and liquid. (b) Schematic diagram showing the forces acting at the solid/liquid/air interface, respectively, and their relationship. Relationship between wetting phenomena and contact angle. (c) Relationship between energy and forces; examples of liquid films stretched on U-shaped and I-shaped wires.

so we introduce the Gibbs free energy, G. For example, considering a gas–liquid interface, the surface area created by the liquid is A, and considering a state of thermal equilibrium with constant temperature T and pressure P, for a liquid surface, the energy required to create a new surface is γ (also known as surface energy). Then the following relationship holds [29, 30]:

$$\gamma = \left(\frac{\partial G}{\partial A} \right)_{T,\ P\ \text{eq.}}, \tag{2.6}$$

where the subscripts denote a state of thermal equilibrium with constant temperature T and pressure P. When described by the Helmholtz free energy, it can also be described as follows:

$$\gamma = \left(\frac{\partial F}{\partial A} \right)_{T,\ V\ \text{eq.}}, \tag{2.7}$$

where the subscripts denote a state of thermal equilibrium with constant temperature T and volume V. Described in terms of distribution function, Z, equation (2.7) are written as follows:

$$\gamma = -kT\left(\frac{\partial \ln Z}{\partial A}\right)_{T,\ V\ \text{eq.}}.$$ (2.8)

The surface of water is the boundary between liquid and gas, i.e. the interface. Outside the interface, the bonds between water molecules are broken and either do not exist or are significantly reduced. On the other hand, the bonds in the inward direction of the liquid are strong, so water molecules on the surface are drawn inwards. This inward force must be counteracted for the internal water molecules to make the surface. There is thus a surface energy at the air–liquid interface as schematically illustrated in figure 2.11(a). As the surface energy tries to be as small as possible, it tries to become a sphere with the smallest surface area. In other words, the surface tension of a liquid is the force that tries to shrink the liquid surface. Surface tension is defined as surface energy per unit area because it is a quantity conceived in this way. Surface tension therefore has an 'energy/area' dimension, but is generally expressed in terms of a 'force/length' dimension. Surface tension considered in terms of 'energy/area' is called surface free energy. The units for surface tension are mN m^{-1} (millinewtons per metre) in the International System of Units (SI) and dyn cm^{-1} (dyne per centimetre) in the CGS system of units. Except for liquid metals such as mercury, water has the highest surface tension of all liquids, approximately 72.7 mN m^{-1} at room temperature. Mercury has a surface tension of about 480 mN m^{-1}. The surface tensions of organic solvents such as n-hexane and acetone are 18.4 and 23.3 mN m^{-1}, respectively. As temperature increases, the surface tension decreases because thermal motion reduces the cohesive energy between molecules. As temperature increases, the surface tension decreases because thermal motion reduces the cohesive energy between molecules. The three tension forces, γ_{ls} (the droplet-substrate interfacial tension), γ_{sa} (the substrate-air interfacial tension), and γ_{la} (the droplet–air interficial tension) start at the points on the solid substrate surface where the three-phase interfacial cross sections of the contact between the droplet and the air are located. γ_{sa} and γ_{ls} are located on the solid substrate surface, and rests are located at the tangent line of the droplet surface at the contact angle of θ, as shown in figure 2.11(b). θ is greater than 90° for a hydrophobic material, and less than 90° for a hydrophilic material. The shape of the droplet on the solid surface is determined by the combination of forces, γ_{sa}, γ_{ls}, and γ_{la}, corresponding to the surface energies per unit area from the interfaces of solid-air, liquid–solid, and liquid-air, respectively. Then, the following relation holds

$$\gamma_{sa} = \gamma_{ls} + \gamma_{la}\cos\theta.$$ (2.9)

This equation (2.9) is an important basic equation for considering wetting phenomena and is called Young's equation.

Let us consider again the meaning and definition of surface tension and surface energy. In preparation for this, the relationship between force and energy is first reviewed. Force can be defined by the gradient of energy. The gradient is obtained from the derivative with respect to position. Defined a force f, energy E and x as a one-dimensional position, the force f is given by

$$f = -\frac{dE}{dx}. \tag{2.10}$$

The direction of the force f is taken in the direction of the x-axis, so the negative sign means that the force acts in the direction of decreasing energy. It can also be said to be the energy stored when a unit length of displacement is applied against the force, i.e. when work is applied. Consider a structure consisting of a U-like-shaped wire and an I-shaped wire with a liquid membrane attached to it, as shown in figure 2.11(c). Imagine a film of soapy water, for example. The energy required to create a unit surface is defined as surface energy and is given by γ.

As the liquid thin film has an area, $2lx$ because surfaces exist on both upper and lower surfaces, the energy of this system is

$$E = \gamma \cdot 2lx \tag{2.11}$$

higher than in the absence of liquid thin film. Here, the state with no liquid thin film is assumed to have zero energy. The force acting on the I-shaped wire due to surface energy is given by

$$f = -\frac{dE}{dx} = -2\gamma l. \tag{2.12}$$

This force is the combined force acting on both the top and bottom surfaces, as seen in the cross-sectional diagram of the liquid thin film, so the force acting on one surface is half of equation (2.12), that is γl. Therefore, the force acting per unit length is therefore γ. This suggests that the magnitude of the surface energy may be considered as the force acting per unit length.

Assume that an object with a freely deforming volume V and surface area A is in equilibrium with the external world at temperature T and pressure P. For the sake of essential understanding, curvature and the like are ignored. Let us define 'the tension acting per unit length of the surface as γ'. Then the tension γ is the energy that can be extracted as work, which is equivalent to defining it using free energy. Let U_{tot} and S_{tot} be, respectively, the total internal energy and total entropy of a system consisting of objects and surfaces of properties, from the first and second laws of thermodynamics, the following relation is obtained:

$$dU_{tot} = TdS_{tot} - pdV + \gamma dA. \tag{2.13}$$

As there are bulk and surface contributions, they are described separately as follows:

$$U_{tot} = U_{bulk} + U_{surface} \tag{2.14a}$$

and

$$S_{tot} = S_{bulk} + S_{surface}. \tag{2.14b}$$

For parts of the bulk, not related to the surface, the following relation:

$$dU_{bulk} = TdS_{bulk} - pdV. \tag{2.15}$$

must hold. Subtracting both sides of equations (2.13) and (2.15), only the surface contribution can be considered, yielding the following equation:

$$dU_{surface} = TdS_{surface} + \gamma dA. \tag{2.16}$$

Since these surface contributions of internal energy and entropy should be proportional to surface area (higher-order proportionality is also possible, but as a zero-order approximation, a proportional relationship is suitable), the following relations can be defined:

$$U_{surface} = U^{\gamma}_{surface} \cdot A \tag{2.17a}$$

and

$$S_{surface} = S^{\gamma}_{surface} \cdot A. \tag{2.17b}$$

$U^{\gamma}_{surface}$ and $S^{\gamma}_{surface}$ are defined here as the proportionality coefficients. From the above equations (2.13)–(2.17), the following relationship can be obtained:

$$\left(U^{\gamma}_{surface} - TS^{\gamma}_{surface} - \gamma\right)dA + A\left(dU^{\gamma}_{surface} - TdS^{\gamma}_{surface}\right) = 0. \tag{2.18}$$

Since this relationship must hold for any surface area, from the first term

$$\gamma = U^{\gamma}_{surface} - TS^{\gamma}_{surface} \tag{2.19}$$

is obtained. This equation (2.19) corresponds to the Helmholtz free energy with respect to the surface. From this relationship (2.19), it can be seen that the surface tension coincides with the surface energy at absolute zero. Next, from the second term, the following relation is obtained:

$$dU^{\gamma}_{surface} = TdS^{\gamma}_{surface}. \tag{2.20}$$

Furthermore, the equation (2.19) is totally differentiated on both sides, we obtain

$$d\gamma = dU^{\gamma}_{surface} - TdS^{\gamma}_{surface} - S^{\gamma}_{surface}dT. \tag{2.21}$$

Combined with equation (2.20) and (2.21), the following relation is obtained

$$S^{\gamma}_{surface} = -\frac{d\gamma}{dT}. \tag{2.22}$$

This is a relation that can be derived by considering only thermodynamics macroscopically. It is also consistent with the equation from a microscopic statistical mechanics point of view. In other words, the definition that surface tension is a force acting per unit length of surface (available for work extraction) is reaffirmed as described in equations (2.7) and (2.8).

2.6 Capillary as an elementary technology for lab-on-a-chip

Below are some of the elemental technologies used to construct the LOC. Although there are many elemental technologies that cannot be introduced here, the basic elements are those related to transportation and mixing. These basic elements are combined to form the LOC. In order to construct a LOC system, it is also necessary to know the unit elements that make it up. Research and development on unit elements has a large scientific aspect. On the other hand, it is also necessary to consider how to control the functionality and performance characteristics of the unit element and how best to incorporate it into the system. In this section, we will learn about the characteristics and structure of unit elements necessary to reach systems thinking.

For transport, there are methods such as capillary action and syringe pumps. A syringe pump is a method in which liquid is mechanically pushed out with a syringe, and is often used in microchemical systems by connecting it to a chip using tubing. The system shown in figure 2.1 also uses a syringe pump to pump liquid. This method is very good in that the control is well established and can be easily systematized. On the other hand, it is not suitable for use in a confined space or in an outside environment because of the wiring caused by tubing. In addition, there is the problem of unused reagents remaining in the tubes. However, it is still the most commonly used method because it is the simplest and quickest to form.

Capillary action is a physical phenomenon in which liquid inside a capillary tube (tubular object) moves through the tube without being given energy from outside. It is explained that when a cloth is immersed in water, the water rises through the cloth to a higher level than the liquid level due to this phenomenon. When a capillary with a small cylindrical cavity is inserted into a puddle of water, the water surface position in the capillary rises as shown in figure 2.12(a). When a capillary is inserted into a mercury (liquid at room temperature) puddle, the free surface position in the capillary is lower than the surrounding liquid surface position, as schematically

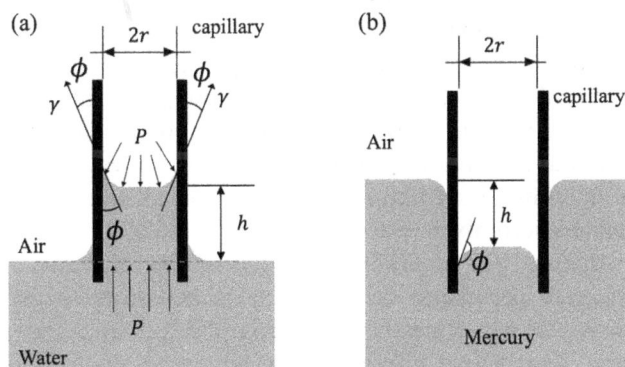

Figure 2.12. (a) Schematic diagram of capillary action. (a) When a pipe with a small cylindrical cavity (capillary) is inserted into a puddle of water, the water surface position in the capillary rises. (b) When a capillary is inserted into a puddle of mercury (liquid at room temperature), the free surface position in the capillary is lower than the surrounding liquid surface position.

illustrated in figure 2.12(b). The height of the liquid rise is determined by the surface tension, the wettability of the wall surface, and the density of the liquid.

Surface tension exerts a force on the liquid surface in the direction of shrinkage. The inclined liquid surface near the wall surface tries to contract, and as a result, the water surface is lifted. In other words, the force of the liquid to rise is equal to the vertical component of the surface tension near the wall. The liquid surface rises until the above two forces and the weight of the lifted liquid are in balance. Assuming the case of a narrow liquid column as shown in figure 2.12(a), the cross-sectional area of the column is very small. The weight of the liquid is determined by ρgrh, where ρ, r, g, and h are the density of liquid, radius of tube, local acceleration due to gravity, and height of a liquid column, respectively. As shown in figure 2.11, the liquid is lifted by the surface tension described above. Estimated capillary rise, h, is expressed by the following equation [29–31]:

$$h = \frac{2\gamma \cos \phi}{\rho gr}, \qquad (2.23)$$

where γ is the surface tension coefficient, ϕ is the contact angle. This is a simplified formula assuming that the free surface profile is spherical. For example, if a capillary tube with a radius of $r = 1$ mm is immersed in water, the capillary rise from the reference water surface position can be estimated using the equation (2.23). Here, in the case of parameters: density of water is $\rho = 1000$ kg m^{-3}, gravitational acceleration $g = 9.8$ m s^{-2}, surface tension $\gamma = 0.072$ N m^{-1} between water and air, and contact angle $\phi = 30°$, the capillary rise h is estimated to be 12.7 mm. These values have been found to be in approximately good agreement in both experiments and simulations. Thus, if the estimates are known in advance by mathematical equations in a simplified model, it is easier to estimate the size of the overall LOC and to consider what functions can be extracted from it.

2.7 Passive micromixer and particle diffusion

When considering microfluidic channels based on capillaries, we have to consider mixing processes to facilitate chemical reactions, which are important in chemical operations. In microfluidic channels, laminar flow dominates due to the low Reynolds number. In laminar flows, liquids are easily transported and weighed, while mixing is difficult. Therefore, various micromixers have been studied based on the properties of microfluidics. Micromixers can be broadly divided into passive and active types. Active micromixers use external energy such as ultrasonic waves and temperature for mixing, but generally have high mixing efficiency. However, it is necessary to integrate peripheral devices such as actuators to generate external energy. Therefore, the structure becomes complicated and the manufacturing process increases. Also, active mixers with mechanisms such as ultrasound and high temperatures can damage biological fluids, so active micromixers are less popular for biochemistry and chemistry applications. Passive micromixers are based on the force that pushes the fluid and the design of the specially shaped flow path. This design is intended to reduce the diffusion distance of the fluid and increase the contact area. Passive mixers are cheaper and easier to manufacture than active micromixers. As examples of active micromixers, examples using surface acoustic waves and those using dielectrophoresis/electroosmotic flow will be introduced in each chapter. Passive micromixers include those that use continuous or parallel layer fluids and those that enhance chaotic advection. In the following, a Y-type micro-mixer using a parallel layer fluid will be described in terms of ease of design and structural adaptation. Figure 2.13(a) shows the outline of the basic design of the Y-type micromixer. The inlet is divided and n sub-channels are recombined into the main channel to form a laminar flow. The fluid is mixed by flowing in the main flow path while still in contact. The shape of this type of mixer enhances the mixing process by reducing the diffusion distance and amplifying the contact area between the two fluids. Therefore, the main flow path needs to have a shape having a narrow width and a long overall length. Also, the fluid needs to flow slowly. Two-dimensional particle diffusion is expressed as follows [29, 30]:

$$x^2 = 2Dt, \tag{2.24}$$

where t is the average time that the particles diffuse over the distance of x, and D is the diffusion coefficient given by

$$D = \frac{kT}{6\pi\mu R}, \tag{2.25}$$

where k is the Boltzmann coefficient, T is the absolute temperature, R is the radius of the particle (or molecule), and μ is the intermediate viscosity. Typical standard value

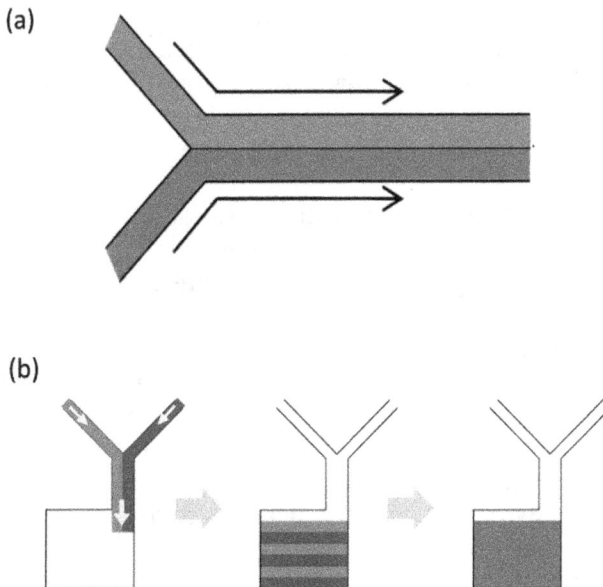

(a)

(b)

Figure 2.13. (a) Schematic diagram of Y-shaped flow channel. (b) Schematics of mixing.

of the diffusion coefficient when the molecule is very small at room temperature is about 10^{-9} m^2 s^{-1}. When this is applied to the flow path of microfluidics, x is the width of the flow path through which the fluid to be mixed flows. Therefore, by narrowing the flow path width, the time required for diffusion can be reduced. Further, according to the equation, since the diffusion time is proportional to the square of x, the time required for mixing can be significantly reduced by narrowing the flow path width. In addition, each fluid in n layers accelerate mixing with the element of n^2, and the mixing time is

$$t = \frac{x^2}{2n^2 D}.$$
(2.26)

When designing a Y-type micromixer, it is necessary to determine the width and length of the flow path in consideration of the time required for particle diffusion. As an example, let us consider the transport and mixer of liquid in a LOD. Since the LOD used here integrates structures other than the mixing structure, there are restrictions on the number of revolutions used and the location where it is installed. Therefore, it is difficult to design the liquid to be completely mixed in the Y-shaped flow path. Therefore, as shown in figure 2.13(b), the diffusion distance is reduced to some extent in the Y-shaped flow path, and the final mixing is performed by leaving the liquid in the reservoir. The widths of the sub-flow path before the introduction of the liquid in the Y-type flow path and the main flow path in which the liquid is layered and mixed are set to 250 and 500 μm, respectively. This value takes into account the ease of flow of the liquid and the time of particle diffusion. With this design value, the liquid is calculated to be mixed in about 30 s.

The device used for the mixing structure experiment is shown in figure 2.14. There are two reservoirs inside the circle and one outside the circle. There is a mixing structure between the reservoir inside the circle and the reservoir outside the circle. The mixing structure is a Y-shaped flow path structure. Below the reservoir inside the circle is a capillary valve. The capillary valve is designed to break at the same number of revolutions. The details of capillary valve in the LOD are described in chapter 5.

Let us examine the mixing of liquids with and without the mixing structure. When using the mixing structure, put the red and blue liquids in the two reservoirs inside the circle, increase the rotation speed above the breaking speed of the capillary valve, and send the two liquids to the mixing structure. When using the mixing structure, the two reservoirs inside the circle are filled with red and blue liquid, respectively, the rotations per minute (rpm) is increased above the breaking rpm of the capillary valve and the two liquids are pumped into the mixing structure. If the mixing structure is not used, the inner circular reservoir is filled with red liquid and the outer circular reservoir with blue liquid, the rpm is increased above the breaking rpm of the capillary valve and the liquid is pumped from the inner circular reservoir into the outer circular reservoir.

Figure 2.14 shows photographs of the pumping behaviour of those without a mixing structure and those with a mixing structure. In the case where the mixing

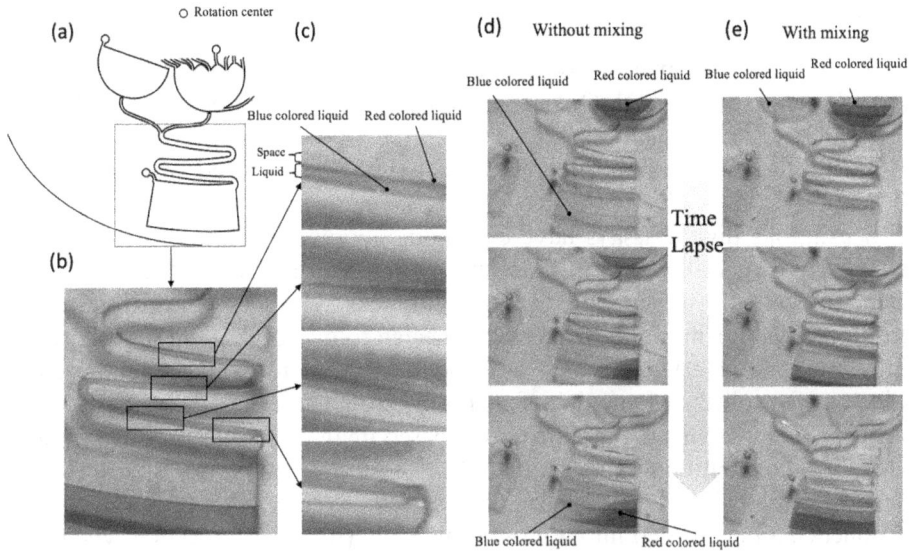

Figure 2.14. (a) Schematic structure for mixing in LOD. (b) Photograph of meandering microchannel for mixing. (b) Magnified image of the meandering channel: near the entrance in the Y-shaped mixer, the red coloured water and blue coloured water in the microfluidic channel do not mix and remain in a laminar flow state. It can be seen that the blue and red coloured water mix as the liquid progresses through the channel and is pumped in the direction of the larger tank in the bottom of the figure. In the other words, it can be seen that diffusion mixing is occurring. (d) Results of centrifugal pumping with blue coloured water in the liquid reservoir tank and only red coloured water in the Y-shaped mixing channel. It can be seen that the blue coloured water and red coloured water are not mixed in the liquid reservoir but remain separated. (e) Centrifugal pumping of blue coloured water and red coloured water into the Y-shaped mixing channel shows that the two liquids mix as they progress through the channel and that a mixed state of the two liquids is formed in the reservoir tank.

structure was not used, the red and blue liquid phases remain separated, indicating that the liquids are not mixed. For those using the mixing structure, no phase separation of the liquids can be seen, indicating that the two liquids have been mixed. Therefore, it was confirmed that the incorporation of the mixing structure ensures that the mixing is efficient. A magnified view of the mixing structure is shown in figure 2.14. It can be seen that the liquid, which was initially divided into blue and red phases, is mixed as the phases disappear as it flows through the mixing structure. It can also be seen that the width of the liquid is not constant and that spaces are opening up in the flow path. In the design, mixing was not completed within the mixing structure and was supposed to be mixed after the liquid entered the reservoir, but it appeared that mixing was completed in the channel. The time for the liquid to pass through the mixing structure was about 0.2 s, which was considerably shorter than the design value of 30 s mixing time.

Although only passive diffusion mixing is presented here, LOD can also utilize active methods of mixing by Euler force by controlling the rotational force. Details are given in chapter 5.

2.8 Recent integrated microsystems

In terms of the embodiment of system thinking, two examples of integrated systems in recent years are presented. One of the expected features of microchemical systems is *in situ* diagnostics and *in situ* analysis. Researchers and engineers around the world are therefore aiming to build microchemical systems combined with smartphones and WiFi.

For example, digital assays have enormous untapped potential for diagnostics, environmental surveillance, and biosafety monitoring. However, due to the instrumentation necessary to generate, control, and measure millions of droplets, normal laboratory experimental microsystems connected to microtubes fail to fulfil requirements for digital assays, in which biological samples are compartmentalized into millions of femtoliter-volume droplets and interrogated individually. Yelleswarapu *et al* succeeded in achieving mobile phone-based imaging technique that was >100x faster than conventional microfluidic droplet detection, which did not require expensive optics [20]. Their microfluidic system is composed primarily for four components, all integrated into a monolithic chip: (i) a bead processor where beads are incubated and washed in successive steps, (ii) droplet generators, (iii) a delay line for the enzymatic amplification reaction, and (iv) the fluorescence detection region. The system is consisting of polycarbonate filter sandwiched between laser-cut layers of adhesive-coated Mylar and PDMS piece which contains the droplet generators fabricated using multilayer soft lithography. Their system can simultaneously measure multiple fluorescent dyes in droplets. By using this time domain modulation with cloud computing, they overcome the low frame rate of digital imaging, and achieve throughputs as high as 1 million droplets per second.

Recently, a number of examples have been reported of disease screening tests in microchemical systems using droplet manipulation. One example is the report by Yan *et al* [19] As shown in figure 2.15, their group focused on the diagnosis of early-stage Parkinson's disease and formed a platform based on microchemical systems. They created a platform that developed on-chip droplet confinement fluorescence methods and used super-resolution direct stochastic optical reconstruction microscopy (dSTORM) to identify a-synuclein on the membrane of L1CAM+ extracellular vesicles (EVs) immunocaptured from human serum. Using this platform, conditioned medium from neuroblastoma cells expressing a-synuclein mutants or patient-derived induced pluripotent stem cell (iPSC) neurons with a-synuclein gene triplets was used to increase the association of a-synuclein with the L1CAM+ EV surface under pathological conditions. This platform facilitates the assessment of EV membrane proteins and has potential application as a diagnostic method for various clinical indications.

Recently, an unconventional approach to common LOD that combines a quadcopter propulsion system, a miniaturized 2.4 GHz WiFi spy camera, 9.74 Watt Qi wireless power, and an Ardiuno into an open-source, miniaturized All-in-one Powered LOD Platform has been demonstrated by Serioli *et al*, as shown in figure 2.16 [21]. What they have shown is that it is possible to build an entire system, including a microchannel and its control system, using only small, low-cost, readily

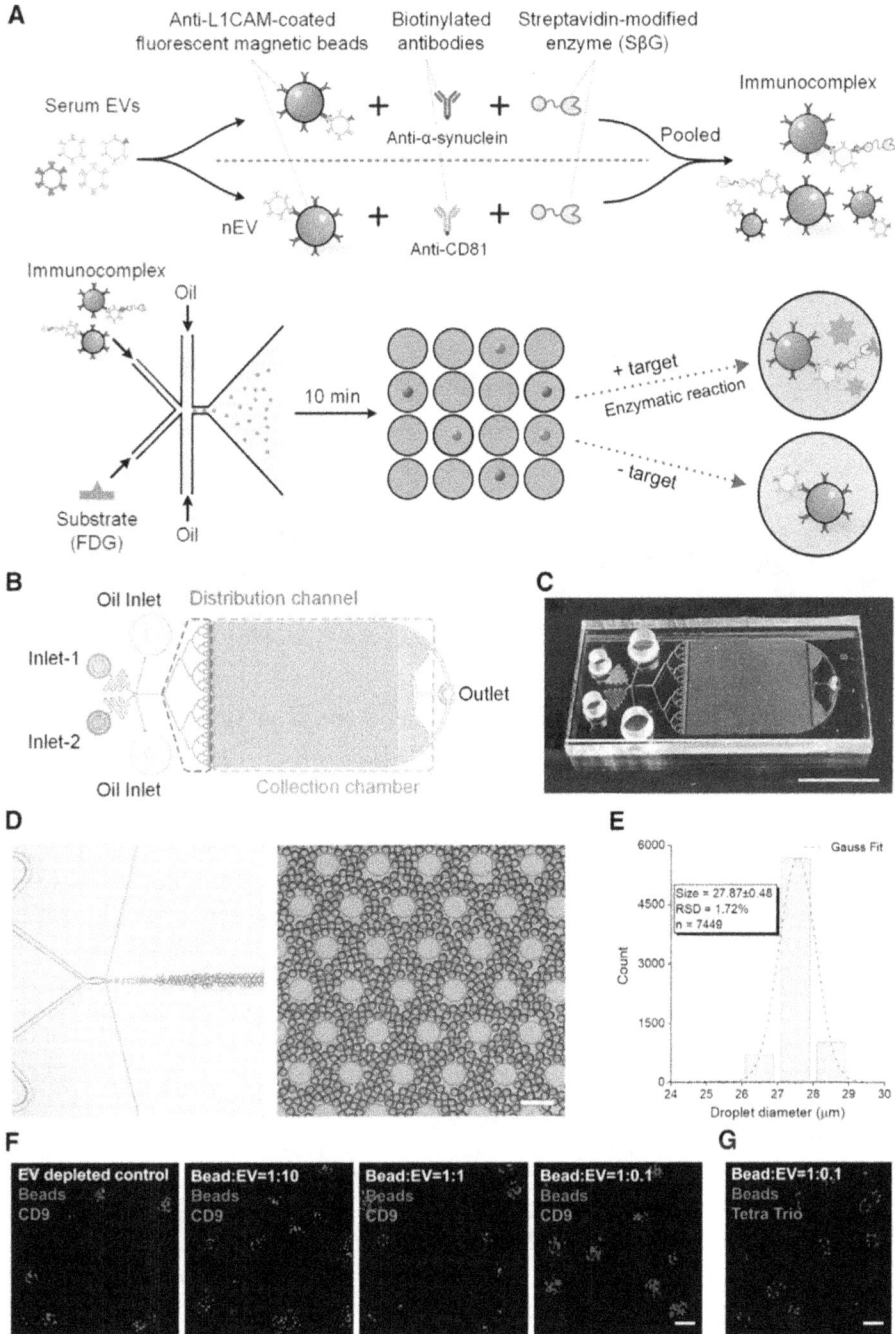

Figure 2.15. Droplet-based microfluidic extracellular vesicle digital immunoassay. (A) Workflow of the assay. (B) Schematic illustration of the droplet microfluidic device. (C) Photograph of the microfluidic device (scale bar, 2 cm). (D) Representative images of the picoliter-sized droplets (scale bar, 100 μm). (E) Size distribution of the droplets averaging 27.9 μm in diameter. (F) Representative fluorescence microscopy image of L1EVs

immunocaptured from serum using DAPI-labelled beads, followed by staining with Alexa Fluor 647-labelled anti-CD9 in EV-depleted serum as a control and bead-to-serum-EV ratio of 1:10, or 1:1, or 1:0.1 (scale bar, 2 μm). (G) Representative fluorescence microscopy image of L1EVs immunocaptured from serum using DAPI-labelled beads, followed by staining with Alexa Fluor 561-labelled anti-Tetraspanin Trio (CD9, CD63, and CD81) using a bead-to-serum-EV ratio of 1:0.1 (scale bar, 2 μm). Reprinted from [19] CC BY 4.0.

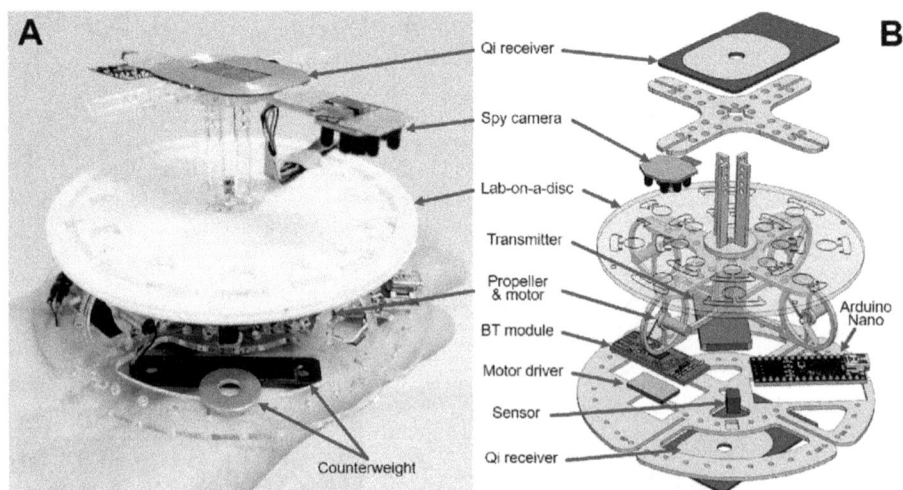

Figure 2.16. (A) Photo and (B) explored diagram of a rotational part of the APELLA. Reprinted from [21] CC BY 4.0.

available items. This is a good example of education that encourages ordinary scientists and engineers if it is possible to design the functions you want to achieve with free ideas and build them with familiar things. Thus, the improved camera functionality of smartphones and the availability of communication systems, combined with cloud computers, have made it possible to publish a low-cost, robust and ultra-sensitive measurement system. In the future, such systems are expected to increase in number, enabling easy on-the-spot diagnosis and analysis, and various analyses and market development using big data.

2.9 Summary

In this chapter, the collection of examples of microfluidic systems shows that the systems idea, design, and demonstration has already been taken up by various groups. The progress of these systems is so rapid that it is difficult to describe all systems and their R&D trends. On the other hand, however, the direction in which researchers and developers are thinking about them is roughly consistent: their use as tools for bio-applications, chemical applications and basic science. As micro-fluidic channels can be used to form precisely controlled fluid interfaces and to perform unit chemical operations, their deployment in combinatorial chemistry and ELISAs tends to be the main R&D direction. The microfluidic systems do not only work on individual components fabricated in diverse technologies. However, the combination of validated fluidic unit operations enables one to build a mechanism that achieves the final process that is desired to be carried out. This approach allows the design and fabrication of application specific systems easily and will lead to a paradigm shift from component and technology-based research to a system-oriented approach. Hence, these systems and platforms will evolve further. For example, they are expected to evolve into devices that are flexible and can be worn like clothing to constantly sense the human body and monitor health, and systems that are networked and matched with statistical data to conduct epidemiological studies. Some of the examples described here are already commercially available as devices and have entered our daily lives. Examples include smartwatches and some medical devices. However, even these smartwatches and medical devices have not yet reached a mechanism that enables chemical sensing at will, and this is expected to be one of the directions of future research and development.

An example of the growing interest in microfluidic platform technology is the fact that a number of spin-off companies have emerged in recent years to commercialize LOC products. For example, ELISA testing in hospitals, which is readily available in larger hospitals but could not easily be done in smaller clinics or pharmacies, is becoming a reality with the development of LOC. If the legal issues are cleared, it could be introduced into clinics and pharmacies as a medical device. In addition, as there are limits to the development of new materials and the investigation of medicinal effects by humans, a system to comprehensively use LOC has been proposed, which can be combined with machine learning and AI, and companies are expected to provide these as a platform.

As described above, it is clear that while research and development of individual elemental technologies will continue to progress in their own right, there will also be a move towards the construction of new systems by combining the elemental technologies that have been created, and the promotion of industrialization as well as research. And when industrial applications are promoted, the need for basic research and development will be realized, and basic research and industrial applications will develop in a complementary manner, like the two wheels of a car.

In this chapter, systems thinking on the development of such individual elemental technologies and the systems that are built by combining them was explained with examples of implementation. It is hoped that this chapter will help the reader to have a multi-scale and flexible mindset that can move between micro and macro perspectives.

References

[1] Primiceri E, Chiriacò M S, Rinaldi R and Maruccio G 2013 Cell chips as new tools for cell biology—results, perspectives and opportunities *Lab Chip* **13** 3789

[2] 2015 *Lab-on-a-Chip Devices and Micro-Total Analysis Systems, A Practical Guide* ed J Castillo-León and W E Svendsen (Switzerland: Springer International Publishing)

[3] Zhu P and Wang L 2017 Passive and active droplet generation with microfluidics: a review *Lab Chip* **17** 34–75

[4] Trantidou T, Friddin M S, Salehi-Reyhani A, Ces O and Elani Y 2018 Droplet microfluidics for the construction of compartmentalised model membranes *Lab Chip* **18** 2488

[5] Kong L X, Perebikovsky A, Moebius J, Kulinsky L and Madou M 2016 Lab-on-a-CD: a fully integrated molecular diagnostic system *SLAS Technol.* **21** 323–55

[6] Zhang Y S, Zhang Y-N and Zhang W 2017 Cancer-on-a-chip systems at the frontier of nanomedicine *Drug Discov. Today* **22** 1392–9

[7] Ingber D E 2022 Human organs-on-chips for disease modelling, drug development and personalized medicine *Nat. Rev. Genet.* **23** 467–91

[8] Zhang Y S *et al* 2016 Bioprinting 3D microfibrous scaffolds for engineering endothelialized myocardium and heart-on-a-chip *Biomaterials* **110** 45–59

[9] Kim S, Lee H, Chung M and Jeon N L 2013 Engineering of functional, perfusable 3D icrovascular networks on a chip *Lab Chip* **13** 1489–500

[10] Kim S, Chung M, Ahn J, Lee S and Jeon N L 2016 Interstitial flow regulates the angiogenic response and phenotype of endothelial cells in a 3D culture model *Lab Chip* **16** 4189–99

[11] Kim H J, Li H, Collins J J and Ingber D E 2016 Contributions of microbiome and mechanical deformation to intestinal bacterial overgrowth and inflammation in a human gut-on-a-chip *Proc. Natl. Acad. Sci. USA* **113** E7–E15

[12] Kasendra M *et al* 2018 Development of a primary human small intestine-on-a-chip using biopsy-derived organoids *Sci. Rep.* **8** 2871

[13] Khetani S R *et al* 2008 Microscale culture of human liver cells for drug development *Nat. Biotechnol.* **26** 120–6

[14] Liu Y, Li H, Yan S, Wei J and Li X 2014 Hepatocyte cocultures with endothelial cells and fibroblasts on micropatterned fibrous mats to promote liver-specific functions and capillary formation capabilities *Biomacromolecules* **15** 1044–54

[15] Lin N Y C, Homan K A, Robinson S S, Kolesky D B, Duarte N, Moisan A and Lewis J A 2019 Renal reabsorption in 3D vascularized proximal tubule models *Proc. Natl Acad. Sci. USA* **116** 5399–404

[16] Homan K A *et al* 2019 Flow-enhanced vascularization and maturation of kidney organoids in vitro *Nat. Methods* **16** 255–62

[17] https://fda.gov/patients/learn-about-drug-and-device-approvals/drug-development-process

[18] Monteduro A G, Rizzato S, Caragnano G, Trapani A, Giannelli G and Maruccio G 2023 Organs-on-chips technologies—a guide from disease models to opportunities for drug development *Biosens. Bioelectron.* **231** 115271

[19] Yan S *et al* 2025 Single extracellular vesicle detection assay identifies membrane-associated a-synuclein as an early-stage biomarker in Parkinson's disease *Cell Rep. Med.* **6** 101999

[20] Yelleswarapu V, Buser J R, Haber M, Baron J, Inapuri E and Issadore D 2019 Mobile platform for rapid sub-picogram-per milliliter, multiplexed, digital droplet detection of proteins *Proc. Natl Acad. Sci. USA* **116** 4489–95

[21] Serioli L, Ishimoto A, Rajendran S T, Yamaguchi A, Zór K, Boisen A and Hwu E -T 2023 APELLA: open-source, miniaturized all-in-one powered lab-on-a-disc platform *Hardware* X **15** e00449

[22] Comsol 1994 Droplet Breakup in a T-Junction https://www.comsol.com/model/droplet-breakup-in-a-t-junction-1994#:~:text=Emulsions%20consist%20of%20small%20liquid,the%20size%20of%20the%20droplets

[23] van der Graaf S, Nisisako T, Schroën C G P H, van der Sman R G M and Boom R M 2006 Lattice Boltzmann simulations of droplet formation in a T-shaped microchannel *Langmuir* **22** 4144–52

[24] Umbanhowar P B, Prasad V and Weitz D A 2000 Monodisperse emulsion generation via drop break off in a coflowing stream *Langmuir* **16** 347–51

[25] Pautot S, Frisken B J and Weitz D 2003 Production of unilamellar vesicles using an inverted emulsion *Langmuir* **19** 2870–9

[26] Matosevic S and Paegel B M 2011 Stepwise syhthesis of giant unilamellar vesicles on a microfluidic assembly line *J. Am. Chem. Soc.* **133** 2798–800

[27] Funakoshi K, Suzuki H and Takeuchi S 2006 Lipid bilayer formation by contacting monolayers in a microfluidic device for membrane protein analysis *Anal. Chem.* **78** 8169–74

[28] Li J and Barrow D A 2017 A new droplet-forming fluidic junction for the generation of highly compartmentalized capsules *Lab Chip* **17** 2873

[29] Landau L D and Lifshitz E M 1959 Fluid mechanics *Volume 6 of Course of Theoretical Physics* translated from the Russian by J B Sykes and W H Reid (Oxford: Pergmon)

[30] Rusanov A I 2005 Surface thermodynamics revisited *Surf. Sci. Rep.* **58** 111–239

[31] Young T 1805 An easy on the cohesion of fluids *Philos. Trans. Royal Soc. London* **95** 65–87

Chapter 3

Nano/microfabrication for creating Nano–Microsystems

The aim of this chapter is to give an overview of the nano/microfabrication for creating Nano–Microsystems. The emphasis will be on fundamental techniques and basic methods, since the goal is primarily to fabricate and prepare the nano/microsystems. Of course, the principles for understanding the technical foundation will also be outlined as necessary to help improve the readers' understanding. Reading this chapter is not necessary either to understand nano/microsystems, or to follow the other chapters of this book. Since readers familiar with nano/microfabrication will be familiar with the content, they may read the chapter or section of interest directly omitting this chapter. For readers and beginners who are not familiar with nano/microfabrication, this chapter describes the manufacturing technology of nano/microsystems. This chapter will be useful for readers who will create Nano–Microsystems in research and development from now on.

3.1 Introduction

A Nano–Microsystem is generally defined as a new type of integrated system by the combination of electronics, fluidics, sensors, optics, mechanics, magnetic, chemistry, biotechnology, and so on. Its development started with the fabrication of microelectromechanical systems (MEMS) and microfluidics. The new concepts on the Nano–Microsystem introduced by the MEMS and microfluidics and, in addition, the potential of the nano/microsystems for both medical and environmental applications boosted the development of an intense activity in a field of research that was later called micro-total analysis system (μTAS) or lab-on-a-chip (LOC) [1–12]. In the demonstrations of these first times of μTAS and LOC, the bioMEMS were often fabricated and demonstrated aiming to realize LOC for prevention and preventive medicine by rapid diagnosis of pandemic by COVID-19 in recent years. To date there has been a lot of research and development related to Nano–Microsystems,

doi:10.1088/978-0-7503-3111-1ch3
3-1

including μTAS and LOC. Their volume is too large to explain all the details in this book. There are several chapters regarding major research fields in this book. Therefore, one can learn how to create and evaluate Nano–Microsystems by choosing them according to one's interest. In this book, each chapter is fully described in detail, including the method of preparing the system to be studied, the evaluations and applications. The important issue is that nano/microfabrication is required to construct a desired Nano–Microsystem. This chapter documents a detailed commentary review on the methods and techniques for fabrication and interconnection of the LOC and Nano–Microsystems; here, the principle is briefly discussed.

In addition, three-dimensional (3D) printing technology has been rapidly developed [13–20]. Recently, 3D printing become a smart additive manufacturing technique which allows the engineering of biomedical devices and microfluidic systems. Nowadays, the 3D-printed microfluidics has gained enormous attention due to their various advantages including rapid production, low-cost, and accurate designing of a range of products even geometrically complex devices. The 3D printing technologies are more suitable candidates for the rapid-prototype fabrication of microfluidic devices that are useful for biomedical applications. 3D printers are used as manufacturing tools for board devices and chips for various applications like diagnostic microfluidic systems to detect biomarkers, chemical analytes such as glucose, lactate, and glutamate. In addition, recently, nano/micro-assisted strategies applied in regenerative medicine are becoming increasingly focused because the biological architectures such as bone, tissue, and other biological structures are composed of nano/microscale functional materials.

In this section, the introduction of fabrication processing of Nano–Microsystems will be outlined. Examples of Nano–Microsystems are μTAS and LOC based on microfluidics, optofluidics, or MEMS. As described above, fundamental studies can be possibly promoted by quantitatively controlling the fluidic properties through fine processing and by creating ideal experimental systems. Nano/microfabrication is essential for basic research and development. In this section, we outline various nano/microfabrication procedures. The 3D printing technologies are also outlined. Some emerging diagnostic technologies using 3D printing as a method for integrating microfluidic chips, living cells or biomaterials are also reviewed.

The basic scaffold structure can be fabricated by a 3D printer or other lithographic methods. On the other hand, surface treatment is also very important. This is because the behaviour of the systems changes greatly depending on whether it is hydrophilic or hydrophobic when constructing microchannels or handling cells. Therefore, surface wettability must always be considered when designing and constructing a microfluidic system. It is necessary to consider the same surface state not only at the solid–liquid interface but also at solid–solid and solid-gas interfaces. Surface treatment alone has a vast history of research and development, and forms an important academic field. It is not suitable for the purpose of this book to explain everything. Therefore, this book will only briefly introduce important surface treatments and their effect. In this chapter, surface modification and nano/microstructure modelling involving novel photo-process such as high-intensity x-rays based on synchrotron radiation, electron beam (EB), and ion beams are also outlined.

3.2 Modelling for nano/microscale structure

The Nano/microfabrication process can provide not only an ideal experimental system for scientific interest but also a device for applications such as Internet of things (IoT). As shown in figure 3.1, the resolution and dimensions of semiconductor micro-fabrication are improving year by year with integration due to higher central processing unit (CPU) functionality and miniaturization due to rapid increase in recording density seen in memory. Semiconductor nano/microfabrication lithography, which started with contact mask exposure, has now become Extended Ultra Violet (EUV) lithography and has achieved a resolution of several nm. In the near future, it is possible that the wavelength of EUV will be halved and further miniaturization will progress. Lithography using EB lithography and atomic force microscopy (AFM) lithography is being considered for microfabrication below that level, but it has not yet been put into practical use as a manufacturing process at this time, and it is thought that it depends on the progress of future research and development.

In this chapter, the nano/microfabrication principles and procedures will be outlined for the sake of preparation of nano/microsystems for both scientific investigations and engineering applications. In systems based on the solid state circuit, usual semiconductor nano/microfabrication is readily available. Compared to solid state device systems, microfluidic devices do not have strict design rule minimum linewidths and accuracy, but they require special structures such as high-aspect ratio machining and tapered shapes. In addition, modelling methods that are not normally used for solid devices, such as modelling for flow path formation, are also required.

In the microfluidic systems, the Reynolds number helps predict flow patterns. It is the ratio of inertial forces to viscous forces within a fluidic which is subjected to relative internal movement due to different fluid velocities. As described in chapters 2 and 4, the Reynolds number is defined as

$$Re = \frac{\rho u L}{\mu} = \frac{uL}{\nu},$$

(3.1)

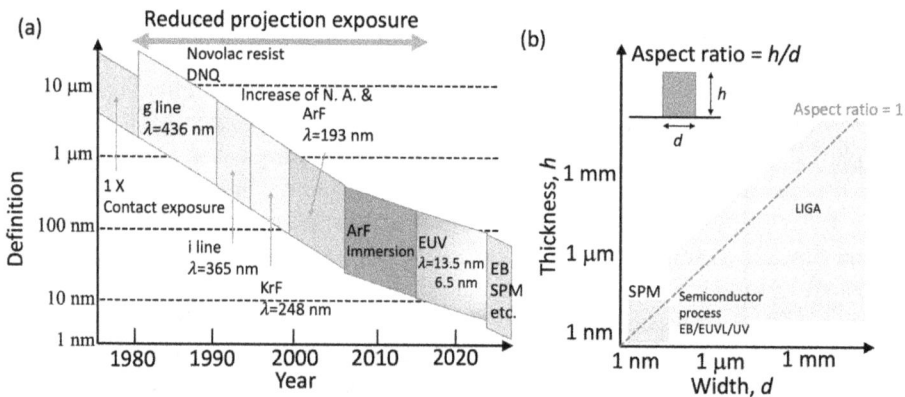

Figure 3.1. Developments in microfabrication: (a) minimum resolution and method evolution; (b) aspect ratios and feasible methods (3D printers are not mentioned here).

where ρ, u, L, μ, and ν denote the density of the fluid (SI units: kg m^{-3}), the flow speed (m s^{-1}), a characteristic linear dimension (m), the dynamic viscosity of the fluid (Pa · s), and the kinematic viscosity of the fluid (m^2 s^{-1}), respectively. This definition (3.1) generally includes the fluid properties of density and viscosity, in addition, a velocity and a characteristic length. Here, this characteristic length L is arbitrary. In the research example conducted by Reynolds, a tube was used, so its characteristic length is the diameter of the tube. At this time, the numerator of the formula corresponds to the amount of movement of the liquid per unit time and unit cross-sectional area, that is, the inertial force, and the denominator corresponds to the viscous force. Since the density and viscosity of a liquid are material parameters, it is difficult to operate once the target liquid or material is determined. Of course, it is possible to adjust the density and viscosity by changing the temperature and dissolving the substance. On the other hand, the flow velocity and the characteristic length are physical quantities that can be manipulated by designing and controlling the device. Especially in microchemical systems and LOCs, the characteristic length is generally small because the flow paths are integrated by nano/microfabrication. That is, it functions as a system in which the Reynolds number is small, turbulence does not occur, and the laminar flow operation is used as the basic unit chemical operation. There are advantages and disadvantages to this point, but the features will be described in detail in another section. Here, the methods and procedures for nano/microfabrication of these Nano–Microsystems such as microfluidic devices, MEMS, solid state devices are outlined. MEMS and Nano–Microsystems based on solid-state devices have been fabricated by ordinary semiconductor lithography. Microfluidics have been created by soft lithography using polydimethylsiloxane (PDMS). Recently, with the development of 3D printers, the number of cases where microfluidic systems are manufactured using 3D printers has increased. We will outline them and add a little more detail about semiconductor lithography and soft lithography so that they are ready for beginners to use.

3.3 Outline of nano/microscale fabrication and procedures

As described in the above, nano/microscale fabrication processes are required for preparing nano/microsystems. Here, these processes and procedures will be outlined below. There are two general categories for fabricating nano/microsystems: top-down and bottom-up. Top-down processes enable one to easily control the shape and to form an ideal system structure, and they can also be combined with bottom-up processes. The top-down procedures often incorporate several techniques such as scanning type beam lithography, ultraviolet (UV) lithography, film formation by sputtering or deposition, and Ar ion and focused ion beam (FIB) etching. Recently, stencil lithography in nanoscale patterning for direct deposition of complex materials and the pattering on non-conventional substrates has been also developed. These nano/microfabrication techniques and processes enable nano/microsystems such as plasmonics, electric devices consisting of 2D materials, biosensing devices, cell and protein pattering, etc. Below, some of these techniques and procedures are discussed.

Process selection depends on Resolution, Throughput, and systems size etc.

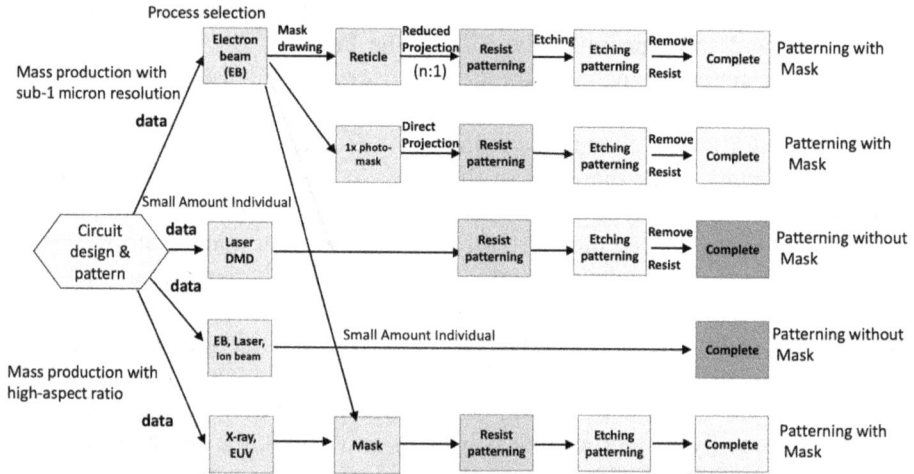

Figure 3.2. Process selection guidelines and process flow.

In lithography including EB, UV and EUV, basically pattern transfer onto a resist is done by shining light through the mask or scanning operation. In UV lithography, one typically uses the g-line (435 nm) or i-line (365 nm) of a mercury lamp. Excepting scanning beam lithography, in general, the smallest feature that can be printed using projection lithography is roughly equal to the wavelength of the exposure source. One should select the appropriate process and manufacturing procedure according to the specifications of the device or system that one would like to fabricate. Figure 3.2 shows the process flow or typical process selection and each fabrication procedure, in particular, for the systems with relatively smaller size such as solid-state based devices.

On the other hand, bottom-up processes have been used extensively in thin film formation, but have been difficult to manipulate. Therefore, in research and development based on solid-state physics such as semiconductor microfabrication and metal artificial lattices, thin-film formation and top-down processes are often used in combination. However, due to the rapid development and technological innovation of 3D printers in recent years, the bottom-up process has come to be generally used for 3D modelling. Looking at the historical development process, we will start with lithography as a practically useful and easy technology, and then move on to thin film formation and 3D printers.

3.4 Lithography

3.4.1 Outline

The word 'lithography' is a coined word. It is originally derived from the Geek word meaning 'stone [lithos]' and 'to write [graphein]'. There are various methods and techniques. The most widely used form of lithography is photolithography. A design pattern is usually transferred to a photosensitive polymer (a photoresist) by exposure to a light source via an optical mask. The pattern formed by the photoresist is further transferred to the underlying substrate by additive or subtractive techniques. The combination of accurate alignment of a successive of masks and exposure of the successive patterns leads to creation of complicated structures. A comparison of various lithographic resolutions and achievable aspect ratios is summarized in figure 3.1. Electron beam lithography (EBL) can create very precise patterns without any masks, while a long process time is required for large-area patterning. EBL is a lithography used for mask fabrication such as UV lithography and EUVL, and photolithography has matured rapidly by continuous improvements in the ability to improve the resolution. The photolithography masks are usually made by EBL. FIB lithography using ions instead of electrons used in EBL enables the usual lithography for patterning a resin or other materials. In addition to subtractive (cutting) process, Focused Electron Beam (FEB) can also be applied to additive processes by combining with chemical-reaction excitation in reactive gas or liquid. The following outlines the process of manufacturing Nano–Microsystems by combining pattern formation and film structure formation by lithography. We will also outline specific methods for creating microchannel structures and 3D printings that have often been used and have become popular recently.

In the case of Nano–Microsystems based on solid-state devices, a nano/micro-fabrication process based on semiconductor fabrication is generally used. That is based on the series process combinations including lithography, thin film formation, milling, ion implantation, and so on. On the other hand, when constructing a microfluidic channel structure in an LOC, it is necessary to introduce or enclose the liquid in the system and perform processes such as liquid feeding, reaction, and extraction etc. In particular, since optical detection and analysis methods have been frequently used to perform analytical chemistry, the LOC is often composed of transparent substrates and materials. Regarding the creation of structures, fine ones are manufactured by semiconductor microfabrication. Larger structures are often implemented in combination with various techniques such as machining, LIGA process [21] and 3D modelling [13–20]. The process for LOC is exhaustively described in some reviews. Here, referring to them, we will describe how to make solid-state devices and LOC for beginners.

3.4.2 Resist processing

Generally, in fine processing, a technique that uses a resist if often employed. This lithography technology proceeds as shown in figure 3.3(a). There are two types of resists: positive-tone and negative-tone. The pattern is formed by exposure to light with positive-tone resists, while the reversal pattern is developed by exposure to light with negative-tone resist. In the positive-tone resists, the molecular chain scission is proceeded with exposure. Since the smaller polymer molecular can be dissolved in an organic liquid solution, the regions are washed away by the developer, leaving the regions that were not exposed due to the presence of a mask in front of the light. In contrast, negative-tone resists are polymerized or strengthened by light exposure. Exposed regions remain after the developing wash.

Here, as an example which is often used, a schematic diagram and procedure of the patterning based on a negative-tone resist process are shown in figure 3.3(b). After the

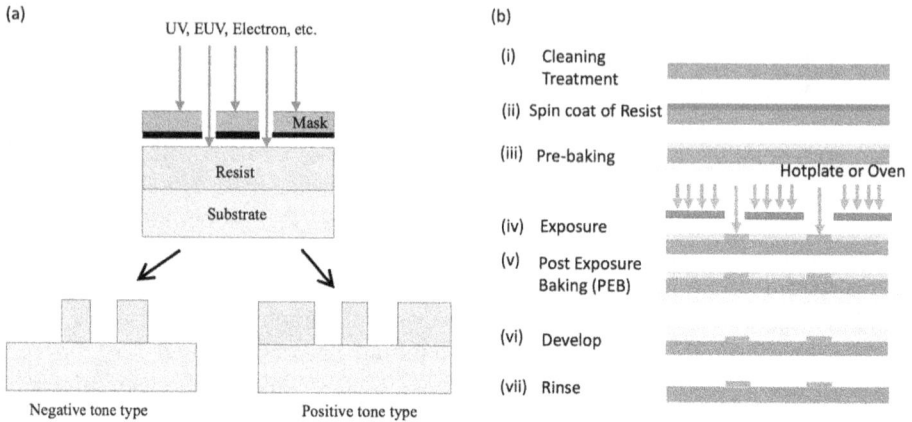

Figure 3.3. (a) Overview of the lithographic process with resist and (b) outline of the process flow.

cleaning and surface treatment (figure 3.3(b)-(i)), the resist is coated using a spin coater (figure 3.3(b)-(ii)). Next, the substrate coated resist is baked using a hotplate or oven to remove the solvent and cure the resist (figure 3.3(b)-(iii)). In lithography processes (figure 3.3(b)-(iv)), several alternatives are available as a light source. For example, UV light, EUV, EB, and a focused ion beam are often used. Herein, the UV lithography, which is the most frequent and convenient to use, will be described below. In the negative-tone resist process, after the exposure process as shown in figure 3.3(b)-(v), post-exposure baking (PEB) is performed. This process is often referred to as 'PEB' because of its acronym. There are two patterns when using PEB:

(1) **When using a chemically amplified resist**

Among the types of resists, there is a chemically amplified resist. These chemically amplified resists have a different reaction mechanism from ordinary resists and are characterized by high sensitivity. As shown in figure 3.3, ordinary resists undergo cross-linking at the same time as exposure. The chemically amplified resist is cross-linked at the same time as PEB. Thanks to the reaction by thermal diffusion of the acid catalyst, the chemically amplified resist can draw a pattern with less wavelength of exposure light. Therefore, the chemically amplified resist has a PEB process. The basic chemical amplification was established by Fréchet, Willson, and Ito [22, 23]. They succeeded in preparing a resist by adding an onium salt to poly(4-[tert-butyloxycarbonlyoxy] styrene) capable of acid-catalyzed deprotection.

(2) **When the pattern is uneven due to the influence of standing waves**

Since light has wave properties, it is possible to use the incident light at the time of exposure and the reflected light from the substrate, etc. It forms a standing wave and may be exposed in a wavy shape. Then, after development, the wall part of the pattern becomes a wavy pattern. In such a case, the PEB process can be used to mitigate the standing wave by thermal diffusion.

Finally, after development (figure 3.3(b)-(vi)) and rinse process (figure 3.3(b)-(vii)), the pattern is formed. Based on the formed resist pattern, a process suitable for the final purpose is performed below to obtain the final structure. In most cases, there are two directions. In the former as schematically illustrated in figure 3.4, modelling is performed through an etching process such as ion milling and wet etch. The latter is a case where lift-off is performed in combination with thin film deposition. In either case, it is necessary to select the optimum process according to the purpose. This process is repeated many times when manufacturing complex semiconductor devices and systems. The following conditions are required to use resist materials for device manufacturing:

- Adhesive to various substrates such as metals, semiconductors, insulators, enabling uniform spin coating of thin films;
- High photosensitivity;
- High resolution (large difference in solubility due to light irradiation);
- Resistance to extremely harsh environments such as high temperature, strong corrosive acids, and plasmas used for post-casting etching, doping, and sputtering operations.

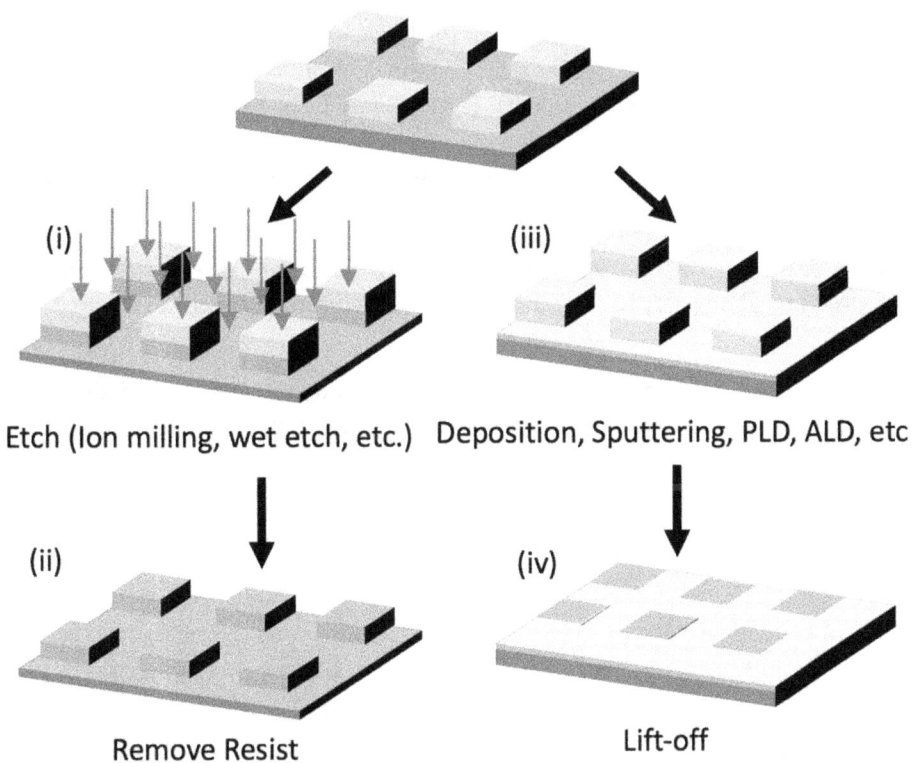

(i)

(iii)

Etch (Ion milling, wet etch, etc.)   Deposition, Sputtering, PLD, ALD, etc

(ii)

(iv)

Remove Resist   Lift-off

Figure 3.4. Nano/microfabrication flow with resist.

3.4.3 Resist properties for lithography

Resist compounds can also be divided into three groups, UV (photo) resists, electron beam resists, and x-ray resists, based on the exposure light source [21–43]. Photolithography using UV light is a technology that has been mainly used in conventional semiconductor manufacturing and continues to be a major technology, as described above. However, the resolution has reached its limit in principle. X-ray lithography including EUV is considered to be the next-generation technology capable of producing high-resolution, high-aspect-ratio (height–width ratio) images, while EBL is used in the production of photomasks. Photolithography can be further divided into near-ultraviolet (350–450 nm), medium-ultraviolet (300–350 nm), and far-ultraviolet (<300 nm) technologies, depending on the exposure wavelength. The resolution (R) is proportional to the exposure wavelength (λ) and inversely proportional to the numerical aperture (NA) of the lens of the exposure device ($R \propto \lambda/NA$). Therefore, the transition from the g-line (436 nm) of a mercury lamp to the i-line (365 nm) with a high NA lens is a 16-megabit (Mbit) dynamic random access memory (DRAM) with a minimum dimension of 0.5 µm. It has made it possible to manufacture elements. Krypton fluoride (KrF, 248 nm), argon fluoride (ArF, 193 nm), and fluorine (F2, 157 nm) excimer laser technology emerged as the minimum shape continued to refine from 0.5 µm to the much shorter 80 nm. As mentioned earlier, there are two main types of resist: positive and negative. The positive-tone resist works by dissolving the molecules in the irradiated area in a solvent as the main chain is cleaved by light irradiation, resulting in a smaller molecular weight. Therefore, poly(methyl methacrylate) (PMMA), which has a simple molecular structure, as shown in figure 3.5(a), is often used. G- and I-line resist structure is shown in figure 3.5(b). On the other hand, in the negative-tone resist, light irradiation causes the molecules to chemically react and increase in molecular weight, resulting in pattern formation when the molecules in the irradiated area are not soluble in the solvent. For this reason, epoxy structures, as shown in figure 3.6, are often used as the main structure. Below, let us take a closer look at commonly used resists.

Positive-tone resist
There are several well-known positive-tone type resists in recent processing. These resist types and their characteristic properties are summarized in table 3.1.

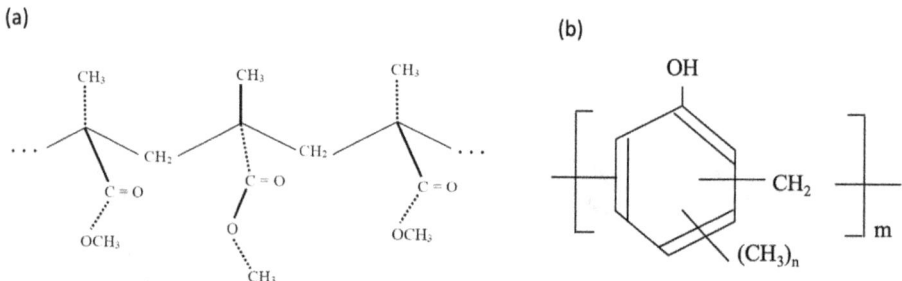

Figure 3.5. Molecular structures of (a) PMMA and (b) G- and I-line resists.

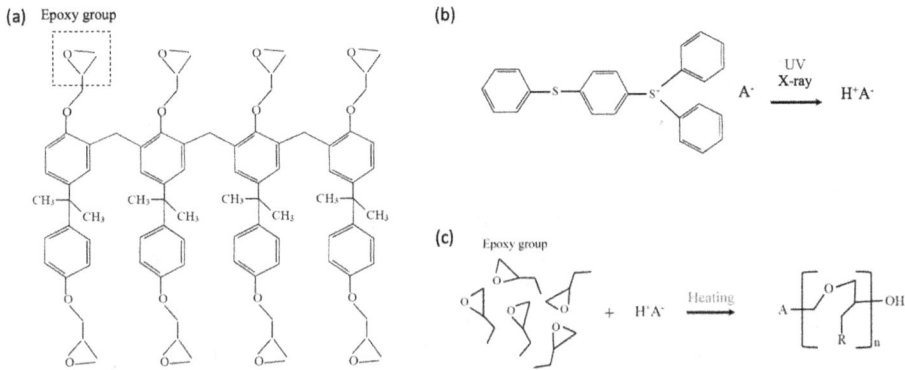

Figure 3.6. Molecular structures of (a) epoxy-based base, (b) acid generator. (c) Reaction mechanism of chemically enhanced resist.

Table 3.1. Summary of positive-tone resist.

Type	Name	Features and typical process	Thickness
UV lithography	AZ P4000 series: [37] 4110 4210 4330-RS 4620 4903	For general photolithography Suitable for electroplating (AZ4903) Developer type: Inorganic (IN) Soft bake: 90 °C–115 °C Expose: 350–450 nm sensitive Develop: spray or immersion	P4110: 1–3 μm P4210: 2–4 μm P4430-RS: 3–5 μm P4400: 4–6 μm P4620: 6– > 20 μm P4903: 8– > 30 μm
	OFPR-800 [38]	For general photolithography Prebake: 65 °C–110 °C Expose: 350–450 nm Develop: spray or immersion Rinse: pure water Option: hard bake	1–3 μm
	TSMR-V90	Diazo-napthoquinone system (DNQ) High resolution Realization of line and space width of 1 μm or less	
EB	ZEP-520A [27]	Resolution: 20 nm (the resolution depends on the thickness, operation at 50 kV) Sensitivity: 100 [μC cm^{-2}]	Typical thickness: 100–500 nm
	PMMA950 [26]	Resolution: 20 nm (the resolution depends on the thickness, operation at 50 kV) Sensitivity: 100 [μC cm^{-2}]	The model number depends on the applicable thickness. Typical thickness range: 200 nm–5 μm

Negative-tone resist

For beginning students, the most popular and frequently used resist, SU-8, will be used as an example. The SU-8 is a negative, epoxy-type, near-UV photoresist based on epoxy resin that was originally developed by IBM [22, 23]. Currently, two companies have now bought the license from IBM to sell the photoresist: Kayaku Advanced Materials [39, 40] and Gersteltec SARL. The photoresist can be as thick as 2 mm and aspect ratio larger than 20 with a standard contact lithography technique. The SU-8 resist is not separated by LIGA process [21]. It is often used in the mould fabricating process for forming microchannels in LOC and for electroplating because of its relatively high thermal stability (glass-transition temperature, $T_g > 200$ °C for the cross-linked resist). On the other hand, this resist is so tightly adhesive that it is difficult to remove. Some techniques and methods are proposed to remove the resist.

Mechanical and physical properties of the SU-8 photoresist are summarized in table 3.2. To increase the adhesion between resist and substrate, some adhesion promoters are commercially distributed. OmniCoat is a commercial product from MicroChem. By using OmniCoat, the bond strength between SU-8 and Au increased. Nordström *et al* have investigated the adhesion properties with various treatments [24]. In particular, they have obtained the results from pull-test experiments between SU-8 and Au. Ge *et al* have performed detailed studies of the adhesion between Cu and photosensitive resists including SU-8 [25]. The important properties of SU-8 are summarized in the web site of references [39, 40]. Mechanical strength is improved by the anchor effect in the case that unevenness on the microscale size is acceptable. On the other hand, another approach is required for chemical bonds. The data obtained by Nordström *et al* show the bond to be of a very weak nature, but they concluded that the value of the bond strength can be increased by up to 75% by using an adhesion promoter and fully optimizing the processing conditions.

3.4.4 General procedure for lithography process

The name of the resist comes from its resistance to chemical corrosion because when applied to a wafer, the applied portion is protected and is not eroded by chemicals such as acids. Photographs are also derived from the application of photographic technology to pattern formation. The process of transferring the mask pattern to the

Table 3.2. Summary of properties of SU-8.

Characteristics	Value	Conditions	References
Modulus of elasticity	4.95 ± 0.42 GPa	Hardbaked at 200 °C	Dellmann [41]
	4.02–4.4 GPa	Postbaked at 95 °C	Lorenz *et al* [42]
Friction coefficient	0.19	Postbaked at t 95 °C	Lorenz *et al* [42]
Bond strength	4.8 ± 1.2 MPa	On Au, pull test	Nordström *et al* [21]
	5.6 ± 2.5 MPa	On Ti, pull test	
	12.1 ± 2.5 MPa	On Al, pull test	
	20.7 ± 4.6 MPa	On Si, pull test	

resist pattern by exposure is shown in figure 3.3. There are two types of photoresists, positive type and negative type. The positive type can obtain a resist pattern having the same pattern as the mask by irradiating with light. On the other hand, it is the negative type that the pattern inverted with the mask can be obtained.

In the positive type of resist, a pattern can be formed by chemically changing the resist material of the portion exposed to light from alkali insoluble to soluble and dissolving it with a developing solution. On the other hand, in the negative type, the portion irradiated with light changes from soluble to insoluble due to the polymer cross-linking reaction. By immersing in a developing solution, soluble sites are removed, and a negative pattern is formed. The photoresist material is a mixture of a resin, a photoactive compound (PAC), and a solvent. Resin is a polymer substance that coats a substrate and serves as a mask for etching. PAC is altered by a photochemical reaction to promote solubility or insolubility. DNQ (Diazoa naphoquinone) is often used as the PAC for positive resists [43].

The solvent uniformly applies the thickness to the wafer by liquefying the resist, but after the coating is completed, it is evaporated and removed to cure the resist. Photoresists are required to have various properties. The most important is the resolution, and the positive type is superior, so the positive type is mainly used in semiconductor microfabrication. In a microsystem such as LOC, microfluidics, and μTAS, it is often used as a mould for creating a flow path structure (microchannel), and a relatively thicker structure is often required. The size is often ranging from several μm to mm. Therefore, in microsystems, especially microfluidic systems, negative resists based on SU-8 are mainly used. The resolution depends on the wavelength of the light used for exposure, and the shorter the wavelength, the higher the resolution. A thick resist has a low transmittance, so the resolution is low, but the resolution can be increased by using x-ray exposure, which will be described later.

Table 3.3 shows a summary of resolution, light used for exposure and the thickness of the resist. The information in this table is simplified information only, so please refer to the process sheets provided by each company for actual usage and conditions. A combination of DNQ/upper rack resin and mercury lamp i-ray was used up to the

Table 3.3. Summary of resist types, lithography methods and spatial resolution.

Type		Name	Resolution	Thickness
Positive tone	EB	ZEP-520A	<20 nm	<500 nm
		PMMA950	<20 nm	<3 μm
Nagative tone	EB	ma-N 2410	30nm	100 nm
				Spin coating @ 300 rpm
		SLA-601	75nm	
Positive tone	UV	AZ P4000 series	~3 μm	1–30 μm
				Depend on series
		OFPR-800	~3 μm	1–3 μm
		TSMR-V90	~0.8 μm	1–2 μm
Positive tone		SU-8-series: 100,	~ 3 μm	5–3000 μm
		2100, 3005 etc.	Depend on thickness	Depend on series

minimum dimension of 0.25 μm. The mercury lamp i-line is visible light with the shortest wavelength, but even with complicated light irradiation techniques such as transfer shift masks and off-axis irradiation, a resolution of 0.25 μm cannot be achieved. For dimensions smaller than that, UV light with a shorter wavelength than visible light is used, and chemically amplified resist (CAR) that chemically reacts with UV light is used in combination. The generation of acid plays an important role in the cross-linking reaction in this CAR. However, if a small amount of ammonia contaminates the resist surface, the acid in the surface layer is neutralized and an insoluble layer is generated, resulting in a poorly developed resist pattern. As a countermeasure, air is cleaned, and pollution is monitored in clean rooms that use CARs. A source of trace amounts of ammonia is humans.

Resist process flow

Resist pattern formation is typically performed in the following order:
  (1) Wafer cleaning.
  (2) Priming: Supports resist adhesion by HMDS surface treatment.
  (3) Applying resist: Forming a resist film of uniform thickness on the entire surface of the wafer.
  (4) Prebake: Evaporation of volatile solvent.
  (5) Exposure: Light irradiation through a mask or reticle or direct drawing in electron beam drawing and laser drawing.
  (6) PEB: Promotes chemically amplified resist reaction. Mainly used for negative resists. Positive type is often unnecessary.
  (7) Development: Development solution immersion or spray treatment. Appearance of resist pattern.
  (8) Post-bake: Resist pattern baking and solidification.

If the resist is not in close contact with the wafer, it will not act as a mask when placed in a chemical solution during the etching process. Priming is performed to convert the surface of the wafer from hydrophilic to hydrophobic so that the hydrophobic resist adheres to the wafer. After the moisture adsorbed on the surface of the wafer is completely removed, the surface is modified by exposing it to the vapour of HDMS (hexamethyledisilazane). Recently, silane coupling agents have also often been used. The figure shows a schematic diagram of the mechanism by which hydrophilicity changes to hydrophobicity. The resist thickness is determined by the combination of the resist viscosity and the spinner rotation speed. Since each manufacturer provides a typical correlation curve between viscosity and rotation speed, the conditions for preparing a resist thickness of a desired thickness are determined with reference to it. However, the recipe provided by the manufacturer is just a typical case. Therefore, it is necessary to adjust the conditions according to the materials and structure of the element that one would like to fabricate.

Next, we will describe some problems that occur during the process. When light is reflected between the silicon wafer interface and the resist surface during exposure, a standing wave is generated inside the resist, and the strength of the photochemical reaction is locally generated, forming an elegant uneven pattern in the resist pattern shape.

The countermeasure is to form an anti-reflective-coating (ARC) and (bottom-ARC) BARC. On the other hand, in the chemically amplified photoresist, the standing wave effect can be extinguished by PEB for an appropriate time. The resist is developed by immersing the wafer in a developing solution, dropping the developing solution onto the wafer, or spraying the wafer. TMAH (tetramethylammonium hydrogen peroxide) is often used in AZ resists as a representative developer. A surfactant is added and used to shorten the development time and dissolve the peeled residue due to development.

Development of positive resist and resolution factors

The transfer from the mask pattern to the resist pattern is directly drawn by an electron beam or a laser in the maskless exposure machine. However, this method is not suitable for mass production because the throughput is slow. This process is suitable for the research and development phase where one wants to change the pattern frequently. On the other hand, when a mask is used, it is difficult to change the pattern, but mass productivity is improved. In each case, it is carried out in three stages: light irradiation, latent image formation by photochemical reaction inside the resist, and development. In the case of an optical process, the resolution of the pattern is determined by the combination of reticle, lens and resist. The resolution factor should be considered from the basic characteristics of light, so it will be outlined below.

Light is a type of electromagnetic wave that propagates in space. The electromagnetic wave propagates while being an oscillator in the plane where the electric field and the magnetic field are orthogonal to each other. In a substance, there is an interaction with the elements and electrons that make up the substance, and its intensity is attenuated. The energy that interacts depends on the electron shell and crystal structure. Electromagnetic waves are classified by frequency or wavelength, and when the frequency is low, that is, when the wavelength is arranged from the longest to the shortest, AM radio wave, FM radio wave, microwave, infrared ray, visible light, ultraviolet light, x-ray, gamma ray. It is called by the name. Visible light is in the wavelength range of 700–350 nm, and ultraviolet light is in the range of 350–10 nm. In ordinary optical processes, light in the short-wavelength visible to ultraviolet region has been used. However, due to recent miniaturization, the process is shifting to a process using extreme ultraviolet light. The propagation speed c_0 of light is $2.998 \times 10^8 \text{ m s}^{-1}$ in a vacuum, and $c = c_0/n$ is the propagation speed of light in a media with the refractive index, n. Here, note that light exhibits characteristic phenomena such as reflection, refraction, diffraction, and interference, and even in the exposure process, it causes problems that affect the resolution and construct the process by fully considering these phenomena.

Resolution factors and depth of focus

The mask pattern is often formed by EBL. One of the reasons why the formed mask pattern cannot be accurately transferred to the resist pattern on the wafer is explained at the boundary between the transparent part and the opaque part of the mask pattern, and the boundary of the transfer pattern is unclear. The effect of dimensional change due to diffraction is greater as the wavelength is longer and can be improved by using a shorter wavelength. To explain this, it is preferable to use the relationship between the image formation by the lens and the magnification. The figure shows the definitions of F-number and numerical aperture (NA), which indicate the optical characteristics of the lens. NA denotes the value which describes the brightness and resolution of an optical system. In general, the lens formula is a formula in geometrical optics. Distance a from the object to the principle point A, distance b from the principle point to the image plane B, and focal length f (distance between principle point and focal length) are considered, as schematically illustrated in figure 3.7(a). The lens formula is ideally given by the following relationships [44]:

$$\frac{1}{a} + \frac{1}{b} = \frac{1}{f}, \tag{3.2}$$

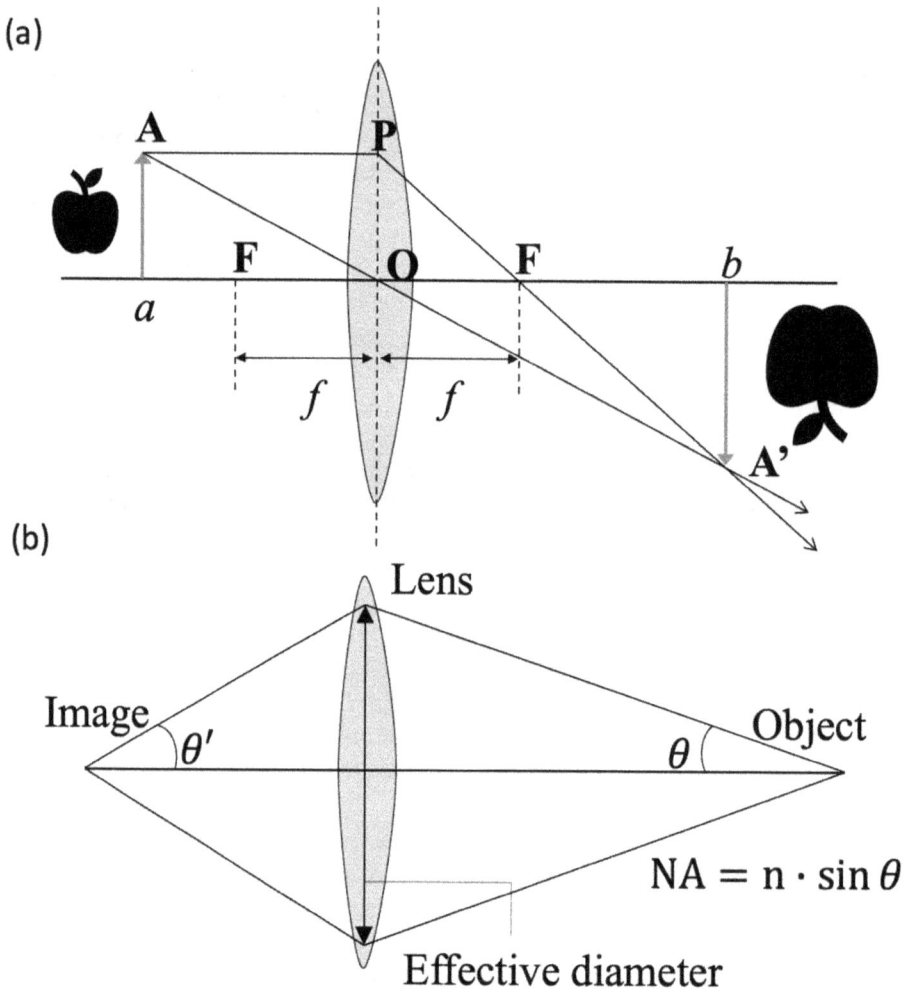

Figure 3.7. (a) On light focusing when the light source is outside. (b) Schematic diagram of magnification with a lens.

Here, the focal length f is negative for a divergent system such as a concave lens, and distance b to the image plane is negative for a virtual image. If the object is at infinity, the first term on the left side is 0, and if the image is a virtual image at infinity, the second term on the left side is 0. This formula can be applied not only to single lenses but also to concave/convex mirrors and optical systems that combine multiple lens/mirrors (if the principal point/focus can be defined).

The larger the diameter of the lens and the shorter the focal point, the brighter the image because the light can be focused on the focal point. Also, the larger the NA, the higher the resolution. The following is the definition of an object on the optical axis as the effective system θ of the lens that is incident on the object, as schematically illustrated in figure 3.7(b).

$$NA = n \cdot \sin \theta, \tag{3.3}$$

where n is refractive index of the medium surrounding the object. $n = 1$ for air. Light has a phenomenon of spreading like a wave, which is called diffraction. This means that even if a high-performance lens with no aberration is used, the image cannot be focused to a single point and spreads out like a disk. The aperture (NA) represents the light gathering limit in the absence of aberration, which is due to the fact that light has the nature of a wave, and this limit value is called the diffraction limit. The wafer is placed at the position of the lens focal point by irradiating the reticle surface with parallel rays perpendicularly. This corresponds to forming an image of an object at infinity at the focal position. In geometrical optics, a perfectly similar image should be able to be formed through a lens, but the image of a point at infinity is not actually an infinitesimal point, but a finite dimension. This will determine the resolution of the lens. These phenomena occur because light has wave-like properties. The figure shows the intensity distribution of the light formed by a point light source at infinity. This pattern is called an Airy disk and is represented by a curve described by the Bessel function. In an Airy disk, a strong emission circle in the centre is surrounded by multiple rings with weak emission intensity. When light of wavelength λ is focused through a lens with NA, the radius δ of its central circle can be written by

$$\delta \approx 0.61 \cdot \frac{\lambda}{NA}. \tag{3.4}$$

Here, δ is the resolution of the lens. According to this formula, the larger the NA, the smaller the radius of the airy disk, i.e., the larger the NA, the sharper the image that will be reproduced. These are common values for evaluating lenses. When an image of a mask pattern having a negligible thickness is projected onto the thick resist layer, the image forming region becomes non-uniform in the thickness direction of the resist layer. The reason is that the irradiation light is focused toward the focal point and emitted from the focal point. Depth of focus (DOF) is defined as the distance within the range where the light intensity is within -20% of the peak value with respect to the light intensity distribution on the focal plane and its peak value, and is described by the following formula:

$$DOF = \pm\frac{\lambda}{2(NA)^2}. \tag{3.5}$$

When exposure is performed under a condition where the depth of focus is small, the non-causal property of the photochemical reaction in the thickness direction of the resist layer becomes large, and when the resist film thickness is non-uniform, the variation in pattern formation becomes large. Therefore, in order to obtain good exposure, it is desirable to increase the depth of focus. However, from the equation, both the resolution and the depth of focus cannot be fully satisfied at the same time.

The developed resist pattern is evaluated for a line width, a side wall angle, a residual resist thickness, and the like, and it is determined whether or not the developed resist pattern is suitable for forming a desired structure. The process is performed by optimizing the parameters that cause resolution, such as the focal plane setting, resist application conditions, and exposure time, until a satisfactory structure is formed.

The exposure device is called an aligner because it aligns the processed pattern of the base with the pattern to be newly developed with high accuracy. An alignment pattern (alignment mark) is incorporated into the exposure pattern to perform alignment drawing. The figure shows some examples of alignment marks. This method is the same not only for UV exposure machines but also for maskless exposure devices such as electron beam drawing and laser drawing. In exposure using a mask, the alignment mark is incorporated in the mask to design and manufacture the mask.

The following is a summary of technical developments in nano/microscale patterning formation mechanisms based on semiconductor nano/microfabrication:
  (1) Imaging method: Adhesion ⇒ Proximity ⇒ Projection.
  (2) Projection method: 1x projection ⇒ reduced projection (stepper) ⇒ reduced projection (step and scan).
  (3) Original version: Mask ⇒ Reticle.
  (4) Original plate production: Master mask or reticle is produced by light exposure ⇒ Reticle production by electron beam writing.
  (5) NExposure antigen: Mercury lamp g-line ⇒ i line ⇒ KrF ⇒ ArF ⇒ F_2 ⇒ EUVL.
  (6) Resist: Negative type ⇒ Positive type ⇒ Chemical amplification type Positive type.

The contact method and proximity method masks are provided with a pattern on the entire surface of the wafer. On the other hand, as microfabrication progresses, the original pattern becomes one-chip field of view in order to cope with alignment and miniaturization and is called a reticle instead of a mask. Originally, the reticle was a master mask used to make 1× exposure masks, but with the miniaturization, it has come to be used in step and repeat reduction projection exposure equipment. The UV light passes through the lens after passing through the reticle, and the reticle pattern is further reduced to about 1/10 and exposed. When the exposure of the one-chip field of view is completed, the reticle remains stationary, the wafer is moved, and then the next one-chip field of view is exposed. The exposure device is called a stepper because it repeats the exposure of one-chip field of view. Furthermore, a step-and-scan reduction projection exposure apparatus has been developed to cope with the progress of miniaturization. This is an exposure device that further advances microfabrication by arranging a concave reflector between the reticle and the wafer and scanning the reticle and the substrate at the same time. Recently, EUVL has been introduced due to the intrinsic limitation depending on the wavelength used.

3.5 Thin film process

In MEMS and solid-state systems, the device structure is often made of semi-conductor by pattering by nano/microfabrication. Electrodes are frequently formed because electrical inputs and responses are essential. Since the electrodes are highly conductive, they are often made of a metallic material. In particular, metals such as gold and copper are used. Also, in LOC it is essential to form electrodes when incorporating dielectrophoresis or MEMS. Therefore, this section outlines the thin film process for creating electrodes and solid-state devices.

There are three types of growth for the film growth, e.g., vapour, liquid and solid phase growth, as summarized in figure 3.8. Thin films are often prepared by chemical vapour deposition (CVD), physical vapour deposition (PVD), and plating. These are used for a variety of different purposes in nano/microstructures. That is, masking materials, structural materials, sacrificial materials, electrical devices, molecular sensors, optical functional materials, and so on. They are formed by each optimal processes based on chemical-reaction driven processes or physical process. For example, the several typical materials and methods are as described below.

When considering thin film growth, one must consider the relationship between system temperature T, mean free path (MFP) λ, and geometric characteristic length L of a structure to be produced. Here, MFP is the average value of the distance that

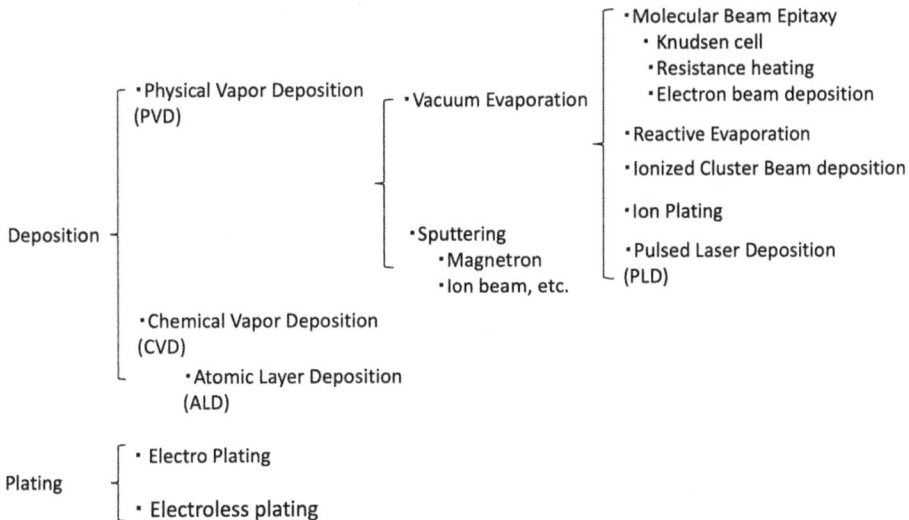

Figure 3.8. Summary of three types of growth for the film growth.

particles such as molecules can travel without being disturbed by scattering by a scattering source in kinetic theory of gas in physics and chemistry. In other word, when a particle moves by the MFP, it collides with other particles once on average. At a finite temperature T, the average speed of a vapour molecule with mass m follows the Maxwell distribution at the equilibrium condition $v = (8k_B T/\pi m)^{1/2}$. Here k_B is the Boltzmann constant. Then, the MFP can be given by $\lambda = (\sqrt{2}\,n\pi\,\sigma^2)^{-1}$, where n is the number of molecules per volume and σ is the molecule diameter. For PVD, $\lambda/L \leqslant 0.01$ and viscous flow is responsible for the molecular motion for deposition. For chemical vapour deposition (CVD), $\lambda/L \geqslant 0.3$ and molecular flow is responsible. Here, laminar flow needs to be avoided.

For PVD, there are many techniques such as thermal evaporation, electron beam evaporation, molecular beam epitaxy (MBE), ion plating, pulse laser deposition (PLD) due to laser ablation and sputtering, as schematically illustrated in figure 3.9. To achieve minimum damage onto a substrate and/or seed layer underneath the film to be grown, thermal evaporation, electron beam evaporation, and MBE is suitable because they can induce kinetic energies of 0.1–1 eV to the evaporating molecules. Sputtering, PLD, and ion plating has molecules with kinetic energies of 1–10, ~10, and 10–5 keV, respectively. For controlling the quality of the film, the

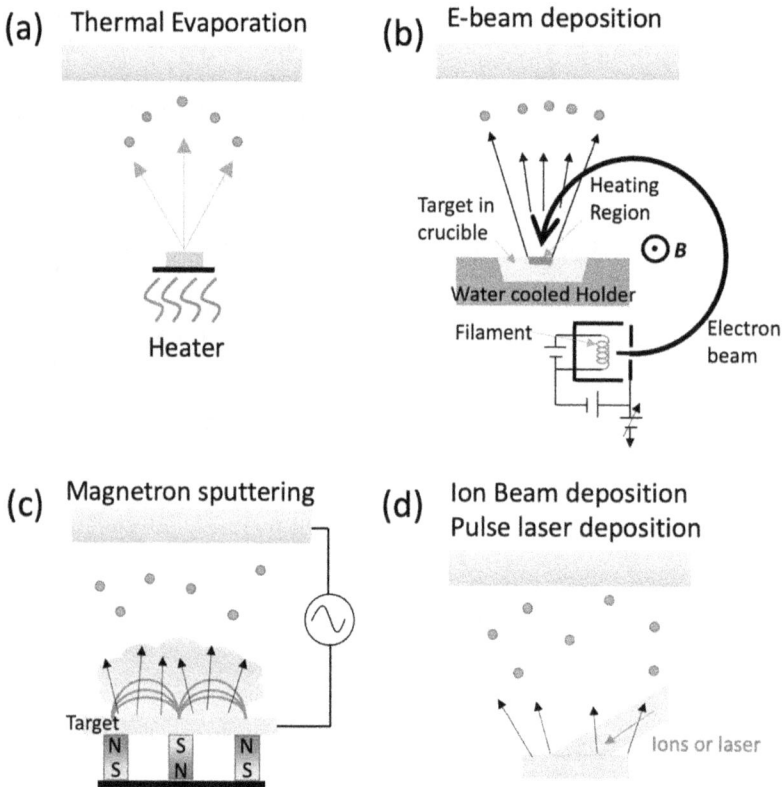

Figure 3.9. Schematics of PVD process: (a) Thermal evaporation, (b) E-beam deposition, (c) magnetron sputtering, and (d) ion beam or pulse laser depositions.

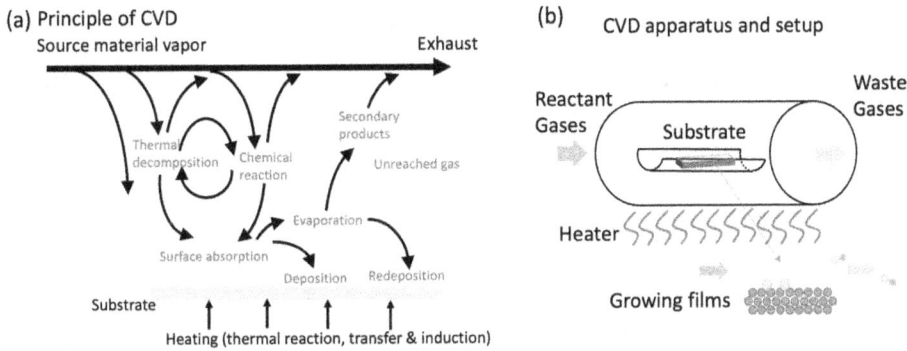

Figure 3.10. (a) Schematic diagram of principle of CVD. (b) Schematic of CVD apparatus and setup.

vacuum level is the most important parameter. By tuning the vacuum conditions, including particle pressure of sputtering gas and/or increasing the deposition rate, the quality of the films can be controlled.

For CVD, a source material is deposited on a substrate by introducing thermal, optical, and electromagnetic energy to the reactive vapour. The films are grown in major modes such as Volmer–Weber, Stranski–Krastanov, Frank–van der Merwe, and Nötzel–Temmyo–Tamamura (self-assembly) modes, as schematically shown in figure 3.10.

Metallic films (Au, Pt, Al, magnetic materials, etc), physically deposited or electroplated, are usually used for formation of electrical interconnects, electrodes, and as replication process masters. They can also be used as surfaces for self-assembled monolayers. One selects individual processes in order to optimize the entire process according to the target system, structure, etc. At that time, it is necessary to prioritize the improvement of consistency and yield with the processes before and after itself, and to create the process or set the process conditions. Therefore, the metallic thin films are frequently deposited by the magnetron sputtering, electron beam deposition, and electric plating. Here, in order to describe the most popular fabrication techniques for metallic thin films, PVDs such as thermal evaporation, electron beam evaporation, ion beam or pulse laser deposition, and sputtering (DC, RF, or magnetron) would be outlined.

3.5.1 Thermal evaporation

Thermal evaporation is one of common methods of PVD. The simplest method in the PVD typically uses a resistive heat source to evaporate a solid material in a vacuum environment to form a thin film, as schematically illustrated in figure 3.9(a). The material is heated in a high vacuum because vapour pressure of material is generally very low in the case of metals and oxide compounds. The evaporated material will become vapour stream or beam, travelling the vacuum chamber with thermal energy and coats the surface of the substrate to form a thin film.

Thermal evaporation deposits both metals and nonmetals such as oxide compound and nitride compounds. This thin film deposition method is often used commonly for applications involving electrical contacts, multilayers, and coating for the mirror and optical elements. More complex applications can be performed. The co-deposition of several components and reactive deposition under the reactive gas pressure like the sputtering method can be achieved by carefully controlling the temperature of individual crucibles and reactive gas pressure.

3.5.2 Electron beam deposition

Electron beam deposition is a method of using electron beams generated from an electron source in a vacuum to irradiate an evaporant material, and heating and evaporating it, as schematically shown in figure 3.9(b). The electron beam is controlled to hit the evaporant material, and the material is used as vapour to form a thin film. This method is often used as the MBE for metallic film depositions such as a lens, mirror, and metallic multilayers. The deflection-type electron source is often installed in a wide range of fields associated with electronics. On the other hand, the direct-type electron sources are large and high powered, enabling high speed evaporation. As a result, they are often used for deposition onto long films and large-area substrates in the manufacturing process for mass production. In the research and development, the deflection-type electron source is generally used. This device consists of two main sections, that are an electron source and a crucible section, as schematically illustrated in figure 3.9(b). The electron source is housed in the vacuum evaporation device. It generates electrons from heating filament and accelerates them by the application of high electronic voltage. Using Lorentz force, they are focused and deflected as an electron beam to irradiate the evaporant material. The crucible section holds the evaporant material to promote evaporation by electron beam irradiation with a heat insulating structure. As the heat source is the kinetic energy of the electrons, the evaporant material is heated directly. This method enables evaporation of various materials including high melting point metals as well as oxide compounds. An electron beam can be precisely controlled using electric and magnetic fields. The electron beam can be scanned at high speed and desired position with the optimal electric density for the evaporant material. It is used in high vacuum (10^{-2}–10^{-5} Pa or so higher). The deposition rate can be precisely controlled. Creation of film as thick as 1 μm or more is easily possible. Of course, the epitaxial ultra-thin film can be formed by adjusting suitable conditions. For example, it is necessary to consider lattice matching, gettability and wettability between the film-forming material and the substrate material.

3.5.3 Sputtering

Sputtering is a phenomenon in which when the surface of a solid is irradiated with high-energy ion particles accelerated by an electric field and collided with each other, the atoms and molecules on the surface of the solid are ejected to the outside and the surface is worn. A metal to be attached as a thin film is installed as a target in a vacuum chamber, and a noble gas element (usually argon is used) or nitrogen (usually derived from air) ionized by applying a high voltage is collided. Then, the atoms on the surface of the target are repelled and reach the substrate to form a film.

Since the principle is simple and there are various types of 'sputtering equipment', it is widely used in various technical fields. Recently, it has been used as a method for manufacturing thin films for semiconductors, liquid crystals, plasma displays, and optical discs, which require high quality thin films. In addition, a reactive sputtering method in which a gas is introduced into a vacuum chamber and reacted with a repelled metal to form a compound is also attracting attention as a new alloy or artificial lattice manufacturing technique. Figure 3.9(c) shows a schematic of magnetron sputtering. A target or a metal precursor, that is required to be deposited, is bombarded with energetic ions of inert gases such as argon, helium, krypton, and xenon etc. The forceful collision of these energetic ions with the target ejects target metal atoms into space. These metal atoms are then deposited on the substrate material forming a metallic film. The target is cooled by water to reduce the little radiation heat. It can deposit metals, alloys, and compounds onto a wide range of materials with thickness up to a millimetre. It also exhibits several important advantages over other vacuum coating technologies, a property that led to the development of a large number of commercial applications from microelectronic fabrication to simple decorative coatings.

3.5.4 Ion beam deposition

Ion beam deposition is the process of applying a material to a target through the application of an ion beam, as schematically illustrated in figure 3.9(d). Ion beam deposition equipment typically consists of an ion source, ion optics, and a vapour deposition target. An optional mass spectrometer can be incorporated. It is a method of creating an ion beam of a thin film material with an ion source and accelerating it to make it thinner. In a broad sense, it is a type of ion plating. Cluster ion beam deposition (ionized cluster beam deposition) is a typical example, and since a film with good crystallinity can be obtained, its application to epitaxial growth of semiconductors is being studied [45, 46].

The features of this method are described below.
  (1) Beam energy and ion density can be controlled independently.
  (2) Accurate control of deposit thickness and thin film characteristics allows for excellent homogeneity and reproducibility.
  (3) High mean free path due to operation in low pressure atmosphere, enabling high-purity, high-density thin film fabrication.
  (4) It is possible to reduce cavities in the deposited thin film and entrained gas.
  (5) Board temperature can be controlled in the range of 15 °C–300 °C pre-cleaning function makes it possible to improve the adhesiveness of thin films.
  (6) Extremely smooth coating with low light scattering and controllable increase in surface roughness below nm scale.

Recently, Terayama *et al* developed and reported that a Penning Ionization Gauge (PIG) ion source that can easily obtain a high-density plasma for plasma CVD enables depositing diamond-like carbon (DLC) layers with a high deposition density and adhesion [47]. In this way, ion beam deposition can be used like CVD by changing the gas in the vacuum chamber. This is one of the interesting methods with a wide range of applications.

3.5.5 Pulsed laser deposition

PLD is an abbreviation for pulsed laser deposition, which is a type of PVD method and is a device for producing thin films. Almost the same device is sometimes called the laser MBE method or laser ablation method. The laser strikes the target multiple time per second. This vaporizes the target material and scatters it onto the substrate, enabling the formation of a thin film.

For example, a high-energy pulsed laser (KrF excimer, wavelength: 248 nm) is applied to a densely sintered target in the chamber to cause instantaneous sublimation (ablation). Sublimated substances (molecules, atoms, ions, clusters, electrons, photons, etc) head toward the substrate while colliding with the reaction gas in the membrane-forming chamber in a plasma state called plume. Substances that reach the substrate settle or re-evaporate on the substrate. This method is often used for deposition of oxide thin films. Basically, oxides have a wide bandgap and a high melting point, but most targets can be ablated by irradiating them with a high-energy laser (energy density: 1–5 J cm^{-2}).

When the deposition process occurs, the inside of the chamber is once evacuated to a high vacuum, and then during film formation, it is controlled to a predetermined pressure with a reaction gas such as oxygen. If the substrate temperature is low, a film with a composition close to the target can be produced. Since there are many film formation and growth parameters, it is possible to form a film under various conditions. Since there are multiple target holders in the chamber, it is easy to make a multilayer film. *In situ* observation by Reflection High Energy Electron Diffraction (RHEED) is possible. RHEED is a popular technique for evaluation of films under the growth, being frequently used in PLD and other PVD processes.

3.6 Plating for nano/Nano–Microsystems

There are various methods of plating such as electroplating and hot-dip plating, and they are used properly according to the purpose and application. The purpose of the plating process is decoration, corrosion protection, surface hardening, and function-alization (mechanical, electrical, magnetic, and optical properties). In addition, typical plated products include galvanized iron and tinplate. Galvanized iron is iron plated with zinc, and tin is iron plated with tin. By plating a base metal such as iron with a metal having a high ionization tendency such as zinc, there is an effect of preventing corrosion of the base metal of a film by plating a metal having a low ionization tendency such as tin due to a potential difference. The conductive material is plated by immersing it in a plating solution and connecting it to the cathode. When plating a non-conductor such as plastic, the surface is subjected to a conductive treatment and then immersed in a plating solution to electrolyze, or plated by vacuum vapour deposition or sputtering. Plating methods include electro-plating, hot-dip plating, and electroless plating [48, 49].

Gold plating and electric field nickel plating are often used for electroplating. Commonly used electric field nickel plating baths are the Watt bath and the nickel sulfamate bath. In addition to electroplating, the technology that utilizes the phenomenon that metal is electrolyzed to the cathode when a conductor is immersed in an electrolytic solution containing metal ions and electrolyzed includes electroplating, electrolysis purification, electrolysis, and electrolytic metal foil. The electroforming is made by layering thick plating with electroplating to give strength, making it look like a cast product. Used for manufacturing records, compact discs, DVD stampers, soft vinyl product moulds, accelerator parts, rocket engine combustion chambers, etc. Since it is suitable for faithful reproduction of details at room temperature, it is suitable for precision processing. On the other hand, it is not suitable for mass production because it takes a long time to process. However, there is a method of mass-producing by moulding using a mould manufactured by electroforming, which is called LIGA process as described below. Electroless plating is a method of depositing metal on the surface of an object to be plated without an external power source. Metals are deposited by the reducing action of chemicals, and the reduced metal itself acts as a catalyst, being also called autocatalytic plating. There are many examples such as electroless silver plating, Electroless copper plating, Electroless nickel plating, Electroless nickel–tungsten alloy plating, Electroless nickel-PTFE composite plating, electroless tin plating, electroless gold plating, etc. Hot-dip plating, where a steel material or the like is immersed in the melted metal for plating, is industrially used. It has the features that a thick film can be obtained, even a large structure can be processed in a short time, continuous plating is easy, and plating can be performed even in places that are out of reach such as inside pipes.

Plating, including electroless deposition (ELD) and electric plating, is a well-known method for forming thin films. They are highly selective methods allowing additive patterning. ELD enables the additive manufacturing of isolated and embedded structures on insulating substrates. A well-organized review was made by Shascham-Diamand et al, showing ELD also has superior advantage due to its simplicity and by having a better step coverage on complex shape substrate [48]. The electroless NiB, CoB, CoW–X (X = P, B), CoSnB, Pd films have superior thermal stability as capping layer or barrier layer. In addition, the insertion of a self-assembled monolayer (SAM) to the formation of these layers is effective to produce conformal and thin film.

Plating is a technology that gives the object to be plated the desired function by precipitating the desired metal on the surface of the object to be plated. In electroplating, electrons are supplied by passing a direct current from an external power source, and metal ions are reduced and deposited on the object to be plated placed on the cathode (−) side, as schematically illustrated in figure 3.11 [49]. The object to be plated is limited to the conductor, and the current density distribution on the object to be plated differs depending on the part, so the plating film thickness will differ. Herein, the typical experimental method and procedure using a Ni-plated sulfamic acid bath are described below. They are often used for LIGA process. An example of the electroplating procedure is described below.

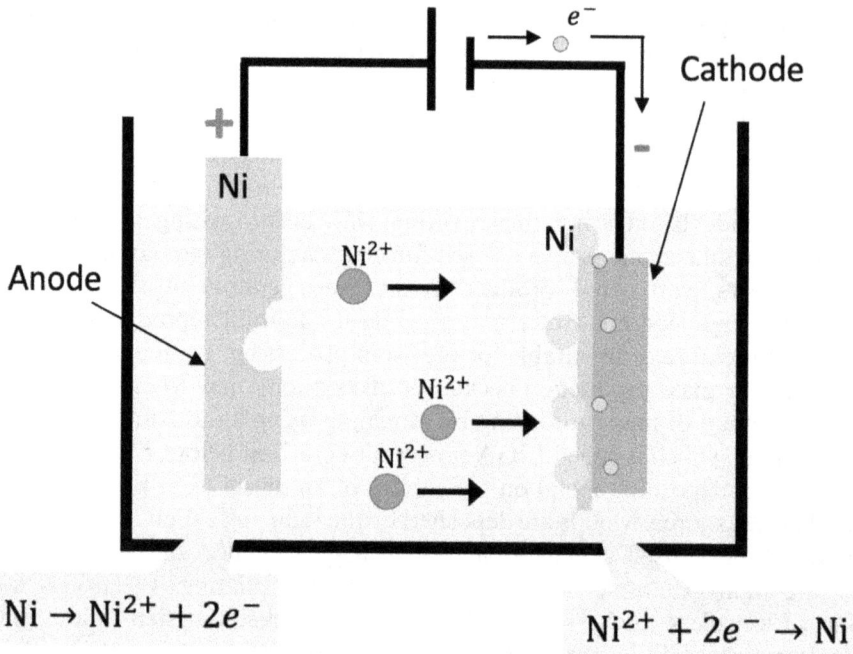

Figure 3.11. Schematic of Ni plating.

Procedure:
  (1) Adjustment of pH meter.
  (2) pH adjustment in the beaker.
  (3) Measurement of metallic nickel concentration.
  (4) Nickel chloride concentration measurement.
  (5) Measurement of boric acid concentration.
  (6) Calculation of surface area, current value, plating time
        I. Accurately determine the area of the part of the board to be plated.
        II. Calculate the current value from the following formula:

Current value [A] = Surface area (cm^2) × Current density (A cm^{-2}).

        III. Determine the plating time.
  (7) Creation of the electrically conductive part of the cathode (base).
  (8) Insulation treatment of the board.
  (9) Substrate surface treatment.
  (10) Start of plating:
        I. Put the anode plate and substrate into the plating bath.
        II. Start plating with the calculated current value.
        III. Leave for the calculated time to get any thickness.
  (11) After plating is completed, take it out, wash it with water, wash it with
        ethanol, and dry it, and observe it. After electroforming, after removing
        the resin, when using it as a mould, it is cut or polished according to the
        nesting of a moulding machine or the like.

3.7 LIGA process

Deep x-ray lithography (DXL) uses x-rays with a wavelength of 0.1–1 nm instead of UV light, which is commonly used as an exposure light source. Due to the short wavelength of x-rays, 'blurring of penumbra' due to diffraction is extremely suppressed, and due to the high transmittance to PMMA, deep digging of 1 mm or more is possible [21]. This makes it possible to fabricate a structure with high accuracy and high aspect ratio (height/width). Range of maximum width and aspect ratio for each process is summarized in figure 3.12. In addition, since it is chemically processed as compared with cutting, its side wall becomes a mirror surface. Figure 3.13 shows the schematic diagram of LIGA process consisting of (a) x-ray or UV lithography, (b) development, (c) electroforming, and (d) mould injection. Specific LIGA process implementation methods are described below.

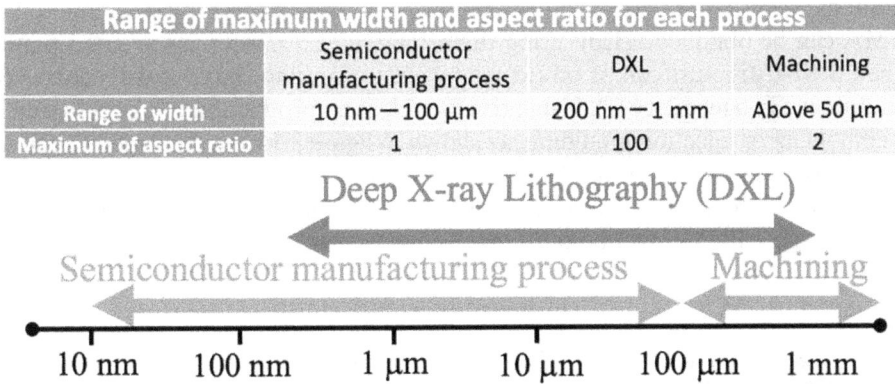

Range of maximum width and aspect ratio for each process			
	Semiconductor manufacturing process	DXL	Machining
Range of width	10 nm – 100 µm	200 nm – 1 mm	Above 50 µm
Maximum of aspect ratio	1	100	2

Figure 3.12. Range of maximum width and aspect ratio for semiconductor manufacturing process, deep x-ray lithography and machining.

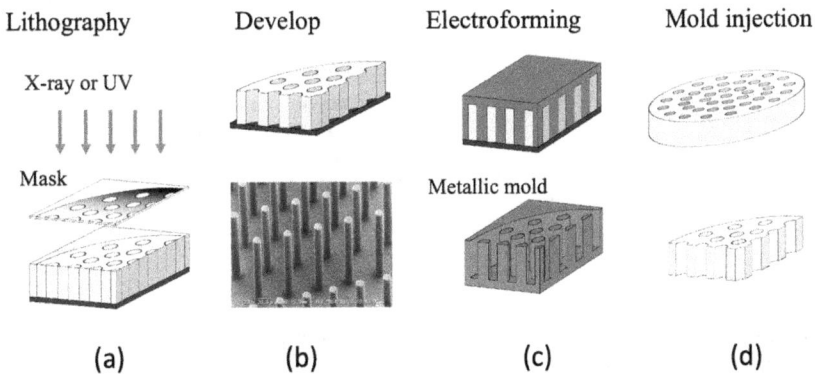

Figure 3.13. Configuration of the LIGA process: (a) lithography, (b) development, (c) electroforming and (d) mould injection.

Figure 3.14 shows the processing process of PMMA substrate by DXL. First, an x-ray mask consisting of Ni electroplating is fabricated. There is also the option of a membrane mask using polyimide, Ti, SiN, and SiC films, but the details of these process are omitted here. Basically, the processes are similar to the one described below. Those interested should refer to the references. In order to ensure conductivity, copper, Cu, is sputtered on the surface of the Si substrate in advance. SU-8 is then applied, and patterning is performed in sequence by UV lithography and developed, as shown in figures 3.14(A) and (B), respectively. After forming the SU-8 pattern, Ni electroforming is carried out as shown in figure 3.14(A)-(c). Next, the Si substrate is dissolved with an aqueous solution of potassium hydroxide, and SU-8 is stripped with a stripping solution (Remover PG, MicroChem Corp.). Then, we obtain the x-ray mask, as shown in figure 3.14(A)-(d). The x-ray mask is placed in contact with the PMMA substrate and exposure is performed using synchrotron radiation, as shown in figure 3.14(A)-(e). After exposure, GG developer (60% diethylene glycol, 20% morpholine, 15% pure water, 5% ethanolamine) is used, and the development conditions were 37 °C for 1 h. Structures with high aspect ratio composed of PMMA can be obtained. Then, using this structure as a master, an inverted pattern of the PMMA structure can be fabricated by electroforming, and this will become the mould shown in figure 3.14(C). Using this mould, mass production can be achieved by hot embossing or injection moulding, as shown in figure 3.15. At this time, this mould should have the same shape as the first x-ray mask.

In addition, the process of developing this technique to soft mould microfluidic structures using PDMS will be outlined. Soft moulding will be presented in section

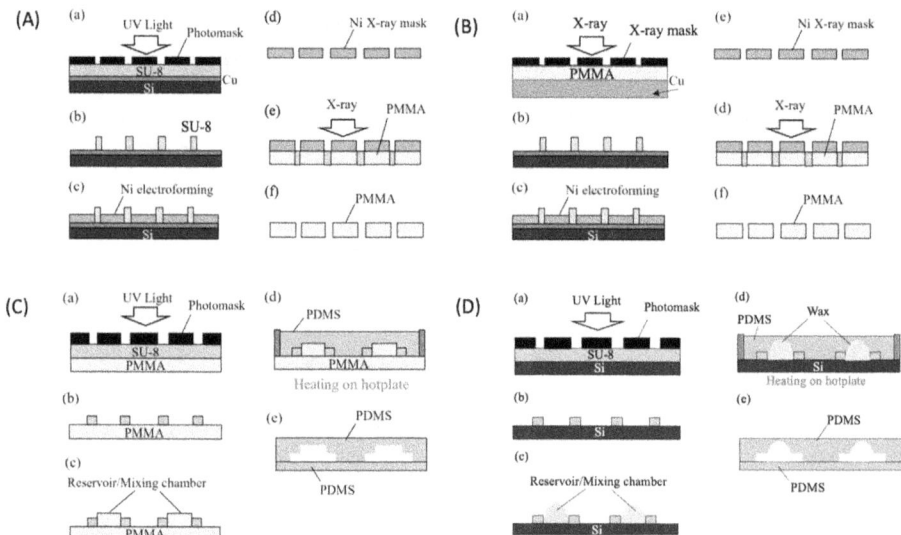

Figure 3.14. (A) X-ray mask fabrication process and process flow of PMMA microstructure fabrication using x-ray lithography. (B) X-ray lithography process for obtaining high-aspect ratio structures with PMMA. (C) Microchannel configuration process combining (A) and (B). (D) Fabrication process of the Lab-on-a-Disk used in an example of chapter 5.

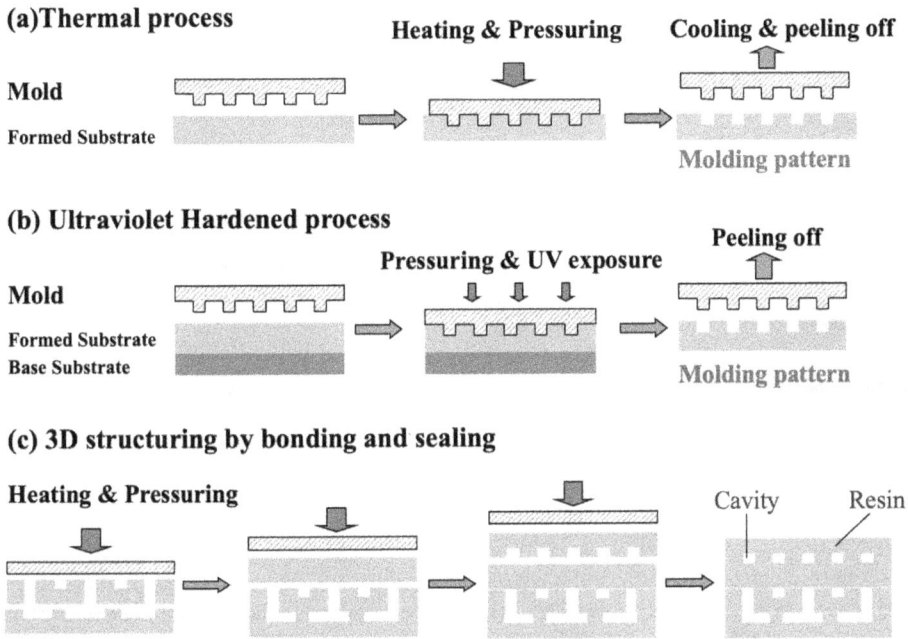

Figure 3.15. Fabrication technology for 3D stacked fluidic devices. Direct stacking enables functional intensification through encapsulation and sealing structures.

3.9. The structure is staircase-shaped in cross-section and requires a two-step process. The process is schematically illustrated in figure 3.14(C). First, SU-8 is applied onto the PMMA substrate and patterned by UV lithography, as shown in figure 3.14(C)-(a). After exposure and development, we obtain the structure shown in figure 3.14(C)-(b). Next, the PMMA reservoir mixing chamber prepared by DXL is joined to the PMMA substrate, as shown in figure 3.14(B). This bonding is performed by adding methyl methacrylate to the bonding surface and heating at 50 °C. for 10 s. After that, a metal frame was attached to the PMMA substrate, and a PDMS (SILPOT 184, Dow Corning Toray Co. Ltd, Japan) precursor (a liquid in which a PDMS monomer and a cross-linking agent were mixed at a ratio of 10:1) was poured into the PMMA substrate for 6 h at 50 °C, then heated on a hotplate, as shown in figure 3.14(C)-(d). Then, the PDMS is peeled off from the mould and adhered to a PMDS substrate having a flat surface as shown in figure 3.14(C)-(e). Here, the depth of the reservoir mixing chamber is determined by the thickness of the PMMA substrate to be machined, and the depth of the flow path of the capillary valve is determined by the film thickness of SU-8.

3.8 Three-dimensional printing technology

3D printing is a smart additive manufacturing process that allows the fabrication of devices and systems that are usually difficult to design and manufacture using conventional methodologies such as semiconductor process, machining or moulding process [13–20]. 3D printers can print various materials including inorganic and

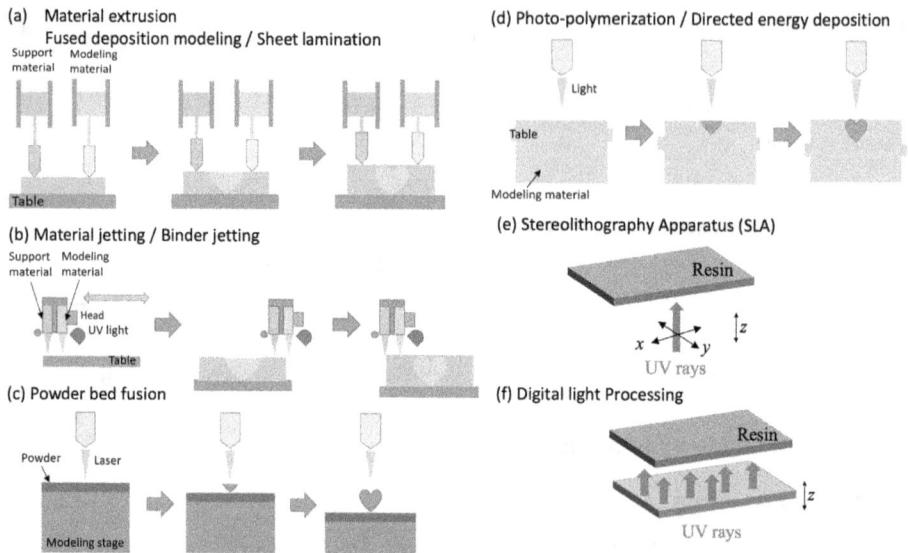

Figure 3.16. List and schematic diagram of 3D printer: (a) Materials extrusion process. Fused deposition modelling (FDM) is an additive manufacturing process in which materials is selectively dispensed through a nozzle or orifice. Sheet lamination (SL) is an additive manufacturing process in which sheets of material are bonded to from an object. (b) Material jetting or binder jetting process. The former is an additive manufacturing process in which droplets of build materials are selectively deposited, corresponding to the inkjet process. The latter is an additive manufacturing process in which a liquid bonding agent is selectively deposited to join powder materials. (c) Powder bed fusion, be an additive manufacturing process in which thermal energy selectively fused regions of a powder bed by laser or electron beam, etc. (d) In a vat (or container) filled with a liquid photopolymer resin, photo-polymerization, being an additive manufacturing process in which liquid photopolymer in a vast is selectively cured by light-activated polymerization. Directed energy deposition is an additive manufacturing process in which focused thermal energy is used to fuse materials by melting as they are being deposited. (e) Stereolithography Apparatus (SLA) is basically the same as (d) Vat photo-polymerization process, but while the process type (c) is larger, it is inverted (suspended) to save space and reduce costs. (f) Digital light processing (DLP). Whereas the suspended SLA method exposes at a point and moves the stage, DLP uses a digital mirror device and exposes in a plane, with stage movement only in the Z-axis. This method saves space and costs even more, and the resolution has improved as the performance of the digital micromirror device (DMD) has improved. For the definition and explanation of additive manufacturing processes, reference is made to American Society for Testing and Materials: https://www.astm.org.

polymers with varying density, strength, and chemical properties, providing users a broad variety of strategic selections to preparing the systems. Below, we have outlined potential 3D printing technologies and process for the fabrication of microfluidic devices that are suitable for engineering applications for biomedical, chemical reaction and analysis, etc. Figure 3.16 shows a summary of 3D printer methods with reference to American Society for Testing and Materials [14].

3.8.1 Fused deposition modelling

Fused Deposition Modelling (FDM) is one of the widely used additive manufacturing processes due to its capability to create complex parts. It has been utilized in the automobile industry, ranging from testing models, lightweight tools to final functional components. It is a method of ejecting heat-melted resin and laminating it. Therefore, FDM technique faces two main obstacles to be developed as an effective processing method in the automobile industry: weak and anisotropic mechanical properties and a limited variety of printing materials. Recently, improving the technique, it can be concluded that the FDM parts used in various fields requiring desired mechanical properties and good dimensional accuracy by adjusting either optimizing printing process or improving materials properties.

3.8.2 Powder method

There are two main types of powder methods: a method in which nylon powder is sintered with a thermal laser, and a method in which gypsum powder is cured and laminated. Nylon has excellent strength and enables the creation of prototype models that can withstand assembly tests on actual machines. This method is often called Selective Laser Sintering (SLS). SLS is one of the 3D printing techniques which belongs to the family of powder bed fusion. As mentioned, in SLS, powder fusion is accomplished utilizing various particle-binding mechanisms, that is, by chemical reactions, by solid-state sintering, or by absolute or partial melting. The resolution is dependent of the particle size of the powder, scanning speed and spacing, laser strength, and quality of the powder. In this SLS process, polymers such as polystyrene (PS), polyamide (PA), and polycaprolactone (PCL), etc are frequently used as laser sintering materials.

On the other hand, plaster is excellent in low-cost and short-time modelling, and can be coloured at the time of output. This method has evolved into metallic 3D modelling using metallic powder. Currently, the metallic 3D printers that use not only lasers but also electron beam are on sale and generally used.

3.8.3 Stereolithography apparatus (SLA) and digital light processing (DLP)

This is a method that often uses 'photo-curing resin' and an ultraviolet laser. Photo-curing resin is a special material that is cured by ultraviolet rays and is shaped by repeating the operation of stacking and irradiating the resin. The strength is that the resin cures immediately after light irradiation, so the moulding speed is fast and complicated modelling can be reproduced smoothly. Cooke et al utilized a biodegradable resin mixture of diethyl fumarate (DEF), poly(propylene fumarate) (PPF) and a photoinitiator, bisacylphosphine oxide (BAPO). They demonstrated the manufacture of biodegradable polymeric scaffolds made of DEF/PPF for tissue engineering utilizing the SLA. Palaganas et al demonstrated 3D printing of cellulose nanocrystal (CNC)-filled biomaterials with significant improvement in mechanical and surface properties [19]. Their findings may potentially pave the way for an alternative option in providing innovative and cost-effective patient-specific solutions to various fields in the medical industry. The 3D printing of CNC nanocomposite hydrogel via SLA forming a complex architecture will be useful for tissue engineering.

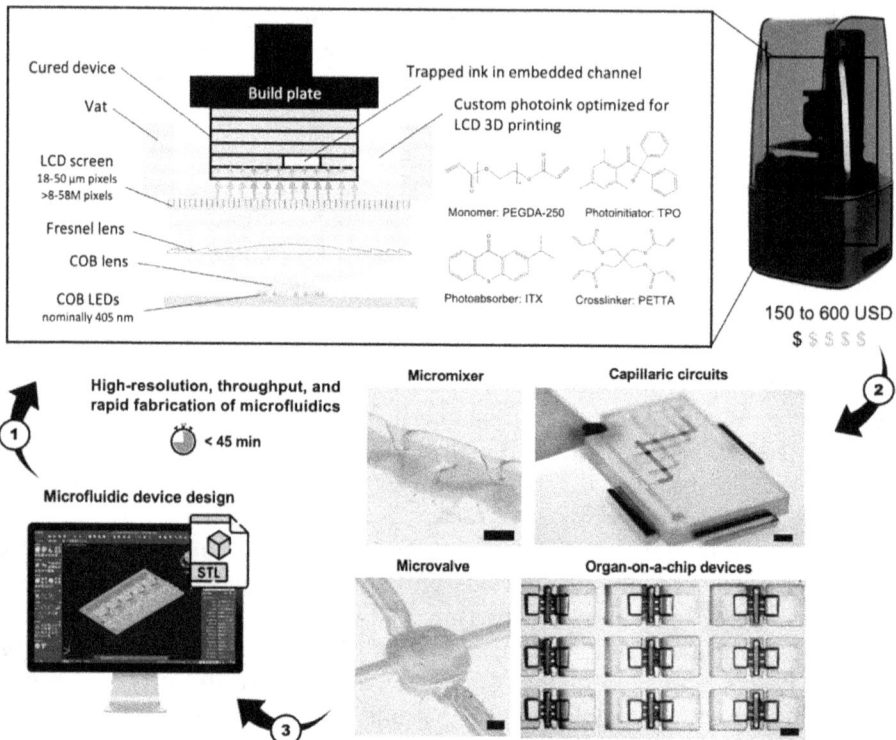

Figure 3.17. Low-cost LCD 3D printing of microfluidic devices and the rapid prototyping cycle. The workflow including (1) manufacturing on low-cost LCD 3D printers using a custom PEGDA-based ink (PLInk) optimized for LCD 3D printing, (2) directly manufactured microfluidic devices that can be tested and characterized, and (3) inform design improvement for the next iteration. Pixel sizes of 18–50 μm and print areas of up to 218 × 122 mm^2 afford high print resolution over large areas. Photoinks optimized for LCD 3D printing with reduced sensitivity to light heterogeneity and low viscosity enable the direct manufacture of microfluidic chips including open and embedded microchannels with a lateral resolution <100 μm and vertical features as thin as 22 μm in <45 min. Scale bars = 500 μm. Reprinted from [20] with permission from the Royal Society of Chemistry.

In the SLA method, small dot-like ultraviolet (UV) lights are applied to the resin for moulding, so even complex and detailed objects can be faithfully output according to the information in the 3D data. However, the disadvantage is that the moulding speed is slightly slower than the DLP method because the range of UV to be irradiated is narrow.

DLP is characterized by the fact that the UV light is applied to an area rather than to a point as in SLA [20]. As a result, DLP is faster than SLA. On the other hand, it has also been said that the wider the moulding area, the coarser the resolution and the more likely distortion is to occur, but this has been improved by technological advances and it is now analyzed that DLP has the highest resolution and accuracy. The DLP exposure mechanism uses DLP technology from TI (Texas Instruments), which projects a concentrated beam directly onto high transmittance glass. The light scattering value is reduced by projecting the beam in a concentrated manner. As a result, the energy light intensity at the bottom of the resin solution reservoir reaches more than 95%. For these reasons, it has a very good performance. This technology is also used in semiconductor maxless exposure machines. Recently, very inexpensive DLP exposure machines have become readily available for sale and use.

The DLP method is a moulding method that hardens the liquid resin by applying ultraviolet rays from below, as shown in figure 3.16(f). A feature of the DLP method is that irradiates planar UV, as opposed to the SLA method, which irradiated point-like UV rays. It will be easier to understand if you imagine how resin is hardened and layered like a mill-feuille. Compared to the SLA method, the DLP method has the advantage of being able to produce objects with a sense of speed because the range of UV rays is larger than that of the SLA method. However, unlike SLA method, where the resolution is consistently the same, it can be said that the disadvantage is that the wider the moulding range, the coarser the resolution becomes. The SLA method takes longer to build than the DLP method, but all figures will be output at the same resolution. In the case of the DLP method, the printing speed is fast, but the resolution is somewhat coarse, and each figure is output with a slightly different size and shape. However, when printing only one small object, it is possible to output speedily while maintaining resolution.

In fact, other 3D printers that melt and solidify filaments made of materials such as plastics and metals (FDM, etc) have the disadvantage of conspicuous layering marks on the finished product during the modelling process. Lamination traces refer to the state in which traces of stacking remain like strata in the process of moulding while melting and solidifying materials. In particular, when making figurines, accessories, and other models whose appearance is important, the traces of lamination impair their appearance. Even in microfluidic systems, it is necessary to pay attention to the fact that fluid behaviour will change if there is surface precision or unevenness. Stereolithography 3D printers also irradiate with ultra-violet rays and gradually build up layers from where the light hits them, leaving traces of lamination. However, by dripping the resin liquid of the material onto the layered part during moulding, it is possible to cover the jaggedness of the layering marks and create a smooth surface as a result. Therefore, the application of stereolithography 3D printers (SLA/DLP) to microfluidic systems is superior to other 3D printers. In addition, one of the features of stereolithography 3D printers is the ability to create highly transparent shapes. The use of highly transparent materials is useful for creating models that allow visualization of the internal structure, as well as for modelling vessels that emphasize transparency. However, it should be noted that it is difficult to maintain the transparency permanently because the colour of the modelled object turns yellow due to deterioration over time.

In recent years, methods for creating microchemical systems using 3D micro-structure fabrication techniques have been established, making it very easy and fast for university laboratories and corporate R&D groups to create them, as shown in figure 3.17. By combining the microchemical system created in this way with a bioprinter, research cases have also been reported in which cells are seeded three-dimensionally in the microchemical system to construct tissues and other structures.

3.9 Moulding and sealing for microfluidics

LOCs and nano/microsystems based on fluidics are typically sealed to confine solvents, samples and reagents defined volumes. They are also necessary to reduce contamination and biohazards. In addition, they prevent uncontrolled spreading of liquids. Processes using PDMS and other resin materials to form microchemical systems are outlined in figures 3.17– 3.19. Here, PDMS is the most popular material in the academic microfluidics community. PDMS is inexpensive and easy to fabricate by replication of moulds. Here, an overview of the manufacturing and fabrication procedure is presented below. The PDMS lithography replication can be divided into eight main steps [50, 51]:

(1) **The preparation of mould**

Before using a mould it has to be made hydrophobic. One can buy a mould, a foundry or do it oneself. The silicon from one's SU-8 mould is normally quite hydrophilic and thus the PDMS will have a good affinity to it, strong enough to make the peeling impossible, as shown in figure 3.18.

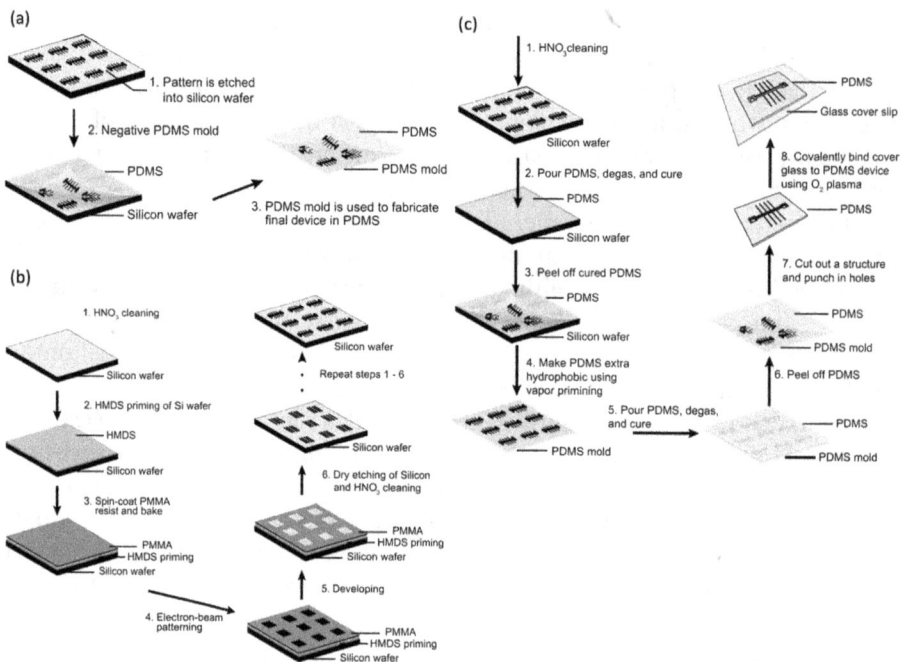

Figure 3.18. (a) Overview of PDMS moulding procedure. Here, etched silicon substrates are used as moulds. (Step 1) Moulds can also be made using SU-8 or with a 3D printer. Next, the PDMS is poured into the mould, moulded and removed. (Step 2) The PDMS can be used as it is for microfluidic channels etc, or another PDMS can be moulded using the PDMS as a mould. (Step 3) (b) Schematic diagram of fabrication process of multiple structure using EBL and dry etching for a Nano–Microsystem with nano/microscale structure. (c) Schematic of the fabrication of multiple structures in PDMS with sub-micron sized channels. Reprinted from [50] CC BY 4.0.

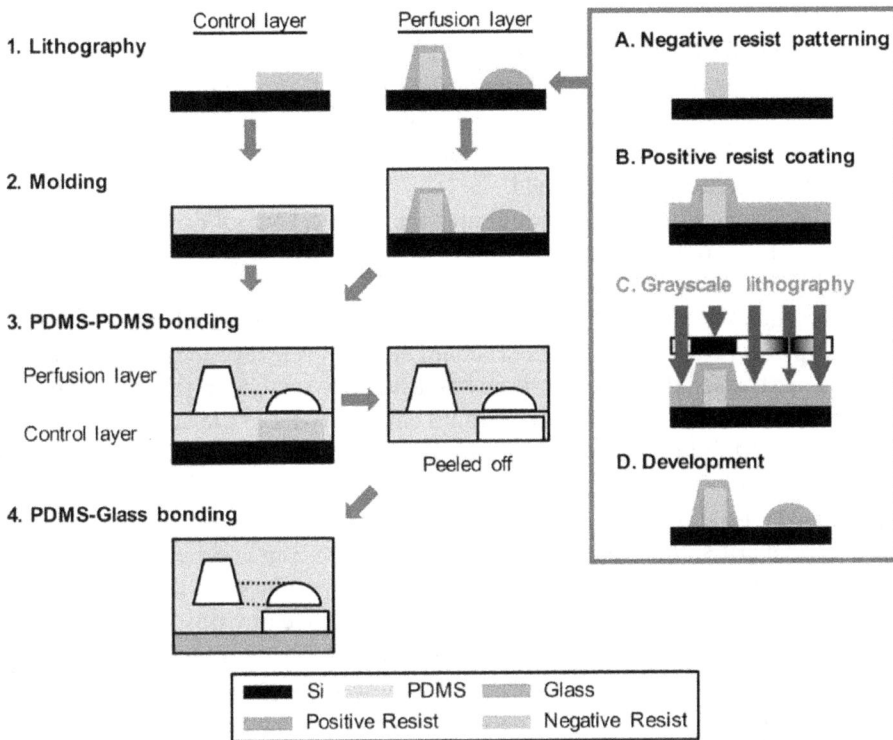

Figure 3.19. Fabrication of the integrated Heart/Cancer-on-a-chip microfluidic device. The device was fabricated as follows: (1) for the control layer, the negative resist was patterned on a Si wafer and was used as a casting mould. A casting mould of the perfusion layer was fabricated by a combination of the negative resist patterning and grayscale lithography using positive resist coating. Following this, DMD-based grayscale lithography combined with the simulation-based fabrication approach was employed. (2) The PDMS pre-polymer was poured onto the mould structure and cured. (3) The perfusion layer was bonded onto the control layer. (4) The assembled structure containing the perfusion and control layers was bonded onto a glass substrate using oxygen plasma treatment. Reprinted from [51] CC BY 4.0.

(2) **Scaling and mixing the PDMS and curing agent**

Once one's mould is ready, one can prepare the PDMS. To make it harder, one has to add a curing agent. The most used PDMS is the Sylgard 184. For the Sylgard 184, the usual ratio between the curing agent and the PDMS is 1:10 (weight).

(3) **Degassing to remove bubbles**

Due to the mixing of PDMS and curing agent, the prepared PDMS is filled with bubbles. These bubbles have to be removed otherwise they will be trapped inside the PDMS chip. There are several ways to degas. The most popular procedure is to create a vacuum using a vacuum pump or a vacuum line.

(4) **PDMS pouring on the mould**

Once the PDMS is degassed, one can pour it on the SU-8 mould. If some bubbles appear during the pouring, one can remove them with a needle or by putting the wafer back under vacuum.

(5) **PDMS baking**

The PDMS baking is performed by hot plates or an oven. The contact angle with water on the surface of the PDMS layer is almost independent of baking condition because the baking will not influence the chemicals of PDMS. The PDMS has been often baked at 80 °C in an oven during 2 h.

(6) **PDMS peeling off the mould**

After waiting for the mould and the PDMS to cool down, the PDMS is peeled off to release devices.

(7) **PDMS cutting and piercing**

When the PDMS is completed, one should cut it to release the different chips. In order to form a microchannel, it is necessary to make an air hole. In that case, the hole is often made with a punch or the like. Since the PDMS is a soft material, one should choose to make the hole a little smaller than the tubing. This way the tubing will be strongly maintained in the chip and it will decrease the leak problems.

(8) **PDMS bonding**

The last step is to bond the PDMS to form a microchannel. The PDMS can be bound on another piece of PDMS or on glass but the protocol is the same. Each part has to be well cleaned to remove any dust or particles from the surface. There are several steps and methods combined with an ultra-sonic bath, cleaning with alcohol, and clean N_2 gas blow etc. To bond the PDMS, the surface must be activated by transforming the Si–CH_3 function of the PDMS to a Si–OH to create strong and permanent Si–O–Si link. To do so the most used tool is a plasma cleaner working with O_2 or air. To functionalize the surface, the treatment should be performed by an appropriate condition. For example, a few minutes is a good time to create strong bonds with glass or PDMS, as shown in figures 3.15 and 3.19.

3.10 Nanoimprint lithography

Nanoimprint lithography is primarily a mechanical process in which a prefabricated mould is pressed into a lithography material such as polymers or resists, like stamping, as shown in figure 3.20 [52, 53]. This lithography method can provide a solution for both high-throughput and low-cost lithographic processes that can easily generate nanoscale and microscale structures in high throughput while EBL has the capability to form a nanoscale structure precisely but it is a fairly expensive and low-throughput process. A nanoscale pattern mould is prepared primarily by using EBL and EUV lithography. After the mould is pressed into the materials, the further hardening of the resist can be achieved using either chemical, optical, or thermal curing. After sufficient curing has been finished, the mould is removed from the resist and further development can be conducted.

3.11 Milling

Ion milling usually refers to a processing device or processing method that irradiates the surface of the sample to be observed with an argon ion beam to polish or etch the

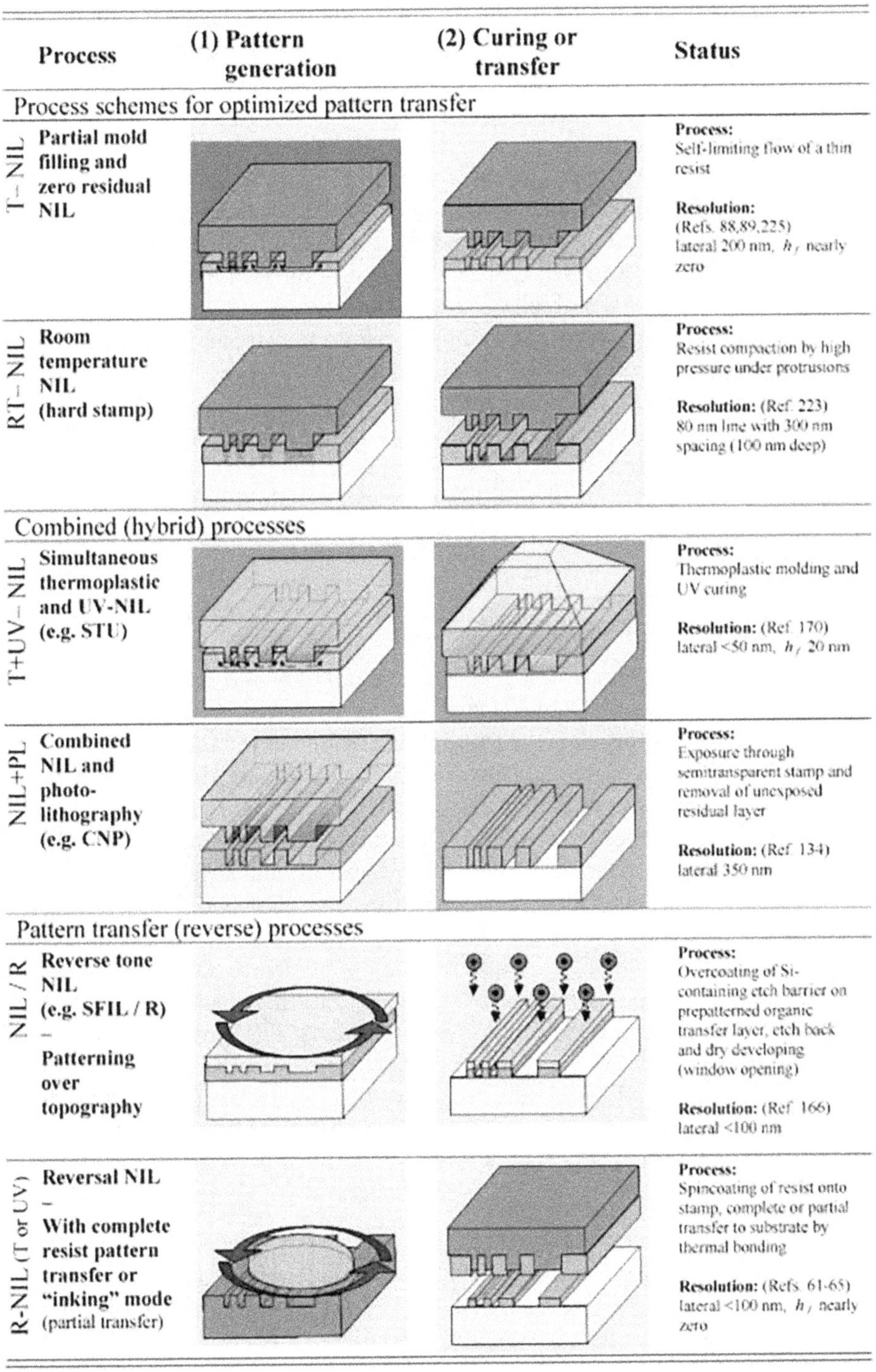

Process	(1) Pattern generation	(2) Curing or transfer	Status
Process schemes for optimized pattern transfer			
T–NIL Partial mold filling and zero residual NIL			**Process:** Self-limiting flow of a thin resist **Resolution:** (Refs. 88,89,225) lateral 200 nm, h_f nearly zero
RT–NIL Room temperature NIL (hard stamp)			**Process:** Resist compaction by high pressure under protrusions **Resolution:** (Ref. 223) 80 nm line with 300 nm spacing (100 nm deep)
Combined (hybrid) processes			
T+UV–NIL Simultaneous thermoplastic and UV-NIL (e.g. STU)			**Process:** Thermoplastic molding and UV curing **Resolution:** (Ref. 170) lateral <50 nm, h_f 20 nm
NIL+PL Combined NIL and photo-lithography (e.g. CNP)			**Process:** Exposure through semitransparent stamp and removal of unexposed residual layer **Resolution:** (Ref. 134) lateral 350 nm
Pattern transfer (reverse) processes			
NIL/R Reverse tone NIL (e.g. SFIL / R) – Patterning over topography			**Process:** Overcoating of Si-containing etch barrier on prepatterned organic transfer layer, etch back and dry developing (window opening) **Resolution:** (Ref. 166) lateral <100 nm
R-NIL (T or UV) Reversal NIL – With complete resist pattern transfer or "inking" mode (partial transfer)			**Process:** Spincoating of resist onto stamp, complete or partial transfer to substrate by thermal bonding **Resolution:** (Refs. 61-65) lateral <100 nm, h_f nearly zero

Figure 3.20. Summary of nanoimprint lithography processes. Reprinted from [52] with permission from AIP Publishing.

surface. In general, mechanical polishing of materials such as copper and aluminum induces crushing and thermal sagging. Further, in the case of resin, plastic, etc, when mechanical processing is performed, deformation occurs, the processing accuracy may not be high, or the element may be destroyed. Even in ceramics and silicon, cracks are likely to occur due to machining and polishing. Furthermore, even with a thin film, conventional mechanical polishing has been difficult for the same reason. However, by processing with an ion-milling device, it became possible to prepare a sample according to the purpose, and it became possible to perform clear observation and develop into a device. If ion milling is used, not only can fine processing be performed with high accuracy, but also processing can be performed while observing an SEM image (scanning electron microscope image) with an FIB observation device. There are two types of ion milling: cross-section milling and planar milling. Since cross-section milling can generate a cross-section without applying external stress, it is suitable for observing the thickness and layer structure of each layer and the cross-section of surface processing. Planar milling is used for final finishing after mechanical polishing because it can generate a surface with reduced unevenness formation by bringing the irradiation angle (Θ) of the ion beam close to 90°.

3.12 Laser micromachining

Femtosecond laser micromachining becomes a popular and major technique that found wide application in LOC fabrication. This technique is based on a nonlinear absorption process. Tailoring of the irradiation parameters enables one to produce a 3D structure. The advantage of this microfabrication technique is the capability to easily produce complex optofluidic devices. In addition, this process also has its unique capability to fabricate 3D optofluidic nano/microstructures taking advantage of the fact that the photochemical reaction induced modification is localized to the focal volume and can be arbitrarily placed anywhere. The additional advantage is the possibility to combine different fabrication processes from material property tuning to selective removal and additive manufacturing. One of the additive manufacturing processes is two-photon polymerization, which can create plastic microstructures embedded in the microfluidic channels. The most important issue of 3D formation technology is the rapid prototyping capability. The direct-formation without any mask can be achieved, a software design can be transferred forthwith onto a working prototype, with the possibility to simply and rapidly adjust the design or to evaluate the properties.

3.13 Deep RIE

Deep RIE is a type of reactive ion etching (RIE), which refers to reactive ion etching with a high aspect ratio (narrow and deep) [53–55]. It is also referred to as high-aspect ratio etching because of its high aspect ratio. In semiconductor devices such as DRAM, it is used to increase the breakdown voltage or to create large capacitors. It is also the main fabrication technique for bulk micromachining in MEMS, and many devices are fabricated using this method [1–7].

The deep drilling technique usually uses high-density plasma and either cools the sample to low temperatures, or uses an etching technique called the Bosch process, or both. The main method for generating high-density plasma is inductively coupled plasma (ICP) RIE; there is also a microwave-based method called ECR-RIE (Electron Cyclotron Resonance-RIE), but ICP-RIE is the predominant method due to the high equipment cost.

The Bosch process 'Bosch-process' is a switching etching method developed by Robert Bosch GmbH, Germany, in 1992, in which etching and passivation (side wall protection film formation) are repeated [53, 54]. This method is often used in MEMS, where alternating flows of SF_6 and C_4F_8 gases are used: in passivation mode, when C_4F_8 is flowing, a Teflon-like substance is deposited by plasma polymerization and the sidewalls are coated with a protective film. In etching mode with SF_6 the protective film at the bottom is scraped away to expose the Si, which is then etched by the F radicals. So, before the protective film on the sides is gone, the next protective film is deposited in turn. This method has the advantage that various things can be done by changing the parameters of the alternating flow of SF_6 and C_4F_8. For example, it is possible to meet various requirements, such as etching at the same speed whether the mask width is narrow or wide, even if the etching rate is slow (countermeasure against micro-loading effect), or etching quickly to create grooves with a large aspect ratio, even if this is sacrificed. The maximum etching speed is about 1.5 times faster than that of the conventional etching speed [55, 56].

3.14 Surface treatment

There are a wide variety of surface treatment methods, each with different principles. It is impossible to introduce them all here, so please refer to the literature according to the readers' needs. In particular, surface treatments used in microfluidics include plasma and ozone treatments to improve the adhesion of PDMS and to control the wettability of glass substrates. Antibody immobilization in bio-microchemical systems is another example of surface treatment. Companies that provide reagents and other materials related to these treatments provide accompanying documents on their recipes and methods.

For example, as a special case, surface treatments such as nano-patterning by polymer phase separation. Block copolymers consisting of flexible, chemically incompatible, heterogeneous blocks (e.g. polystyrene and polyisoprene) are capable of microphase separation with nanoscale structural morphology. Surfactants may be used in these polymer blends to highlight the differences in surface tension of each pattern. Other commonly used methods include surface atomic monolayer (SAM). Polyelectrolytes mainly have a large number of charged groups in the side chains, and the constituent monomer units dissociate into ions, and the characteristics of the solution in a high dielectric constant solvent range over a wider range than the typical molecular scale. These substances can be used as dispersants in water-soluble solvents, coagulants for coagulating slurries and industrial wastes, sizing agents in the textile and paper industries (controlling liquid absorption properties). In addition, it is widely used in industrial applications as a conditioning additive to prevent polishing damage. SAM technology is a simple but highly sophisticated method, an important nanoscale method for making functional supramolecular assemblies that can be applied to a variety of devices [57]. With this technique, anionic and cationic polyelectrolytes are alternately adsorbed (alternately laminated) on a suitable substrate. Fluoropolymer coating on the mould surface or coating with diamond-like carbon to make it easier to remove the mould after moulding by nanoimprinting are other uses of surface treatment. As mentioned above, this technology is important as a basic science and is actually used in the industrial field. It is also a technology that is not easily disclosed and must be kept secret by companies that earn profits from product development. In fact, when it is used in research and development at universities and other institutions, it is sometimes used by itself, and sometimes the mechanism itself is studied. In this case, it is mainly used for the fabrication of microchemical systems. As mentioned above, plasma treatment, UV and ozone treatment, and antibody immobilization are particularly important treatments. For example, the mechanism of surface modification of fluoropolymers by plasma is that radicals generated in plasma pull F atoms from the $-CF_2$ skeleton, and carbon atoms on the skeleton become radicals. The highly reactive radicals generated on the surface form C–C, C–H, C–O, C–O, and other chemical bondings through the reaction between radicals in the framework, cross-linking reactions, or surface addition reactions with radicals in the plasma. Okubo et al performed surface modification of PTFE by thermal assisting plasma treatment utilizing graft polymerization followed by Cu plating. As a result, their process increases the adhesive strength between Cu plating film and PTFE substrate. Ikeda and Kobayashi has recently showed that the plasma treatment with using ammonia gas improves the bonding strength between Cu and fluorinated ethylene propylene (FEP) or perfluoroalkoxyalkane (PFA) films [58]. Based on microprobe x-ray fluorescence spectra and microprobe near-edge x-ray absorption fine structure spectra using synchrotron radiation SPring-8, it is considered that the delamination of the FEP/Cu piece is mainly caused by resin failure, while the delamination of PFA/Cu is caused by interfacial delamination in addition to resin failure, as shown in figure 3.21 [59]. The hard x-ray photoelectron spectroscopy was also performed to confirm that the bonding via nitrogen is formed. Thus, the plasma treatment enables the achievement of high adhesion by N-mediated bonding at the interface of the Cu substrates with fluoropolymer FEP and PFA. These results are expected to provide significant insight into the bonding mechanism between fluoropolymers and metals, which is essential for the future development of electronic devices, microchemical systems, and so on.

3.15 Example of microsystem fabrication

The fabrication method of the micro hot stirrer is shown in figure 3.22(a). This device was fabricated using lithography and etching techniques used in semiconductor fabrication processes [60, 61]. First, a piezoelectric substrate (LiNbO$_3$: 128-degree rotation Y-plate X propagation) is used as a substrate. The thickness of the piezoelectric substrate is 500 μm and its surface roughness is less than 0.3 nm.

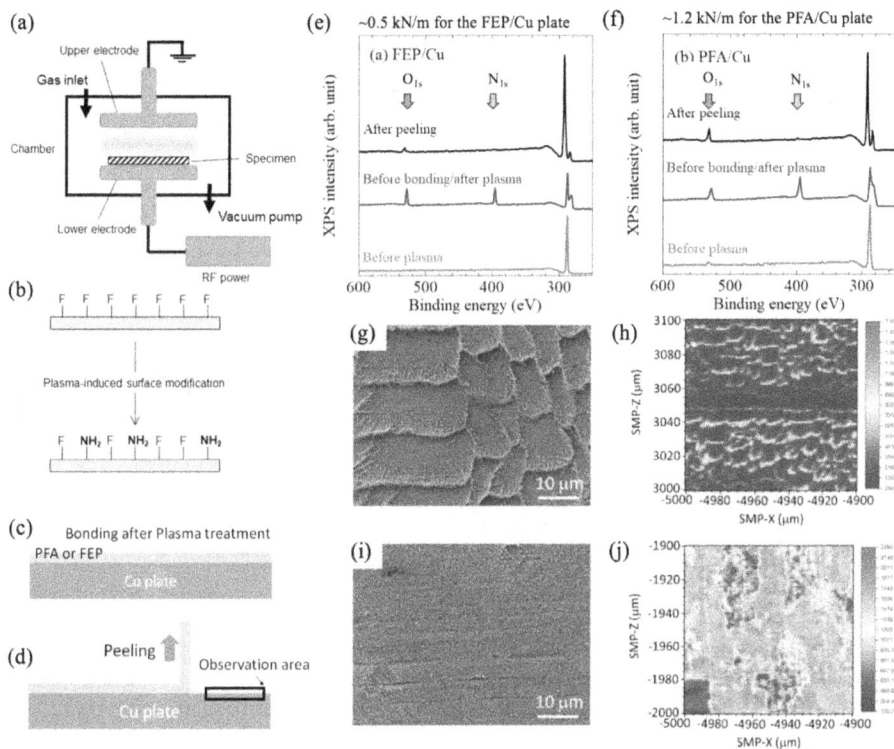

Figure 3.21. (a) Examples of physical and chemical SAM film treatments by carrying out a plasma treatment containing amino acids, (b) by replacing amino acids on the fluoropolymer surface and bonding copper substrates. (c) Peel test the bonded fluoropolymer from the copper substrate. XPS spectra before plasma treatment, after plasma treatment before bonding, after bonding and after peeling of samples (e) FEP/Cu and (f) PFA/Cu, respectively. (g), (i) SEM observations and (h), (j) synchrotron absorption spectroscopy mapping results on the copper substrate surface after peel test. In the (j) PFA/Cu substrate, there are red areas, i.e. areas corresponding to fluorine, and not only resin fracture coexisting with interfacial delamination as well as resin fracture. In (j) FEP/Cu substrate, only resin fracture occurs, so the peel strength is apparently small.

On this piezoelectric substrate, Cr and Au are deposited by RF sputter (CFS-4EP-LL, Shibaura Mechatronics) in this order. The thicknesses of the Cr and Au films are 10 and 100 nm, respectively. Next, UV exposure is performed using a positive resist (OFPR800LB-54cP, Tokyo Ohka Kogyo), and the IDT pattern on the glass mask is transferred using a mask aligner (MA6, SUSS Micro Tec). Then, based on this pattern, etching of Au and Cr is performed and the resist is peeled off to fabricate the IDT and heater that will be the source of surface acoustic wave (SAW). The device is then completed by attaching a liquid reservoir wall (Stratasys, Vero White) fabricated by a 3D printer (Stratasys, Connex 500) to this piezoelectric substrate using adhesive (Toagosei, Aron Alpha).

Next, an image of the fabricated micro hot stirrer device is shown in figure 3.22(a). Here, a batch-type SAW device was fabricated to investigate its basic characteristics as a stirrer. The device was fabricated on a piezoelectric substrate (LiNbO$_3$), and a lidless reservoir was placed in the centre. The reservoir consists of a hollow cylinder

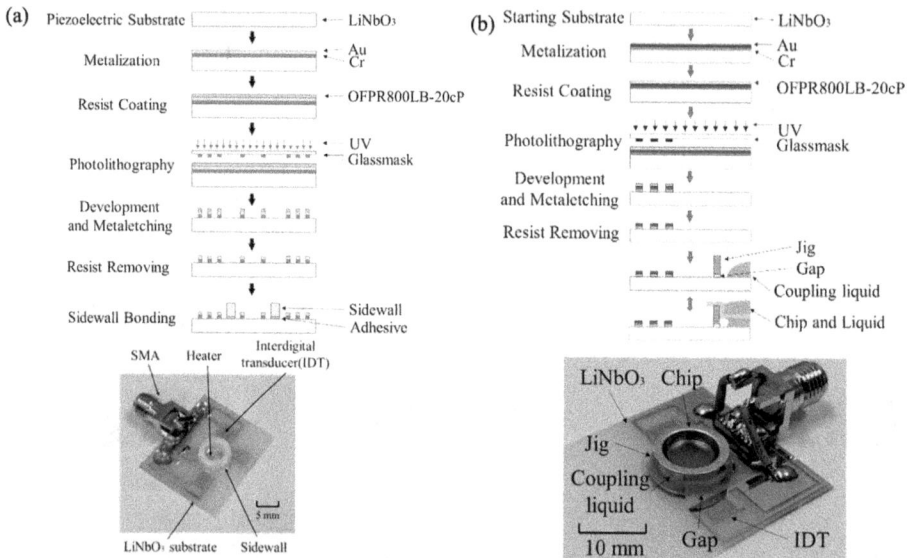

Figure 3.22. Fabrication procedures related to micro-mixing systems with surface acoustic waves presented in chapter 6. (a) Example where a heater is also built into the mixing vessel. The mixing vessel is directly bonded on the LiNbO₃ substrate. (b) System in which a coupling liquid is introduced between the piezoelectric substrate and the mixing vessel to propagate the sound waves, thus separating the piezoelectric substrate pump function from the microsystem for chemical processing and preventing contamination.

with an outer diameter of 8 mm, an inner diameter of 5 mm, and a depth of 2 mm, and can hold 40 µl of solution. SAWs generated from two IDTs (pitch 200 µm, aperture width 4.0 mm, log 20) are injected parallel to the central axis of the reservoir and rotate the solution in the reservoir. In addition, heaters (width 100 µm, resistance 55 Ω) are installed at two locations at the bottom of the reservoir to heat the solution while agitating it [60]. Figure 3.22(b) shows an example of a removable and replaceable microsystem with SAW actuator [61]. The details are described in chapter 7.

3.16 Summary

Here we have outlined the advances in lithographic technology and described the minimum technical elements required for the novice to construct Nano–Microsystems; for MEMS, a sophisticated combination of semiconductor processes is required, including the use of techniques such as sacrificial layer etching and surface-activated bonding. It also requires the use of semiconductor packaging technology, and there is much more that cannot be covered in this book alone, so we have omitted it here and focused on microfluidic channel formation and microchip fabrication. Specific examples are given on how to construct microfluidic channels and microfluidic systems, how to fabricate solid-state devices, and how to fabricate surface acoustic wave systems, so that the reader can prepare the necessary materials and equipment to enable the creation of systems. Recently, bioprinters that print cells have begun to become more widespread, making them readily available to researchers in the life sciences as well as in engineering research. Methods and procedures for creating Nano–Microsystems are also developing day by day, and it is recommended to keep an eye on technological trends and continue to take advantage of and use any new ones as they emerge.

References

[1] Korvink J G and Paul O (ed) 2006 *MEMS: A Practical Guide of Design, Analysis, and Applications* (New York: Springer)

[2] Yang Z (ed) 2022 *Advanced MEMS/NEMS Fabrication and Sensors* (Cham: Springer)

[3] Esashi M (ed) 2021 *3D and Circuit Integration of MEMS* (Weinheim: Wiley-VCH)

[4] Castillo-León J and Svendsen W E (ed) 2015 *Lab-on-a-Chip Devices and Micro-Total Analysis Systems–A Practical Guide* (Cham: Springer)

[5] Manz A, Neužil P, O'connor J S and Simone G 2020 *Microfluidics and Lab-on-a-chip* (London: Royal Society of Chemistry)

[6] Yamaguchi A, Hirohata A and Stadler B (ed) 2021 *Nanomagnetic Materials: Fabrication, Characterization and application* (Amsterdam: Elsevier)

[7] Judy J W 2001 Microelectomechanical systems (MEMS): fabrication, design and applications *Smart Mater. Struct.* **10** 1115–34

[8] Trantidou T, Fruddin M S, Salehi-Reyhani A, Ces O and Elani Y 2018 Droplet microfluidics for the construction of compartmentalized model membranes *Lab Chip* **18** 2488

[9] Rotem A, Abate A R, Utada A S, Steijn V V and Weitz D A 2012 Drop formation in non-planar microfluidic devices *Lab Chip* **12** 4263

[10] Kong L X, Perebikovsky A, Moebius J, Kulinsky L and Madou M 2016 Lab-on-a-CD: a fully integrated molecular diagnostic system *J Lab Autom* **21** 323–55

[11] Maguire I, O'Kennedy R, Ducrée J and Regan F 2018 A review of centrifugal microfluidics in environmental monitoring *Anal. Methods* **10** 1497–515

[12] Strohmeier O, Keller M, Schwemmer F, Zehnle S, Mark D, Von Stetten F, Zengerle R and Paust N 2015 Centrifugal microfluidic platforms: advanced unit operations and applications *Chem. Soc. Rev.* **44** 6187–229

[13] Jyothish Kumar L, Pandey P M and Wimpenny D I (ed) 2019 *3D Printing and Additive Manufacturing Technologies* (Singapore: Springer)

[14] *American Society of Testing and Materials* https://astm.org/

[15] Gebhardt A 2011 *Understanding Additive Manufacturing—Rapid Prototyping, Rapid Tooling, Rapid Manufacturing* (Munich: Carl Hanser Verlag)

[16] Cooke M N, Fisher J P, Dean D, Rimnac C and Mikos A G 2003 Use of stereolithography to manufacture critical sized 3D biodegradable scaffold for bone ingrowth *J. Biomed. Mater. Res.* B **64B** 65

[17] Rosen D W 2008 Stereolithography and rapid prototyping *BioNanoFluidic MEMS* ed P J Hesken (Boston, MA: Springer) p 175

[18] Guo N and Leu M C 2013 Additive manufacturing; technology, applications and research needs *Front. Mech. Eng.* **8** 215

[19] Palaganas N B, Mangadlao J D, de Leon A C C, Palaganas J O, Pangilinan K D, Lee Y J and Advincula R C 2017 3D printing of photocurable cellulose nanocrystal composite for fabrication of complex architectures via stereolithograph *ACS Appl. Mater. Interfaces* **9** 34314

[20] Shafique H, Karamzadeh V, Kim G, Shen M L, Morocz Y, Sohrabi-Kashaniab A and Juncker D 2024 High-resolution low-cost LCD 3D printing for microfluidics and organ-on-a-chip devices *Lab Chip* **24** 2774

[21] Saile V, Wallrabe U and Tabata O (ed) 2009 *LIGA and its Applications (Advanced Micro and Nanosystems)* (Weinheim: Wiley-VCH)

[22] Ito H and Willson C G 1984 Chemical amplification in the design of dry developing resist matetials *Polym. Eng. Sci.* **23** 1012

[23] Ito H 2008 Development of new advanced resist materials for microlithography *J. Photopolym. Sci. Technol.* **21** 475–91

[24] Nordström M, Johansson A, Noguerón E S, Clausen B, Calleja M and Boisen A 2005 Investigation of the bond strength between the photo-sensitive polymer SU-8 and gold *Microelectron. Eng.* **78–79** 152–7

[25] Ge J, Tuominen R and Kivilahti J K 2001 Adhesion of electrolessly-deposited copper to photosensitive epoxy *J. Adhes. Sci. Technol.* **15** 1133–43

[26] Kayaku Advanced Materials, PMMA 950 https://kayakuam.com/

[27] ZEON Corporation Electronic Materials Devision https://zeonchemicals.com/

[28] Nishida T, Notomi M, Iga R and Tamamura T 1992 Quantum wire fabrication by E-beam elithography using high-resolution and high-sensitivity e-beam resist ZEP-520 *Jpn. J. Appl. Phys.* **31** 4508–14

[29] Oyama T G, Oshima A, Yamamoto H, Tagawa S and Washio M 2011 Study on positive–negative inversion of chlorinated resist materials *Appl. Phys. Exp.* **4** 076501

[30] Mohammad M A, Koshelev K, Fito T, Zheng D A Z, Stepanova M and Dew S 2012 Study of development processes for ZEP-520 as a high-resolution positive and negative tone electron beam lithography resist *Jpn. J. Appl. Phys.* **52** 06FC05

[31] Micro Resist Technology, ma-N 2410, https://microresist.de/en/

[32] Kim Y and Jeong H 2008 Characteristics of negative electron beam resists, ma-N2410 and ma-N2405 *Microelectron. Eng.* **85** 582–6

[33] Chen Y 2015 Nanofabrication by electron beam lithographys and its applications: a review *Microelectron. Eng.* **135** 57–72

[34] Dobisz E A 1999 Resist materials and nanolithography *MRS Online Proc. Libr.* **584** 85–96

[35] Gentili M, Gerardino A and Fabrizio E D 1998 Electron-beam study of nanometer performances of the SAL 601 chemically amplified resist *Jpn. J. Appl. Phys.* **37** 4632

[36] Grigorescu A E, van der Krogt M C, Hagen C W and Kruit P 2007 10 nm lines and spaces written in HSQ, using electron beam lithography *Microelectron. Eng.* **84** 822–4

[37] AZ4000, MicroChemicals GmbH, https://microchemicals.com/products/photoresists/

[38] OFPR & TSMR-V90 & OMR-83, https://tok.co.jp/eng/products/semiconductor/list

[39] SU-8 & KMPR, http://nkc-mems.com/index_e.html

[40] https://kayakuam.com/wp-content/uploads/2019/09/SU-8-table-of-properties.pdf

[41] Dellmann L, Roth S, Beuret C, Paratte L, Racine G A, Lorenz H, Despont M, Renaud P, Vettiger P and de Rooij N F 1998 Two steps micromoulding and photopolymer high-aspect ratio structuring for applications in piezoelectric motor components *Microsyst. Technol.* **4** 147

[42] Lorenz H, Despont M, Fahrni N, LaBianca N, Renaud P and Vettiger P 1997 SU-8: a low-cost negative resist for MEMS *J. Micromech. Microeng.* **7** 121–4

[43] THB-430N, https://jsr.co.jp/jsr_e/products/

[44] Rowlands A 2017 Fundamental optical formulae *Physics of Digital Photography* (Bristol: IOP Publishing Ltd) ch 1

[45] Aisenberg S and Chabot R 1971 Ion-beam deposition of thin films of diamondlike carbon *J. Appl, Phys.* **42** 2953

[46] Fallon P J, Veerasamy V S, Davis C A, Robertson J, Amaratunaga G A J, Milne W I and Koskinen J 1993 Properties of filtered-ion-beam-deposited diamondlike carbon as a function of ion energy *Phys. Rev.* B **48** 4777

[47] Terayama N 2017 Preparation of Diamond Like Carbon (DLC) films by hot cathode penning ionization gauge type plasma-enhanced chemical vapor deposition method *J. Vac. Soc. Japan* **60** 85–91

[48] Shascham-Diamand Y, Osaka T, Okinaka Y, Sugiyama A and Dubin V 2015 30 Years of electroless plating for semiconductor and polymer micro-systems *Microelectron. Eng.* **132** 35

[49] Schlesinger M and Paunovic M (ed) 2011 *Modern Electroplating* 5th edn (New York: Wiley)

[50] Moolman M C, Huang Z, Krishnan S T, Kerssemakers J W J and Dekker N H 2013 Electron beam fabrication of a microfluidic device for studying submicron-scale bacteria *J. Nanobiotechnol.* **11** 12

[51] Kamei K-I, Oka A, Tsuchiya T, Kato Y, Hirai Y, Ito S, Satoh J, Chenad Y and Tabata O 2017 Integrated heart/cancer on a chip to reproduce the side effects of anti cancer drug *in vitro RSC Adv.* **7** 36777

[52] Schift H 2008 Nanoimprint lithography: an old story in modern tiems? A review *J. Vac. Sci. Technol.* B **26** 458–80

[53] Laermer F and Schilp (Robert Bosch GmbH) A 1993 Method of anisotropically etching silison *US-Patent* USA US5501893A

[54] Laermer F, Schilp A, Funk K and Offenberg M 1999 Bosch deep silicon etching: improving uniformity and etch rate for advanced MEMS applications *Proc. of the IEEE Int. Conf. on Micro Electro Mechanical Systems (MEMS)*

[55] https://psi.ch/en/lnq/nanoimprint-lithography

[56] https://spp.co.jp/infinity/reason/si/

[57] Ulman A 1996 Formation and structure of self-assembled monolayers *Chem. Rev.* **96** 1533–54

[58] Ikeda S and Kobayashi Y 2020 Plasma-induced surface modification of fluorine resin films for direct plating *J. Surf. Soc. Jpn.* **71** 775–80
Kobayashi Y and Ikeda S 2021 Plasm-induced surface modification of fluoropolymer films for direct adhesion with copper *J. Surf. Soc. Jpn.* **72** 333–9

[59] Yamaguchi A, Ikeida S, Nakaya M, Kobayashi Y, Haruyama Y, Suzuki S, Kanda K, Utsumi Y, Sumida H and Oura M 2023 Soft X-ray microspectroscopic imaging studies of exfoliated surface between fluoropolymer and Cu plate directly bonded by plasm irradiation with nitrogen-based gas *J. Electron. Spectrosc. Relat. Phenom.* **267** 147385

[60] Takahashi M, Saegusa S, Saiki T, Amano S, Utsumi Y and Yamaguchi A 2024 Micro hot stirrer driven by surface acouctic wave (SAW) for agitation of high viscosity liquids *IEEJ Trans. Sens. Micromach.* **144** 106–10

[61] Yamaguchi A, Takahashi M, Saegusa S, Utusmi Y and Saiki T 2024 Removable and replaceable micro-mixing system with surface acoustic wave actuators *Jpn. J. Appl. Phys.* **63** 030902

Chapter 4

Nano/microfluidics

This chapter covers the fundamentals of fluid mechanics for describing fluid dynamics at the nanoscale or microscale. In practice, if readers study fluid mechanics at university, fluid behaviour in microfluidic channels is treated as a special case with a small Reynolds number. In other words, laminar flows are mainly formed in microfluidic channels. Laminar flow is often used in lab-on-a-chip (LOC) because it facilitates unit chemical operations except mixing and is suitable for miniaturization and high density. Therefore, it is only a review for students and researchers who have already mastered fluid mechanics, so it can be omitted and they can move on to chapters on research examples and experiments on real systems. In this chapter, the handling of fluid behaviour is described for beginners, followed by an introduction to the handling of fluids in microfluidic channels and the policies and methods to be deployed in actual research. Examples of microfluidic channels may be included in other chapters, for example when they are actually used. The aim of this chapter is to learn the basic equations and descriptions of fluids.

4.1 Introduction

Here, readers will learn the basics about fluid mechanics. Since microchemical systems mainly deal with laminar flow conditions, only special conditions in fluid mechanics need to be considered. However, knowing why such treatment is possible is always necessary when designing, fabricating, and using the system to evaluate the results. In addition, while laminar flow in microchemical systems facilitates unit chemical operations, it does not allow for the mixing necessary to speed up chemical reactions. It is very important to know the concepts and methods to solve these problems. Therefore, it is necessary to study fluid mechanics itself. There are many excellent textbooks on fluid mechanics [1–3], and we hope that you will refer to them as needed. This section will provide the basic fluid mechanics knowledge that is considered necessary for dealing with microchemical systems.

doi:10.1088/978-0-7503-3111-1ch4

4.2 Basic of fluidics

Microchemical systems and microfluidic chips handle fluids. As mentioned in chapters 1 and 2, these systems have succeeded in systematically integrating chemical analysis by performing unit chemical operations by intentionally creating laminar flows [4–10]. Laminar flow is one characteristic of fluids. In the following sections, we will consider fluid mechanics, going back to how to create a laminar flow.

4.2.1 Fluid dynamics

In Nano–Microsystems, especially LOCs called microchemical systems and microfluidic systems, fluids are often handled. The basic dynamics of fluids are described in fluid mechanics. In LOC, it is possible to perform unit chemical operations by processing the flow path to micro/nanoscale. The unit chemical operation can be realized because the laminar flow state is actively controlled and used. As will be described later, the laminar flow is related to the Reynolds number. Laminar flow occurs when the Reynolds number decreases. Conversely, if the Reynolds number is reduced, a laminar flow can be created. Utilizing this fact, a microfluidic system can be formed by creating a microchannel having a small Reynolds number and wiring the channels in combination like an electronic circuit. Therefore, the microchannel basically needs to handle only laminar flow. However, in order to create LOC that realizes chemical analysis and chemical synthesis, the process of mixing is indispensable in LOC. This is because the microchannel composed of laminar flow is not good for the mixing process due to its nature. Therefore, proposals and demonstration experiments on various methods and flow path structures have been reported. This point will be described in detail in another chapter and section. Since we are dealing with fluids, we will describe the necessary knowledge about fluid mechanics. By reading through this book, beginners will be familiar with Nano–Microsystems and microfluidic systems to some extent, and the following will explain fluid mechanics with the aim of providing sufficient knowledge for experiments. For those who want to learn more about fluid mechanics itself, there are many excellent textbooks published [1–4], so please refer to the textbooks listed in the references. The following descriptions will be given for the readers, i.e. beginning students, according to the excellent textbooks [1–4].

4.2.2 Continuity equation

When a small thin plate is placed in a stationary fluid, its surface receives pressure from the fluid. That is, the force F exerted by the fluid on one surface is (i) perpendicular to that surface, (ii) proportional to the area of the surface, and (iii) in which direction this minute surface is at the same location. Now, let dS be a vector whose magnitude is equal to the area ΔS of this surface and whose direction is normal, then F is given by

$$F = -P(r)\Delta S, \qquad (4.1)$$

where the normal direction is defined as the (−) direction from the plane to the inside of the fluid. On the other hand, the (+) direction is defined as the direction from the plane to the outside of the fluid. The coefficient $P(r)$ is the pressure, which is generally a function that varies from place to place.

Now let us consider a small rectangular parallelepiped element with one vertex at position $r = (x, y, z)$. This minute rectangular parallelepiped is submerged in a fluid, and the sum of the forces exerted on the fluid in it by the fluid around it, that is, the forces acting on the six surfaces, is calculated. First, for simplicity, consider the forces acting on two planes perpendicular to the x-axis. From the assumption that the magnitude of each surface is very small if the value at the centre is used, the magnitude of pressure is given by

$$
-\left[P\left(x + dx, y + \frac{1}{2}dy, \quad z + \frac{1}{2}dz\right) - P\left(x, y + \frac{1}{2}dy, z + \frac{1}{2}dz\right)\right]
$$
$$
dydz \cong -\frac{\partial P(x, y, z)}{\partial x}dxdydz.
$$

(4.2)

Here, since we considered a tiny space, we ignored the power of higher-order terms. Since the same calculation can be performed for the y-axis and z-axis, the force acting on the six planes of the small rectangular parallelepiped is described by

$$
-\left(\frac{\partial P(r)}{\partial x}, \frac{\partial P(r)}{\partial y}, \frac{\partial P(r)}{\partial z} \right)dxdydz = -\nabla P(r)dV,
$$

(4.3)

However, the volume element is given by $dV \equiv dxdydz$. In general, if the external force per unit mass acting on the fluid at point r is $f(r)$ and the density at the element is $\rho(r)$, the equilibrium equation is given by

$$
-\nabla P(r) + \rho(r)f(r) = 0,
$$

(4.4a)

or

$$
-\frac{1}{\rho(r)}\nabla P(r) + f(r) = 0.
$$

(4.4b)

This equation (4.4a) or (4.4b) shows the equilibrium relationship in the fluid by considering the tiny volume elements are in the fluid. Equations (4.4a) or (4.4b) alone cannot determine both pressure and density. That is, another relationship between pressure and density is required. Two special examples are shown below. (1) If the fluid is a uniform fluid in which the spatial change in density is negligible, the spatial change in pressure in the fluid that is stationary under an external force $f(r)$ can be determined. (2) If there is no external force, it can be said that the pressure gradient in the stationary fluid is zero, that is, the pressure is constant everywhere.

4.2.3 Description of flow

How should we express the motion of a fluid? For example, considering the flow of a river or the flow around the wings of an airplane, we recognize the flow as a pattern or pattern of flow. What is the flow pattern? It recognizes where the flow velocity is high, where the vortex is generated, and how the flow changes over time. That is, in order to express the flow pattern, it is nothing but the velocity distribution of the fluid shown at each point in the space. In general, velocity varies in magnitude and direction with position and time. That is, it is a velocity field stretched by a velocity vector. When there is a flow, the density $\rho(r,\ t)$ of the fluid at each point also generally changes. Since the density is a scalar quantity, the density field is a scalar field. In the following, we consider the equation of motion that defines $\rho(r,\ t)$ and $v(r,\ t)$ when each element of the fluid exerts pressure on each other and moves under external force.

4.2.4 Continuity equation

The motion of fluid also follows the basic laws of mechanics. That is, they are all expressed in the form of conservation law. Only non-relativistic movements are dealt with here. First, let us consider the law of conservation of mass. Please consider an arbitrary region V in the space where the liquid is. We define S as the closed phase surrounding V. When there is a flow, there is fluid in and out through this surface S. If there are no fluid outlets or inlets in this area, what is the relationship between the inflowing mass and the outflowing mass? This will be considered below. The mass, $M(t)$, in the region V at time t is given by

$$M(t) = \int_V \rho(r,\ t)dV, \tag{4.5}$$

where the density is $\rho(r,\ t)$. Next, we consider the mass that flows in a unit time through the surface S. The volume that flows in a unit time through the area element dS at the position r on the surface S is given by $v(r,\ t) \cdot ndS = v(r,\ t) \cdot dS = |v(r,\ t)|dS \cos\theta$, where n is the unit vector in the normal direction of the dS. The sign of n is defined as positive when the direction of the unit vector is outward from the surface S. Its volume is $dS \equiv ndS$, and θ is the angle between n and v. Therefore, the mass in this pillar is equal to $\rho(r,\ t)v(r,\ t) \cdot dS$. Here, if $j(r,\ t) \equiv \rho(r,\ t)v(r,\ t)$ is defined, $j(r,\ t)$ is the flow rate (or momentum density) at the position r at time t. Therefore, the mass flowing out through the closed curved surface S in a unit time is given as follows by taking the sum over the entire surface S:

$$J_S = \int_S j(r,\ t) \cdot dS. \tag{4.6}$$

Here, the direction of dS was outward, that is, the outflow was taken positively. That is, in the case of inflow, the sign of J_S is negative. If there is no gushing or sucking in the region V, then J_S is equal to the change in mass M_V in V, time evolution of M_V can be described by

$$\frac{dM_V}{dt} = -J_S. \tag{4.7}$$

This equation (4.7) shows the law of conservation of mass for any region V that has neither gushing nor sucking. Since this equation holds in any region, we consider the minute region ΔV including the point r_0 and calculate the limit of that when $\Delta V \to 0$.

$$\lim_{\Delta V \to 0} \frac{d}{dt} \int_V \rho(r, t) dV = - \lim_{\Delta V \to 0} \frac{d}{dt} \int_{\Delta S} j(r, t) dS, \qquad (4.8)$$

ΔS is the surrounding closed surface of the minute region ΔV. Considering the Gauss's theorem: $\int_S A(r) \cdot dS = \int_V \mathrm{div} A(r) dV$ (Here, S is the surrounding surface of V), the equation (4.8) can be calculated as the following:

$$\frac{\partial \rho(r_0, t)}{\partial t} \Delta V = -\mathrm{div} j(r_0, t) \Delta V. \qquad (4.9)$$

Here, the point r_0 is an arbitrary point in the region where there is no outlet or suction port. Therefore, the equation (4.9) is generalized as follows:

$$\frac{\partial \rho(r, t)}{\partial t} + \mathrm{div} j(r, t) = 0, \qquad (4.10a)$$

or

$$\frac{\partial \rho}{\partial t} + \mathrm{div}(\rho v) = 0. \qquad (4.10b)$$

This equation (4.10) is called the equation of continuity and is one of the basic equations that govern the motion of liquids. This corresponds to the charge conservation law of electromagnetism. This equation holds in the area where there is neither springing nor suction.

From the correspondence with electromagnetism, it is possible to consider the case where there is a spout or a suction port. For example, considering the velocity field when a liquid flows out (or sucks) in all directions from a sphere with radius a at a constant rate, the mass J of the fluid flowing out in a unit time across the sphere at a distance r from the origin is described by

$$4\pi r^2 j(r) = J, \qquad (4.11)$$

where $j(r) = \rho(r) \cdot v(r)$ is the mass $\rho(r) \cdot v(r)$ flowing out in a unit time through a unit area of a sphere with radius r.

We consider Newton's second law for a small element of fluid at position r at time t. That is, the ratio of the time change of momentum, that is, the relationship that the product of mass and acceleration is equal to the external force is considered for the minute element of the fluid. When the volume of the minute element is ΔV, its mass is equal to $\rho(r, t) \Delta V$. Next, we consider the acceleration acting on this minute element. Since the microelements of the fluid are in the velocity field $v(r, t)$, the velocity at time t is $v(r, t)$. At time $t + \Delta t$, the minute element moves to position $r + \Delta r = r + v(r, t) \Delta t$, so the velocity at that point is $v(r + v(r, t)\Delta t, t + \Delta t)$. From this, the acceleration of the minute elements is

$$\frac{\Delta v}{\Delta t} = \frac{v(r + v(r, t)\Delta t, t + \Delta t) - v(r, t)}{\Delta t} \cong \frac{\partial v}{\partial t} + (v \cdot \nabla)v. \qquad (4.12)$$

The derivation of this equation can be considered as follows. The x component of $v(r + \Delta r, t + \Delta t)$ is $v_x(x + \Delta x, y + \Delta y, z + \Delta z, t + \Delta t)$, and when expanded to the first order by Taylor expansion,

$$v_x(x + \Delta x, y + \Delta y, z + \Delta z, t + \Delta t) = v_x(x, y, z, t)$$

$$+ \frac{\partial v_x}{\partial x}\Delta x + \frac{\partial v_x}{\partial y}\Delta y + \frac{\partial v_x}{\partial z}\Delta z + \frac{\partial v_x}{\partial t}\Delta t \qquad (4.13)$$

$$= v_x + (\Delta r \cdot \nabla)v_x + \frac{\partial v_x}{\partial t}\Delta t$$

Here, the relation $\Delta r = v(r, t)\Delta t$ holding, the equation (4.13) can be calculated as follows:

$$v_x(x + \Delta x, y + \Delta y, z + \Delta z, t + \Delta t) = v_x + \left\{ (v \cdot \nabla)v_x + \frac{\partial v_x}{\partial t} \right\}\Delta t. \qquad (4.14)$$

The acceleration of the fluid element is

$$\lim_{\Delta t \to 0} \frac{\Delta v}{\Delta t} \equiv \frac{Dv(r, t)}{Dt} = \frac{\partial v(r, t)}{\partial t} + [(v(r, t) \cdot \nabla)]v(r, t). \qquad (4.15)$$

The force acting on the fluid element is the external force and the pressure and viscous force exerted by the fluid around it. The viscous force will be described later, but if ignored here, the equation of motion for the fluid element per unit volume is

$$\rho \left\{ \frac{\partial v}{\partial t} + (v \cdot \nabla)v \right\} = -\nabla P + \rho f, \qquad (4.16)$$

Where ρ, v, P and f are the density, velocity field, pressure and external force at (r, t), respectively. Here, a fluid having no viscous force is called a perfect fluid or an ideal fluid, and the above equation of motion is called an equation of motion for a perfect fluid.

4.2.5 Viscous fluid

It is considered that the fluid freely deforms without changing the volume. However, when the deformation progresses at a finite speed like a flow, the actual fluid shows resistance. The origin of this resistance is viscous force, the property of viscous force acting is called viscosity, and the viscous fluid is called viscous fluid. The fluid that actually exists is basically a viscous fluid, and the equation of motion of the viscous fluid will be considered below. Even when viscosity works, the basic equation of motion of fluid is based on Euler's equation of motion derived earlier. That is, from the conclusion, when viscous force works, the term of viscous force is added to Euler's equation, which is the equation of motion of fluid. This is because the left side of the Euler equation is the inertial force of a fluid of a unit volume, and the right side is each term of the force acting on the fluid. Therefore, adding the viscous force $\eta \nabla^2 v$ to the equation (4.16), the general equation of the viscous fluid is given by

$$\rho \left\{ \frac{\partial v}{\partial t} + (v \cdot \nabla)v \right\} = -\nabla P + \eta \nabla^2 v + \rho f. \qquad (4.17)$$

Here, η is a coefficient that determines the viscous force and is called a viscous coefficient or a viscous coefficient. It is a value peculiar to each liquid. The table shows examples of typical viscosity values. (The example of water shows that viscosity strongly depends on temperature.) This equation is the equation of motion of an uncompressed viscous fluid and is called the Navier–Stokes equation. Usually, a standardized formula by dividing both sides by ρ is often described by

$$\frac{\partial v}{\partial t} + (v \cdot \nabla)v = -\frac{1}{\rho} \nabla P + \nu \nabla^2 v + f, \tag{4.18}$$

where $\nu = \eta/\rho$ is the kinematic viscosity.

Now, let us consider the microscopic origin of viscous force below. Based on the analogy with the kinetic theory of gas, the viscous force of molecular motion was mainly due to the change in momentum caused by the collision of molecules. In other words, since the movement changes through the momentum transition, the momentum transition is interpreted as the action of force. The change in impulse is the change in momentum, and the change in impulse is the change in angular momentum. In dense liquids, it is necessary to consider not only molecular collisions but also cases where they are connected to each other and move in unison via strong intramolecular forces, and it is necessary to combine macroscopic and microscopic scales. In the following, the origin of viscous force will be considered. When there is a flow $v_x(z)$ in the z-direction, consider the force exerted by the upper liquid on the lower liquid through the minute surface $dxdy$ perpendicular to the z-axis. The pressure acts in the direction perpendicular to the plane, so it is in the z-direction. The viscous force corresponds to the frictional force handled by the dynamics of mass points or rigid bodies, so it is a force in the x-direction. Like pressure, this force is proportional to the area of the surface, so we write it as $F_{xz} dxdy$. Here, F_{xz} is the force in the x-direction acting on the unit area perpendicular to the z-axis. Newton considered that this force was proportional to the velocity gradient $\partial v_x/\partial z$ of the fluid, and assumed a relationship of $F_{xz} \propto \partial v_x/\partial z$. This assumption makes sense because it can be explained from this assumption that no matter how high the velocity is, the viscous force does not work in a uniform flow, and the greater the slip of the fluid above and below, the greater the viscous force. Extending this relation to the case of general flow, the gradients of viscous force and fluid velocity are given in the same form, indistinguishable from any direction, in that the fluid is isotropic. Therefore, if the following subscripts i and j can be any of x, y, and z, the following relation can be described:

$$F_{ik} = \eta \frac{\partial v_i}{\partial x_k}. \tag{4.19}$$

Viscous forces work when there is relative motion between the elements of the fluid. On the other hand, the viscous force does not work for the movement in which the relationship between the positions does not change. Motions that do not change their positions include uniform flow and rotation as a whole, that is, rigid body rotation. In the case of uniform flow, as mentioned above, the viscous force does not work. However, in fact, in this equation, a viscous force acts when the rigid body is rotating. As an example, consider a system in which a rigid body rotates at an angular velocity ω with the z-axis as the rotation axis. The speed field at that time is

$$v = (-\omega y, \omega x, 0) \tag{4.20}$$

If we introduce the formula into the formula equation (4.19) and calculate it,

$$F_{ik} = \begin{pmatrix} 0 & -\eta\omega & 0 \\ \eta\omega & 0 & 0 \\ 0 & 0 & 0 \end{pmatrix}. \tag{4.21}$$

That is, the viscous force works. Therefore, the equation needs to be rewritten as follows, taking into account the relationship $\frac{\partial v_i}{\partial x_k} = -\frac{\partial v_k}{\partial x_i}$ to the rigid body rotation around an arbitrary axis. Therefore, the corrected equation can be given by

$$F_{ik} = \eta \left(\frac{\partial v_i}{\partial x_k} + \frac{\partial v_k}{\partial x_i} \right). \tag{4.22}$$

The viscous force acting on the small elements of a rectangular parallelepiped whose sides are dx, dy, and dz is the resultant force acting on each surface. For example, if you describe only the x component,

$$
\begin{aligned}
&[F_{xx}(x + dx, y, z) - F_{xx}(x, y, z)]dydz \\
&+[F_{xy}(x, y + dy, z) - F_{xy}(x, y, z)]dxdz \\
&-[F_{xz}(x, y, z + dz) - F_{xz}(x, y, z)]dxdy \\
&\cong \left[\frac{\partial F_{xx}}{\partial x} + \frac{\partial F_{xy}}{\partial y} + \frac{\partial F_{xz}}{\partial z} \right]dxdydz \equiv \frac{\partial F_{xk}}{\partial x_k}dV
\end{aligned}
\tag{4.23}
$$

Here, when expressing arbitrary components x, y, z of the vector $\boldsymbol{r}$, it is defined as x_i. Also, like F_{xk}, when the same subscript k appears in opposition, we promise to take the sum for $k = x, y, z$. Therefore, the i component of the viscous force acting on a unit volume is

$$\frac{\partial F_{ik}}{\partial x_k} = \eta \frac{\partial}{\partial x_k} \left(\frac{\partial v_i}{\partial x_k} + \frac{\partial v_k}{\partial x_i} \right) = \eta \left(\nabla^2 v_i + \frac{\partial}{\partial x_i} \mathrm{div} v \right). \tag{4.24}$$

Here, when considering only a fluid that does not shrink, the term of $\mathrm{div} v$ becomes unnecessary, so the Navier–Stokes equation, which is the equation of motion of an uncompressed viscous fluid, is given.

The flow of a fluid according to the equation of motion with viscous force is qualitatively different from that of a perfect fluid. One of them is that when viscous, the kinetic energy of the flow is dissipated, that is, converted to thermal energy, and the kinetic energy is attenuated. This also applies to the problem of mechanics. In the motion of an object when there is a frictional force, there is energy dissipation due to the frictional force, and the motion of the object is attenuated, which corresponds to stopping if no external force is applied. Regarding heat generation due to frictional force, it can be well understood that the dissipation of kinetic energy is caused by the conversion to thermal energy. Here, since a fluid that does not shrink is considered, it is not necessary to consider the change in the internal energy of the fluid due to the change in fluid density. Energy changes also need to be considered. Now, considering the case of describing the motion of a viscous fluid that does not shrink, it is sufficient to handle the continuity equation and the Navier–Stokes equation in a simultaneous manner. At this time, what is required is a boundary condition on the solid surface. Fluid molecules collide with the surface of a solid, are adsorbed by intramolecular forces, and separated due to flow. That is, molecular desorption at the solid–liquid interface occurs repeatedly and continuously. At this time, it is reasonable to consider that the velocities of the fluid and the solid on the solid surface are equal. It is a daily experience that the liquid on the surface does not easily flow down, and it is thought that there are many people who have experienced shaking off the liquid by accelerating the solid to remove the liquid on the surface. In a perfect fluid, only the components in the normal direction of the velocity should be equal, but the boundary condition in a viscous fluid is the surface (solid–liquid interface), where u is the velocity on the surface of the object. The relation, $v = u$, must be established. This relationship is the same as the condition that the rigid body does not slip on the surface with frictional force in the dynamics of the rigid body.

The momentum equation is well described by the well-known Navier–Stokes equation of incompressible Newtonian fluids:

$$\rho \frac{\partial v}{\partial t} + \rho(v \cdot \nabla)v = - \nabla P + \rho \nu \nabla^2 v + f. \tag{4.25}$$

This equation is rewritten by rewriting equation (4.17) and applying an external force, f. Here, let us reconsider the meaning of each term in this formula. The left side of equation describes inertial acceleration composed of time-dependent acceleration $\rho \frac{\partial v}{\partial t}$ and convective acceleration $\rho(v \cdot \nabla)v$ derived from the spatial effect. The right side represents force densities, that is, forces per unit volume. The first term of the right side is induced by the spatial pressure changes. The diffusion term $\rho \nu \nabla^2 v$ is the divergence of viscous stress $\sigma_s = \eta_s [\nabla v + (\nabla v)^T]$. The body force, f, can cover effects from external fields, such as electrical, magnetic, gravity, and centrifugal forces.

4.2.6 Reynolds number

The problem of fluid mechanics is that it is necessary to deal with the motion of fluids with different densities and viscosities such as water and air. In addition, the scale of flow is very wide, from the problems of the atmosphere and ocean currents in outer space to the global environment, to the handling of fluid behaviour in thin pipes and microchemical systems. However, there are many similarities in their flow. For example, whirlpools when mixing coffee with a cup and whirlpools that occur in ocean currents look similar. In other words, there is a law of similarity. Here, consider the Navier–Stokes equation of viscous fluid and the continuity equation.

However, for the sake of simplicity, if there is no external force and the fluid does not shrink ($\partial \rho / \partial t = 0$), the following equations

$$\frac{\partial v}{\partial t} + (v \cdot \nabla)v = -\frac{1}{\rho} \nabla P + \nu \nabla^2 v \qquad (4.26)$$

and

$$\mathrm{div} v = 0 \qquad (4.27)$$

will be solved simultaneously. When an object is in a uniform flow, consider the flow that occurs around it. Let u be the velocity of the uniform flow and l be the magnitude of the object. Here, we consider a normalized length and time to confirm the law of similarity. That is, the following relations $\tilde{r} = \frac{r}{l}$ and $\tilde{t} = \frac{t}{(\frac{l}{u})} = t\frac{u}{l}$ hold. If we try to measure the velocity with v as a variable and u as a unit, then $\tilde{v} = v/u$. All of these quantities are non-dimensional quantities. To rewrite the Navier–Stokes equation in quantities that do not have these dimensions, divide each term by u^2/l. Since $\widetilde{\nabla} = l\nabla$, the formula is

$$\frac{\partial \tilde{v}}{\partial \tilde{t}} + (\tilde{v} \cdot \widetilde{\nabla})\tilde{v} = -\widetilde{\nabla} \widetilde{P} + \frac{1}{Re} \widetilde{\nabla}^2 \tilde{v} \qquad (4.28a)$$

$$\mathrm{div} \tilde{v} = 0 \qquad (4.28b)$$

$$Re = \frac{ul}{\nu} = \frac{\rho ul}{\eta} \qquad (4.28c)$$

is a dimensionless parameter called Reynolds number, Re. $\widetilde{P} = P/\rho u^2$, which is a dimensionless pressure. Only one parameter appears in the dimensionless basic equation (4.28). This means that the law of similarity holds, and it shows that all kinds of problems can be described by this basic equation, and if the Reynolds number is changed according to the physical phenomenon, the physical phenomenon can be reproduced. Conversely, if you want to reproduce a remarkable physical phenomenon, you should select a Reynolds number that matches the physical phenomenon. That is, the flows in which the boundary conditions are geometrically similar and the Reynolds numbers are the same or similar, and this law is called the similarity law. The reason why the law of similarity holds is that the fluid can be regarded as a continuum. The idealized continuum is a continuum on any scale. If the size of the molecules that make up the fluid becomes an issue, the law of similarity does not hold.

The coefficient η that determines the viscous force is the viscous coefficient, and has a value peculiar to each fluid. The unit is $\mathrm{N \cdot s \ m^{-2}}$ in the MKSA system of units. The dimension of kinematic viscosity $\nu \equiv \eta/\rho$ is $\mathrm{m^2 \ s^{-1}}$. As you can see from the unit and the derivation process, the viscosity η is the pressure acting per unit area multiplied by time, so from a microscopic origin, the molecules that make up the fluid collide with the object. Therefore, it is considered to be the impulse exerted per unit area. Assuming that the average velocity of the random motion of fluid molecules is v_m and the mean free path is l_m, from the definition of Reynolds number, the following relation holds:

$$\frac{\eta}{l_m} \propto \rho v_m \text{ or } \upsilon \propto v_m l_m. \tag{4.29}$$

The left side of this equation corresponds to the impulse that the fluid numerator acts on a unit area per unit time. On the other hand, the right side of this equation shows the amount corresponding to the change in momentum of the fluid molecule per time. Therefore, microscopically, it can be understood again that the viscous force is generated by the fluid molecules randomly colliding with the surface of the object and repeatedly adsorbing and desorbing. Therefore, as long as u and l given in the flow problem are taken so that the parameters R are the same, the similarity rule holds as long as it can be regarded as a continuum.

4.2.7 Dimensionless numbers

Fluid motions are strongly dependent on competition with physical effects in fluidic channels. The force balance normally determines the behaviours of fluid motions. As described in Reynolds number, dimensionless numbers can characterize the relative predominance of different effects and can contrast flows in parameter space, thereby unifying flowing features between different systems. The typical key dimensionless numbers characterizing the droplet generation and microfluidic flow are summarized in table 4.1. Here, the above-mentioned governing equations such as Navier–Stokes equation indicate four types of forces: inertial force, viscous force, gravity (for the case without other external force fields) and capillary force. Considering a volume of liquid fluid flowing at velocity u_s in a microfluidic device of characteristic length L, inertial stress scales as $f_i \sim \rho_s u_s^2$, viscous stress $f_v \sim \frac{\eta_s u_s}{L}$, gravity $f_g \sim \rho_s gL$, and capillary pressure $f_\gamma \sim \frac{\gamma}{L}$. Base on this scaling, the ratio of any two stresses of the four defines a dimensionless number as described in table 4.2. The Reynolds number represents the relative importance of inertia to viscous force, that is, $Re = \frac{f_i}{f_v}$. Re usually ranges between 10^{-6} and 10 in microfluidic flows. This means that the viscous stress dominates over fluid inertia, yielding a laminar flow in microchannels. Second, the capillary

Table 4.1. Dimensionless parameters in microfluidic systems.

Symbol	Name	Formula	Physical meaning (Ratio = A/B)
Re	Reynold number	$Re = \frac{\rho u L}{\eta}$	Inertial force/viscous force
Ca	Capillary number	$Ca = \frac{\eta u}{\gamma}$	Viscous force/interfacial tension
We	Weber number	$We = \frac{\rho u^2 L}{\gamma}$	Inertial force/interfacial tension
Bo	Bond number	$Bo = \frac{\Delta \rho g L^2}{\gamma}$	Buoyancy/interfacial force

Table 4.2. Examples of typical viscosity values [11].

Substance	Chemical formula	Temperature (°C)	Viscosity (mPa · s)
Water	H_2O	0	1.788
		10	1.3059
		20	1.0016
		25	0.89002
		50	0.54652
		70	0.40355
		80	0.35405
		100	0.2825
Ethanol	C_2H_6O	25	1.074
Glycerol	$C_3H_8O_3$	25	934

number is the ratio of viscous stress to capillary pressure, that is, $Ca = \frac{f_v}{f_\gamma}$. The inertial stress is independent of the device length-scale L. The smaller the devices length-scale L is, the smaller the gravitational effect because the gravity is proportional to L. In contrast, the viscous stress and capillary pressure enlarge with reducing the device length-scale L. Therefore, viscous and capillary would become dominant over the other two if L is sufficiently small. Third, we consider the competition between inertia and capillary pressure, that yields the Weber number $We = \frac{f_i}{f_\gamma}$ when we treat the nonlinear bubble formation at high flow velocity or in the vicinity of droplet and bubble pinch-off. For most microfluidic flows, $We < 1$. Finally, the relative importance of gravity to capillary pressure, $\frac{f_g}{f_\gamma}$ yields the Bond number Bo.

4.2.8 Development to microchemical systems

Researchers and engineers often use test tubes to manually discover drugs and synthesize functional substances. Also, in blood tests performed in hospitals, clinical laboratory technicians often perform various biochemical treatments manually while using large machines. Screening in the initial drug discovery stage does not require much material synthesis. In addition, although there is a problem of detection sensitivity in medical analysis and environmental analysis, the detection sensitivity has been improved due to recent advances in science and technology, and not all the blood that has been taken is used. From the above viewpoints, it is expected that various applications can be developed if there is a mechanism that can automatically perform chemical synthesis and analysis with a small amount. Therefore, a microfluidic chip was devised, LOC. LOC can perform continuous unit chemistry operations using fine channels with a width of tens to hundreds of microns. This makes the best use of laminar flow when the Reynolds number mentioned above is small. In fluids, vortices are formed or liquids stick to the walls and cannot be controlled by chemical operations in many cases. Such a fluid control unit or system can be constructed. Taking advantage of this, research and development of LOC has been promoted under the concept of putting a chemistry laboratory in the palm of your hand. These features are given as follows.

(1) Micro space:
- The molecular diffusion time can be expected to be shortened.
- It is possible to dramatically improve the reaction speed.
- It is possible to reduce reagents and samples.

(2) Large specific surface area:
- In the immobilization of antibodies and catalysts, it is possible to improve the sensitivity and reaction efficiency by increasing the specific surface area.

(3) Small heat capacity:
- High-speed thermal reaction and high temperature controllability are possible.
- From this, by-products can be suppressed.

(4) Low Reynolds number:
- It is possible to form a stable interface by laminar flow, and it is possible to improve the controllability of unit chemical operations. It enables control of chemical reactions and high integration of chemical operations. However, there is a demerit that mixing becomes difficult, but there is a possibility that it can be solved by installing a mixing mechanism in the system.

(5) Continuous flow operation:
- Continuous and parallel unit chemical operations are possible. A fully automatic system can be realized.

Therefore, it has the above characteristics and can be expected to be applied to the following fields.

(1) Inspection/analysis:
- Environmental analysis, clinical tests, pharmaceutical screening, bio-chemical research, food quarantine, etc.

(2) Tissue/cell culture:
- Pharmacokinetic analysis, cell screening, protein function analysis.

(3) Chemical synthesis:
- Organic synthesis, combinatorial chemical synthesis, biochemical synthesis.

(4) Various pretreatments:
- Amplification, separation, extraction, purification, concentration, etc.

With the above points in mind, it is possible to develop microfluidic and solid systems having various functions by utilizing the characteristics of fluids, solids, liquids, gases, and fine particles of micron size or smaller. Expected to expand into fields such as medical care, environment, energy, material synthesis, and planetary exploration. By constructing highly efficient function-intensive microdevices and Nano–Microsystems that integrate them, LOC can be also applied to actual medical and testing fields. In addition, we will also expand into scientific research such as solid–liquid interface reaction dynamics.

4.3 Droplet based microfluidics

As described above, fluid flow is generally classified into two types: laminar flow and turbulent flow. The characteristic of laminar flow is smooth and uniform fluid operation, and the characteristic of turbulent flow is vortex and flow rate fluctuation. The difference in physical properties between the two types of laminar and turbulent flows is represented by the Reynolds number, as described above. The Reynolds number is determined by the inertial force and viscosity of the liquid. If the Reynolds number is lower than a certain value, the viscous force becomes dominant over the complete force, and the fluid becomes laminar. If the relationship is reversed, turbulence will occur. In microfluidics, the Reynolds number basically shows a low value, resulting in laminar flow. Since the laminar flow is a flow in which the fluid moves regularly, the fluid layers do not mix in parallel with each other even if a plurality of flows are mixed by bundling the microchannels. At that time, mass transfer between the fluid layers occurs only in the direction in which the fluid flows. That is, mixing occurs solely by particle diffusion. Therefore, the micromixer that mixes microfluidics aims to improve the efficiency of particle diffusion. Therefore, it is necessary to design so that the width of the mixing flow path is reduced and the contact area is increased. A system that performs mixing operations while considering the design of the flow path is called a passive flow path. Various shapes of passive flow path have been proposed, including three-dimensional structures and zigzag flow paths. Recent advancements in 3D printing technology mean it is now possible to create complex shapes with ease. However, simpler structures that can be easily fabricated are preferable for initial prototyping, as they provide clearer experimental results. Therefore, combinations with active mixing mechanisms, such as ultrasonic agitating or mechanical stirrer mixing, have also been proposed. These mixers based on the surface acoustic wave agitation are introduced in the chapter 6.

Nowadays, digital fluidic systems using droplets have attracted much attention in microfluidics for precise control of fluids and combinatorial analysis of small quantities and varieties. In order to create droplets, T- and Y-shaped channels, as shown in figures 4.1 and 2.2, are often used; more examples are given in chapter 2. Droplets are created by mixing two different liquids, e.g. oil and water. The example in figure 4.1 shows the results of a study on droplet formation and behaviour when water and cyclohexane are mixed in the flow path shown in figures 4.1(a)–(e). Figure 4.1(f) shows the effect of junction diameter on the flow patterns. This result suggests that the flow patterns can be controlled by the junction pattern and volumetric flow rate of liquid. The plug formation process is generally described by considering the balance among three forces. One is the surface tension force, Second is the shear stress forces and third is the resistance forces on the plug tip as it enters into the main channel.

Therefore, these parameters should be designed with deep consideration. The most significant parameter affecting the flow pattern in a microchannel is the material from which the channel is made. The wetting behaviour of the fluids on the channel well is strongly dependent on the contact angle. The microchannel is

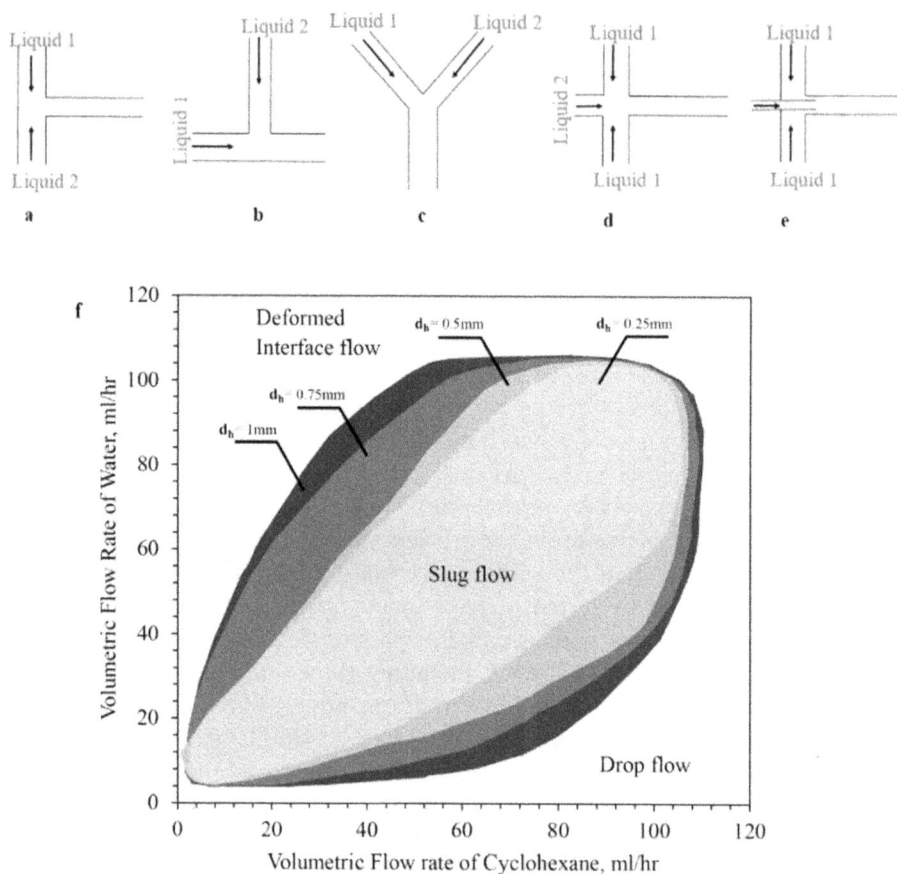

Figure 4.1. Different junctions used as inlet in the literature: (a) T, (b) perpendicular T, (c) Y, (d) cross, and (e) cross-dripping configurations. (f) Effect of junction diameter on the flow patterns. Reprinted with permission from reference [6], copyright (2021) American Chemical Society. The original data is from [9].

commonly made of glass, polydimethylsiloxane (PDMS), poly(methyl methacrylate) (PMMA), quartz, or polytetrafluoroethylene (PTFE).

Recently, there has been an increase in the number of studies dealing with interface consisting of the liquid–liquid two-phase flow, particle–liquid flow, electrode–liquid interface, and gas–liquid in a microchannel. The progress of microdevices can provide the ideal experimental environment with the combination of microchannels whose dimensions vary from submicrometres to submillimetres, because researchers desire to obtain knowledge of the flow patterns of various flows in microchannels to make it applicable on an industrial scale. However, thus far, we have obtained no universal map that takes into consideration all factors that influence the flow regimes and flow structure. In actual cases, various parameters such as channel size and geometry, wettability, liquids properties, temperature, and flow conditions of liquids heavily affect the flow patterns. As a result, further studies are necessary. Ultimately, it is a matter of repeated simulation and experimentation to find the flow channel fabrication and control conditions that perform the liquid manipulation we desire. Recently, the numerical studies have shown good agreement with the experimental data. Therefore, further studies based on the combination with the numerical simulations and experimental investigations open the door to get the deep understanding and knowledge to precisely control flow pattern in LOCs. The chemical operations and processes on the LOCs can be optimized for use on a high-throughput industrial scale, maximizing the rate at which energy resources can be utilized.

4.4 Summary

This chapter describes the basic knowledge of fluids required for the creation of Nano–Microsystems, in particular microfluidic systems. There are already excellent textbooks on fluid mechanics and published software to simulate them. Therefore, the simplest basics of fluid mechanics are described here for beginners. Fluid behaviour described by the Navier–Stokes equations cannot generally be solved analytically. However, simulations are to be used, but before using simulations, it is important to understand what are the original equations intended to describe physical phenomena. The reader is encouraged to further utilize and develop the knowledge learned here and develop it into the creation of new functions, such as analysis and chemical synthesis, using fluid systems designed and created by themselves.

References

[1] Landau L D and Lifshitz E M 1959 *Fluid Mechanics* Course of Theoretical Physics vol 6 *(Oxford: Pergamon) translated from the Russian by J B Sykes and W H Reid*
[2] Imai I 1993 *Fluid Mechanics* (Tokyo: Iwanami)
[3] Tsuneto T 2017 *Elastic Bodies and Fluids* (Tokyo: Iwanami)
[4] Gravesen P, Branebjerg J and Jensen O S 1993 Microfluidics—a review *J. Micromech. Microeng.* **3** 168–82
[5] Castillo-León J and Svendsen W E (ed) *Lab-on-a-Chip Devices and Micro-Total Analysis Systems—A Practical Guide* (Berlin: Springer)
[6] Al-Azzawi M, Mjalli F S, Husain A and Al-Dahhan M 2021 A review on the hydrodynamics of the liquid-liquid two-phase flow in the microchannels *Ind. End. Chem. Res.* **60** 5049–75
[7] Saiffet S and Thiele J 2020 *Microfluidics–Theory and Practice for Beginners* (Leck: CPI Books GmbH)
[8] Seemann R, Brinkmann M, Pfohl T and Herminghaus S 2012 Droplet based microfluidics *Rep. Prog. Phys.* **75** 016601
[9] Kashid M N and Agar D W 2007 Hydrodynamics of liquid–liquid slug flow capillary microreactor: flow regimes, slug size and pressure drop *Chem. Eng. J.* **131** 1–13
[10] Wu B and Sundén B 2019 Liquid–liquid two-phase flow patterns in ultra-shallow straight and serpentine microchannels *Heat Mass Transfer* **55** 1095–108
[11] https://omnicalculator.com/physics/water-viscosity

IOP Publishing

Nano–Microsystems
Science and applications
Akinobu Yamaguchi

Chapter 5

Lab-on-a-disk

Microchemical systems play an important role in enabling many novel applications in chemical synthesis, environmental analysis and life sciences. Liquid manipulation and pre-treatment are essential unit chemical manipulation process steps to be incorporated into microchemical systems. Microchemical systems that can combine these unit chemical operations to perform chemical analyses on minute quantities of specimens are considered a promising key technology in the field of medical diagnostics. Microfluidic technology is characterized by minimal consumption of samples and reagents, short time to results and the possibility of process integration and parallelization. These market requirements are met by so-called 'lab-on-a-chip' (LOC) systems. These systems integrate basic analytical steps such as sample injection, separation, weighing, mixing, reaction and detection, and perform complex diagnostic tasks on a chip.

A conventional LOC often uses syringe pumps and capillary pumps as pumping mechanisms. Capillary pumps are liquid pumping mechanisms that utilize capillary action due to the microstructure formed on the chip, but their slow pumping speed increases the measurement time as a system, and throughput in *in situ* diagnosis is not increased. Syringe pumps, on the other hand, are convenient and easy to use, but have the problem of residual fluid remaining in the tube or syringe. LOC is aimed at microanalysis, but the fact that residual fluid remains is a major problem. In addition, syringes and tubes are an obstacle in *in situ* diagnosis and analysis, and a more compact and user-friendly mechanism is desired.

This is where the microstructured disc called the 'Lab-on-a-disk' (LOD) platform came in: first studied by Madou's group of University of California in 1998. LOD is a system that allows sequential unit chemical operations to be carried out by transporting liquid sealed in a disk using centrifugal force. Basically, a designed structure is created on the disc and unit chemical operations are controlled and executed by rotation control alone. For example, processes such as blood cell separation, aliquoting and mixing of blood by centrifugation can be realized, e.g. enzyme immunoassay. As the basic policy of this book is to guide novice users in the creation and fabrication of microchemical

doi:10.1088/978-0-7503-3111-1ch5

systems, the principles and design principles of the system and examples of fabrication methods will be introduced along with examples. The results of implementation using the actual fabricated systems are also presented as examples. As the book is written with the intention that beginners will implement it on the basis of this book, it does not necessarily describe the latest research trends and systems. Therefore, it is recommended that interested readers, after understanding the basic principles and learning the approximate method of creation, access the latest research papers, etc, and proceed with research and development.

5.1 Introduction

LOC is attracting attention in point-of-care-testing (POCT) and pandemic counter-measure applications in that they enable drug screening and trace molecule and virus detection to be carried out rapidly and easily by anyone [1–6]. Among them, in particular, LOD, which is based on centrifugal pumping, is the most promising system for social implementation, as it enables POCT to be performed on multiple samples using only trace amounts of specimens, incorporating complex unit chemistry operations [7–22]. This chapter discusses the fabrication of an LOD device for POCT, which is a clinical test performed rapidly in the presence of the patient using simple equipment. The advantages of POCT are, for example, that if the patient's DNA and blood information can be obtained on the spot, the best medical treatment for the patient's individuality (personalized medicine) can be performed quickly. In addition, the physical and mental burden on the patient can be reduced by using a smaller amount of blood sampling.

LOD, which control liquids by centrifugal force due to the rotation of the device, has attracted much attention in the field of microchemical analysis systems, as schematically illustrated in figure 5.1 [7–22]. Compared to electrophoretic liquid control, this device is more stable against chemical changes in the solution. Also, compared to syringe pumps, the required volume of liquid can be reduced to several tens of microlitres and the dead volume can be extremely low because no pump tubing is required. Furthermore, the device is basically driven solely by a rotary mechanism, which means that it can be performed with a small, simple and inexpensive device. On the other hand, the device can be adapted to various

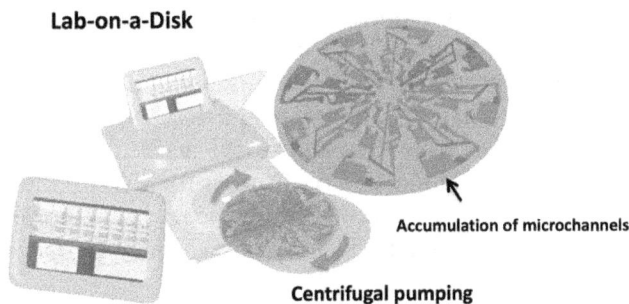

Figure 5.1. Conceptual diagram of LOD.

solutions by appropriately designing the flow path in consideration of the properties of the liquid, such as viscosity, conductivity and surface tension, and can also perform multiple unit chemical operations such as washing, mixing and volume preparation required for chemical reaction systems in a single operation. Due to the above advantages, the system is expected to be applied to various fields, including clinical immediate diagnosis (POCT), where highly reliable and rapid detection of multiple specimens is required, as well as biochemical and environmental analysis. Due to the above advantages, the system is expected to be applied to various fields, including POCT, which requires highly reliable and rapid detection of multiple specimens, as well as biochemical and environmental analysis. In fact, some practical instruments have already been manufactured and sold by various companies. As a POCT instrument in microchemical systems, it is one of the most promising mechanisms for practical application.

Immunoassay is an effective method for quantitative detection of protein concentrations. This method uses antigen–antibody reactions that specifically bind to a certain substance, and in particular, the method that uses enzyme reactions by binding an enzyme to an antibody is called Enzyme-Linked Immunosorbent Assay (ELISA) and is the most commonly used method. Most of the LOD under development around the world are aimed at systems for performing ELISA. For more information, see references [17, 19, 22].

This chapter presents a laboratory level example procedure of adapting the ELISA to LOD, designing a flow path structure for simple and rapid biomarker detection and sequential feeding operation of each solution. Examples of mixing for higher sensitivity and shorter time of detection are also presented. Mixing solutions in microspace is difficult because the effect of viscous forces on the inertial forces of the liquid is greater (lower Reynolds number), making it difficult to generate turbulence. Several mixing methods have been proposed on LOD, but the method using Euler force is simple in its flow structure and can be performed only by rotational movements of the device. Here, an example of the effectiveness of the mixing method is also presented, for reagents used in ELISA.

5.2 Physics of centrifugal microfluidics

In centrifugal microfluidics, let us consider the forces that are exploited on this system to comprehend the unit operations used in centrifugal microfluidics. As schematically illustrated in figure 5.2, we distinguish intrinsic and extrinsic forces. The former forces are induced merely by the presence or absence of centrifugation. The later forces are caused by the external means. In centrifugal microfluidics, the following three forces arise from the centrifugal acceleration of the rotor and controllable: (i) the centrifugal force (F_c), (ii) Coriolis force (F_{Co}), and (iii) Euler force (F_E). These forces exerted on a point-like body (mass m) at position r in a disk rotating with an angular rotational frequency ω are given by the following equations (5.1)–(5.3) [7, 8]:

$$F_c = -m\omega \times (\omega \times r), \tag{5.1}$$

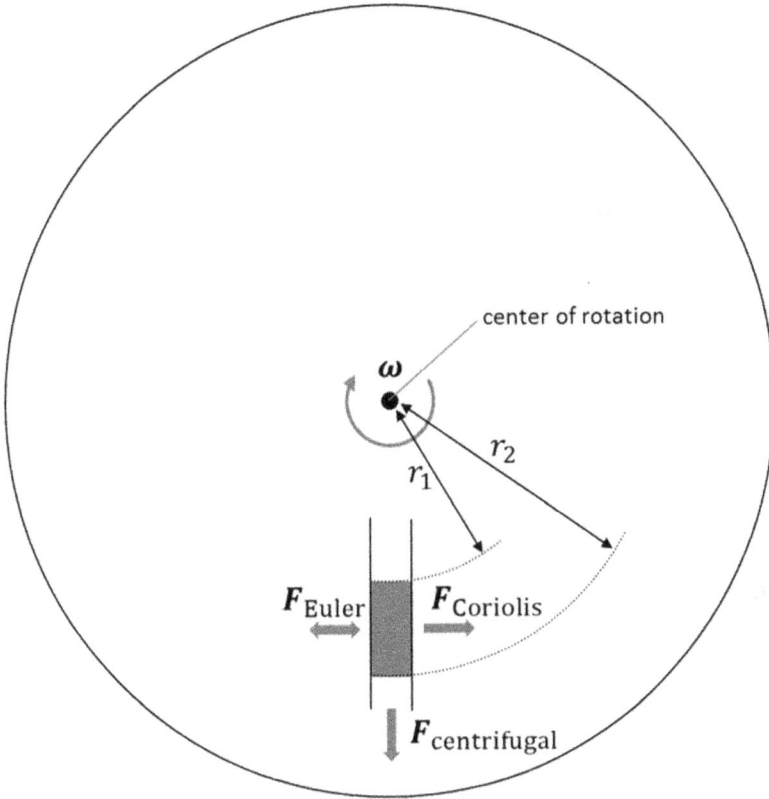

Figure 5.2. While the centrifugal force always acts radially outward, the Coriolis force acts perpendicular to both ω and the fluid velocity and the Euler force is proportional to the angular acceleration. Reprinted from [9] CC BY 3.0.

$$F_{Co} = -2m\omega \times \frac{d\mathbf{r}}{dt}, \tag{5.2}$$

$$F_E = -m\frac{d}{dt}\omega \times \mathbf{r}, \tag{5.3}$$

where ω and $\mathbf{r}$ are the angular velocity vector and position vector, respectively. The disk rotates with angular rotational frequency ω about the centre of the disk, resulting $\omega = (0,\ 0,\ \omega)$ in the coordinate system shown in figure 5.2. Forces that do not result from rotation, i.e. forces acting in the microchannel, such as viscous resistance and capillary forces, exist in both rotating and non-rotating systems. Therefore, these forces also need to be taken into account. These forces are discussed in chapters 2 and 4 and are therefore omitted here. For more information, please refer to chapters 2 and 4.

5.3 Valve mechanism

The basic valve mechanisms used in LOD are capillary valves and syphon valves [7–9, 11, 12, 23, 24], as described below. In capillary valves, the liquid is held in a state where the centrifugal force F_c applied to the liquid surface at the liquid outlet is balanced by the force F_s applied by surface tension. By disrupting this balance through changes in centrifugal force, liquid can be pumped at any desired timing.

5.3.1 Principle of capillary valves

Capillary valves have an enlarged section in the flow path, as shown in figure 5.3. The enlarged section holds the liquid in a state where the centrifugal force F_c exerted on the liquid surface in the channel is balanced by the force F_s exerted by the surface tension, thus acting as a valve. The principle is explained below.

The microfluidic channel is rotating at an angular velocity ω [rad s^{-1}]. The centrifugal force dF_c acting on the liquid per unit volume in the channel at a radius r [m] can then be expressed by the following equation:

$$dF_c = \rho\omega^2 r = \rho\frac{v^2}{r}. \tag{5.4}$$

Here, ρ is the density [kg m^{-3}] and $v = rw$ is the rotational velocity of a droplet in the microchannel at a radius r. As the liquid fills the microchannel between the outlet end $r = R_1$ and $r = R_2$, as shown in figure 5.3, the pressure P_c [N m^{-2}] at the microchannel outlet end can be calculated as follows:

$$P_c = \int_{R_1}^{R_2} \rho\omega^2 r\,dr = \rho\omega^2 (R_2 - R_1)\left(\frac{R_1 + R_2}{2}\right) = \rho\omega^2 \Delta R\overline{R}$$
$$= \frac{1}{2}\rho\omega^2\left(R_2^2 - R_1^2\right) = \frac{1}{2}\rho\left(v_2^2 - v_1^2\right), \tag{5.5}$$

where, $\Delta R = R_2 - R_1$ and $\overline{R} = \frac{R_1 + R_2}{2}$. $v_i = R_i\omega$ $(i = 1, 2)$ is the rotational velocity at radius of R_i, where i is an index indicating the position.

Here, let us consider the capillary pressure for any solid–liquid–air system. Based on the analysis of Madou et al the theoretical description is developed below. The capillary pressure can be derived from the change of total interfacial energy of

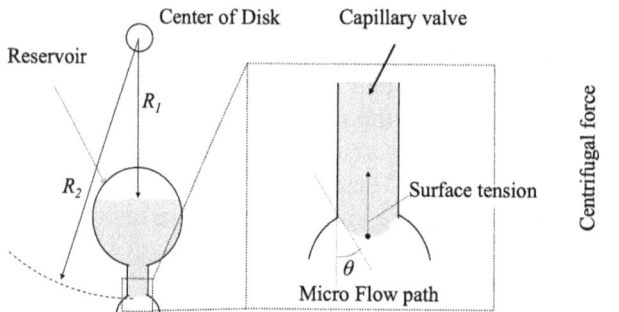

Figure 5.3. Schematic of capillary valve.

the solid–liquid–air system, U_T, with respect to the injected liquid volume, V, according to references [12, 23, 24],

$$P_{capillary} = -\frac{dU_T}{dV}. \tag{5.6}$$

The total interfacial energy of the system can be given by

$$U_T = A_{sl}\gamma_{sl} + A_{sa}\gamma_{sa} + A_{la}\gamma_{la}, \tag{5.7}$$

where A_{sl}, A_{sa}, and A_{la} are the solid–liquid, solid–air, and liquid–air interface areas, respectively [12, 23, 24]. γ_{sl}, γ_{sa}, and γ_{la} denote the corresponding surface energies per unit area. By applying equation (2.8), Young's equation, that is $\gamma_{sa} = \gamma_{ls} + \gamma_{la}\cos\theta$, equation (5.7) can be simplified as follows:

$$U_T = U_0 + (A_{la} - A_{sl}\cos\theta)\gamma_{la}, \tag{5.8}$$

where U_0 is a constant term given by $(A_{sl} + A_{sa})\gamma_{sa}$, being a constant energy component due to the sum of the solid–liquid and solid–air interfaces. θ is the contact angle of the liquid with the solid microchannel wall. The capillary pressure $P_{capillary}$ can be then described as

$$P_{capillary} = -\left[\frac{d}{dV}U_0 + \left(\frac{d}{dV}A_{la} - \frac{d}{dV}A_{sl}\cos\theta\right)\gamma_{la}\right]. \tag{5.9}$$

Since $dU_0/dV = 0$ and $dA_{la}/dV = 0$ due to that U_0 is a constant and A_{la}, the surface boundary between liquid and air in a capillary channel, does not vary as long as the liquid is still travelling within the channel. the surface boundary. Equation (5.6) simplifies to

$$P_{capillary} = \frac{dA_{sl}}{dV}\cos\theta\gamma_{la}, \tag{5.10}$$

For a circular capillary with diameter D, the surface boundary between the solid and liquid, and the volume of the liquid are given as

$$A_{sl} = \pi Dx, \tag{5.11}$$

and

$$V = \frac{\pi D^2 x}{4}, \tag{5.12}$$

where x is the length of liquid progression in the channel with respect to an arbitrary reference point. Similarly for a rectangular capillary of width w, and height h, the surface boundary and the volume of the liquid are given as

$$A_{sl} = 2(w + h)x \tag{5.13}$$

and

$$V = whx. \tag{5.14}$$

For a rectangular channel,

$$\frac{dA_{\text{sl}}}{dV} = \frac{4}{D_{\text{h}}}, \tag{5.15}$$

where D_{h} is the hydraulic diameter and is equal to D for a circular capillary, being described as the following:

$$D_{\text{h}} = \frac{4 \times \text{Area}}{\text{Perimeter}} = \frac{4wh}{2(w + h)} = \frac{2wh}{w + h}. \tag{5.16}$$

Inserting relations described by equations (5.15) and (5.16) to equation (5.10), the general expression for the capillary pressure is given as

$$P_{\text{capillary}} = \frac{4\gamma_{\text{la}} \cos \theta}{D_{\text{h}}}. \tag{5.17}$$

The burst speed of the capillary valve, f_{b} ([1/s] or [Hz]), can be calculated as follows. The capillary valve is ruptured and the liquid is pumped when the pressure applied by centrifugal force exceeds the pressure $P_{\text{capillary}}$. Therefore, f_{b} can be calculated using the relation $P_{\text{c}} = P_{\text{capillary}}$,

$$\rho\omega^2 \Delta R\overline{R} = \frac{4\gamma_{\text{la}} \cos \theta}{D_{\text{h}}}. \tag{5.18}$$

Substituting the relationship $\omega = 2\pi f_{\text{b}}$ into equation (5.7), the following relation is obtained:

$$\rho(2\pi f_{\text{b}})^2 \Delta R\overline{R} = \frac{4\gamma_{\text{la}} \cos \theta}{D_{\text{h}}}. \tag{5.19}$$

Solving for f_{b}, we obtain the burst frequency as

$$f_{\text{b}} = \sqrt{\frac{\gamma_{\text{la}} \cos \theta}{\pi^2 \rho \Delta R\overline{R} D_{\text{h}}}}. \tag{5.20}$$

Based on the above relationship given by equation (5.20), once the liquid is determined and the structural material of the microfluidic channel is determined, it is possible to break the valve at any rotation speed by adjusting the diameter d, R_1 and R_2 of the capillary valve. In other words, once the device is fabricated, the pumping and loading of liquid can be controlled only by the rotational speed. Readers interested in more detailed analytical formulae are referred to the literature.

On the other hand, the syphon valve shown in figure 5.4 is a valve using the syphon principle, where liquid is pumped when the liquid in the chamber reaches a position higher than the channel bend. The characteristic feature of this valve is that the break is determined solely by the height of the liquid surface, independent of the rotational speed of the tip. This allows the valve to be incorporated at the same time as a capillary valve, which breaks only dependent on the rotational speed, and allows a high degree of freedom in design. However, it should be noted that the use of syphons is explained below.

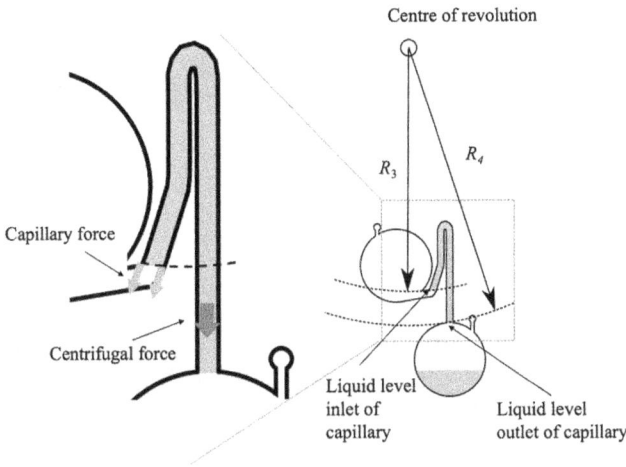

Figure 5.4. Schematic diagram of the liquid discharge of a syphon valve integrated in the LOD.

5.3.2 Theoretical formula for draining fluid in a syphon

When the liquid in the reservoir is drained, the inside of the syphon channel is as shown in figure 5.4. When the syphon channel is filled with liquid and the next liquid flows into the reservoir, liquid remains before and after the syphon fold, as shown in figure 5.4. When rotation is stopped in this state, the capillary forces of the liquid before and after the turn of the syphon channel are balanced and an air space is created in the syphon. If rotation is applied in this state, the centrifugal forces of the liquid before and after the fold of the syphon channel are balanced and the liquid stops moving. Therefore, the liquid in the reservoir will not be discharged. To avoid this situation, all liquid in the syphon must be drained before the next liquid is injected.

Here, we consider the equilibrium between the forces acting on the liquid surface in the syphon channel connecting the turnaround point of the syphon channel to the bottom of the reservoir inside the circle (liquid level in) and the forces acting on the liquid surface in the subsequent syphon channel (liquid level outlet), as shown in figure 5.4. Then, the forces acting on each liquid surface are shown in figure 5.4. Centrifugal and capillary forces act on the liquid surface 'in' and centrifugal forces act on the liquid surface 'ex'. Liquid surface 'ex' is the end of the capillary tube, so no capillary force acts on it. If the position from the centre of the circle to the liquid surface 'in' is R_3 and the position to the liquid surface 'ex' is R_4, the centrifugal force at the liquid surface 'in' is balanced by the centrifugal force of the liquid at the position to R_3 acting on the liquid surface 'ex'. Therefore, the liquid in the syphon channel is discharged when the centrifugal force at the liquid surface 'in' and the centrifugal force of the liquid from R_3 to R_4 at the liquid surface 'ex' exceed the capillary force. The capillary force at the liquid surface 'in' is $C\gamma \cos \theta$, where C is the circumference of the capillary valve, γ is the surface tension, θ is the contact angle. The centrifugal force at the liquid surface 'ex' is derived in the same way as in equation (5.5) by changing the integral of r from R_3 to R_4. If the same derivation is used as for the syphon valve, the rotational speed f_{ex} [1/s] or [Hz] at which the liquid in the syphon channel is discharged is

$$f_{ex} = \sqrt{\frac{\gamma_{la} \cos\theta}{\pi^2 \rho \Delta R \overline{R} D_H}}, \qquad (5.21)$$

where $\Delta R = R_4 - R_3$ and $\overline{R} = \frac{R_3 + R_4}{2}$.

Thus, if a syphon valve is incorporated and used in the LOD, the height of the liquid level and the number of revolutions to drain the liquid inside the syphon valve should also be calculated, considering the height of the liquid level and the liquid that accumulates inside the syphon.

A typical example is illustrated in figure 5.5. This example is shown as an example only. The reader should configure the liquid arrangements and flow paths according to the chemical operations they wish to perform. By combining a capillary valve and a syphon valve, chemical operations such as those shown in figure 5.5 can be carried out using rotary operation alone. However, because capillary valves and syphon valves can have variations in the valve drive rotation frequency depending on the processing accuracy and environmental temperature control, several methods have been proposed, including a method using magnetic nanoparticles and a method in which the rotation frequency is constant and the timing is timed.

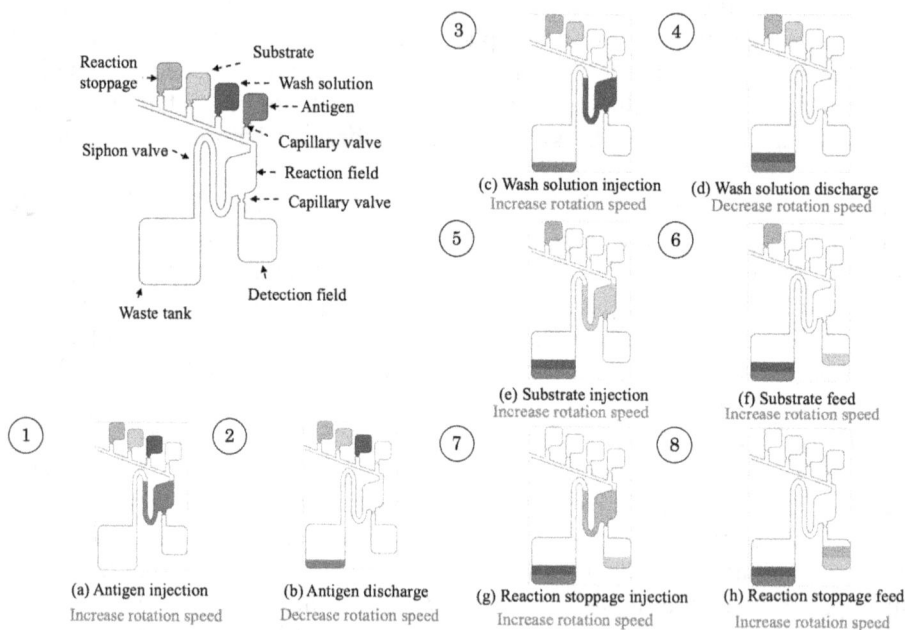

Figure 5.5. Example of LOD for ELISA using a combination of capillary and syphon valves.

5.4 Fabrication and experimentation

The drawing is made using CAD and the LOD device containing the PDMS syphon pump is fabricated from the drawing using the rapid prototyping method (figures 3.14, 3.15, 3.18 and 3.19). The rapid prototyping method involves the fabrication of a mould (template) by UV photolithography and the transfer of the microfluidic structure by applying and curing the PDMS on the mould.

This produces two PDMS disk-shaped plates: one PDMS plate has the proposed and designed LOD structure on one side. The other PDMS plate has a smooth surface. The LOD device is assembled by laminating the two disks. The two PDMS plates are laminated together due to the self-adhesive properties of the PDMS. A pipette is used to inject the coloured liquid into the LOD device and the LOD device is rotated on a rotating table to pump the liquid. The pumping is observed by stitching together snapshots of the device at a fixed angle of rotation.

Figure 5.6 shows the typical LOD structure for a syphon valve evaluation experiment. There is a fluid supply reservoir inside the circle, an upstream reservoir in the middle of the circle and two downstream reservoirs outside the circle. The lower part of the upstream reservoir has a syphon pump and capillary valve, each leading to a downstream reservoir. The upper part of the upstream reservoir is connected to the liquid supply reservoir.

Four types of syphon valves are used, with widths of 100, 150, 200 and 250 μm. The depth of each syphon pump and the reservoir to which the syphon pump is connected is 250 μm. All syphon valves and reservoirs are arranged identically. The capillary valve has to be held to the highest rotational speed when adapted to the proposed LOD. Therefore, a width of 20 μm is applied, which is the narrowest width known to be possible to process.

In the experiment, the same syphon is used to transfer liquid twice to verify the feasibility of repeated use. We verify whether the capillary valve could hold when

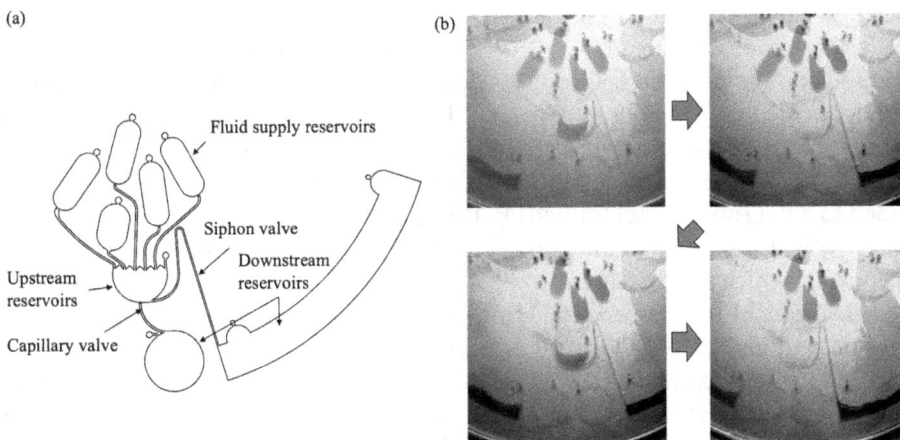

Figure 5.6. (a) LOD for syphon valve evaluation experiments. (b) Syphon bubble and flow-selective valve pumping.

liquid was introduced from the liquid supply reservoir into the upstream reservoir. Also, we measure the break and discharge rotational speeds of the syphon and compare them with the design values. A photograph of the sequential pumping behaviour is shown in figure 5.6, where the reservoir inside the disk is filled with liquid. (i) At high rotational per minute, the liquid is stopped in the middle of the syphon flow path and no pumping takes place. At this time, the capillary valve is holding the liquid and is not ruptured. (ii) When the rotational speed is reduced, the liquid crosses the syphon turnaround point and is pumped towards the downstream reservoir on the right. Pumping continues until the liquid inside the semicircular reservoir placed on the middle of photograph is gone. (iii) Next, when the rotational speed is increased again, the capillary valve in the second reservoir ruptures, pumping begins and fluid accumulates in the central semicircular reservoir. Figure 5.6(b) shows that the liquid in the syphon channel was also pumped. The second liquid is then pumped into the reservoir inside the semicircular reservoir at a higher speed than when the first liquid was pumped. It can be seen that the second liquid is also pumped into the reservoir on the right side in the same way as the first liquid. In addition, it is found that the other capillary valves have also maintained their holding state. (iv) When the rotation speed is reduced, the syphon valve operates to transport the liquid in the semicircular reservoir to the waste reservoir. This procedure was repeated. The last liquid supplied was pumped to the downstream reservoir on the left side due to the capillary valve rupturing as a result of increased rotation speed. Thus, the operation of the syphon valve and the flow-selective valve structure achieved the selective pumping of liquid through the flow path.

5.4.1 Example of lab-on-a-disk for competitive ELISA

An example of LOD was designed for competitive ELISA (figure 5.7) [25]. PDMS is used as the structural material for this device. As indicated in this book, PDMS is one of the commonly used materials for microfluidic chips. The device fabrication process is shown in chapter 3. First, SU-8 is applied to a Si substrate and patterned by UV lithography (figure 3.14(D)). Wax is then added to the chambers that store the liquid and the reaction field. This is done to ensure the volume of each chamber. A PDMS (SILPOT 184, Dow Corning Toray Co. Ltd, Japan) precursor (liquid mixture of PDMS monomer and crosslinker at 10:1) is then poured and heated on a hot plate at 80 °C for 2 h. The final PDMS is peeled off from the substrate and bonded to a flat (no channel formed) PDMS plate. The reaction field was previously coated with a solid-phase antigen; capillary valves were installed in reservoirs, respectively, which were designed to break in numerical order as the device rotation speed increased. First, reservoir (1), into which a solution containing a mixture of target protein and enzyme-labelled antigen is inserted, is ruptured and pumped to the reaction field. Subsequently, reservoir number (2), into which the washing solution is inserted, is ruptured and pumped to the reaction field in the same way. This raises the liquid level in the reaction field above the bend in the syphon valve, the syphon valve operates and the liquid in the reaction field is pumped to the waste tank. Next, reservoirs (3) and (4), into which the washing liquid is inserted, are

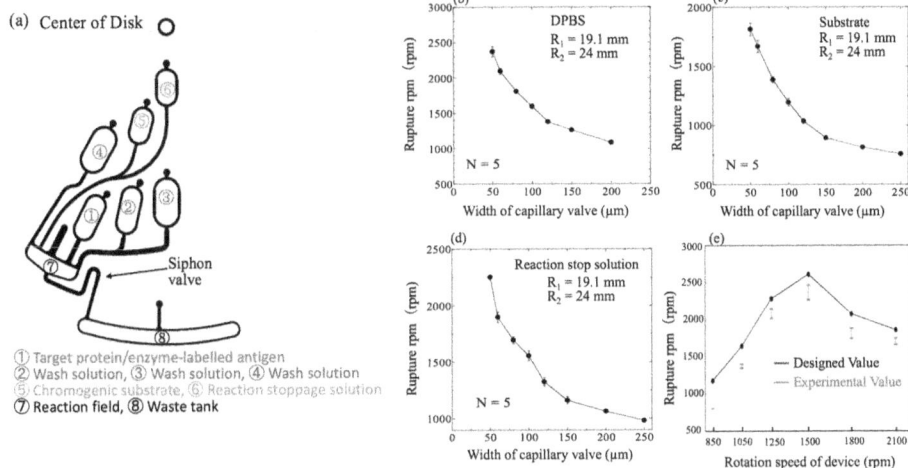

Figure 5.7. (a) Examples of LOD. Valve characteristics of this LOD: rupture rpm dependence on width of capillary valve with various solutions: (b) DPBS, (c) substrate, and (d) reaction stop solution. (e) Comparison of design and actual measurements on actual device operation. DPBS stands for Dulbecco's phosphate-buffered saline (PBS). Here, the original DPBS developed by Dulbecco contains $CaCl_2$ and $MgCl_2$. In general, the solution without $CaCl_2$ and $MgCl_2$ is sold as DPBS because it inhibits cell dispersion by trypsin. Reproduced with permission from [25].

Table 5.1. Composition of each solution [25].

Solution	Composition
Antigen·Antibody	
Washing solution 1, 2, 3	DPBS (137 mM NaCl/2.7 mM KCl/10 mM Na_2HPO_4/1.76 mM KH_2PO_4)
Substrate	
Reaction stop solution	1.0 M H_2SO_4

ruptured in turn and the liquid is pumped through the reaction field to the waste tank in the same way. Subsequently, reservoir (5), into which the colouring substrate is inserted, is ruptured and pumped to the reaction field. Afterwards, reservoir (6), in which the reaction stoppage liquid is inserted, is ruptured and the reaction stoppage liquid is inserted into the reaction field. The sample antigen concentration can be measured by absorbance measurement in this detection tank.

The composition of the solutions used in this experiment is shown in table 5.1. The rupture rpm of each solution is shown in figure 5.7. Based on these results, the rotational speed at which each solution breaks is calculated and sequential pumping is carried out. Table 5.2 shows the rupture rotation speed, R_1, R_2 and capillary valve width for each solution. The black line in figure 5.7 shows the rupture pressure at the set value of the breaking speed for each solution. The red lines are the maximum and

Table 5.2. Rupture rpm and capillary valve width of each solution [25].

	Rupture rpm (rpm)	Valve width (μm)	R_1 (mm)	R_2 (mm)	$R_2 - R_1$ (mm)
Antigen·Antibody	800	–	21.9	27.4	5.5
Washing solution 1	1050	120	22	27.5	5.5
Washing solution 2	1250	100	21.5	27	5.5
Washing solution 3	1500	100	16.5	22	5.5
Substrate	1800	100	12.7	16.7	4.0
Reaction stop solution	2100	150	7.2	11.2	4.0

Table 5.3. Wight of wax and liquid volume inserted to each solution chamber [25].

Reservoir	Wax (mg)	Liquid volume (μl)	$R_2 - R_1$ (mm)
Antigen·Antibody	10	11	5.5
Washing solution 1, 2, 3	19	17.5	5.5
Substrate	8	8	4.0
Reaction stop solution	6	7	4.0
Reaction chamber	12.5	–	–

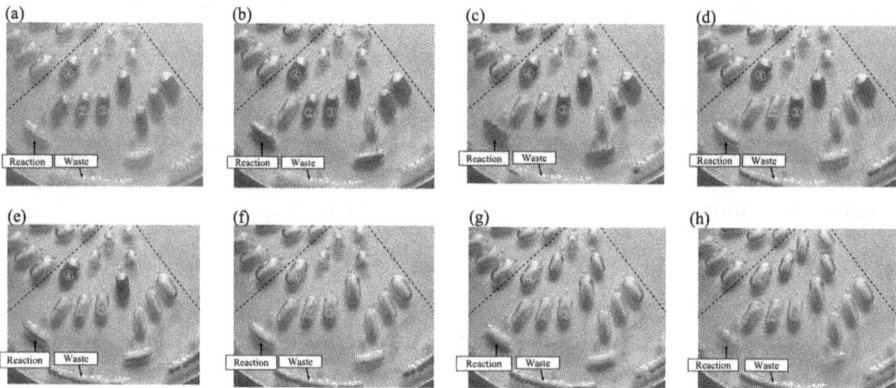

Figure 5.8. An example of sequential solution sending on LOD for competitive ELISA [25].

minimum rupture rpm applied to the capillary valve calculated from the values of each solution in table 5.2 and figure 5.7 and equation (5.20). This ensures that the capillary valve can be ruptured. The volume of each chamber is controlled by the mass of the wax. Table 5.3 shows the mass of wax added to each solution chamber and the volume of liquid injected by using fabrication process shown in figure 3.14(D). Figure 5.8 shows an image of the sequential pumping of the fabricated device shown in figure 5.7. First, a liquid mixture of target protein and antigen is injected into the reaction field (figure (5.8(a)). Next, washing solutions No. 2, 3 and 4 are added to the reaction field and pumped through a syphon valve to the waste tank. Substrate from reservoir No. 5 is then injected into the reaction field (figure 5.8(g)) and finally the reaction stopping solution from reservoir No. 6 is injected into the reaction field (figure 5.8(h)). From the above, this device has successfully achieved sequential pumping in LOD assuming a competitive ELISA. What is shown here is just one example. More sophisticated LODs have been reported by several research groups. Examples are given here that can be designed, fabricated and evaluated immediately in the laboratory by the novice learner.

5.4.2 Mixing methods using Euler forces

Mixing of liquids is one of the important unit chemical operations in the construction of chemical reaction systems using LOD. However, mixing in micro-space is difficult because the viscous force of the liquid is dominant (low Reynolds number), making it difficult to generate turbulence. Several methods have been proposed for mixing liquids using the rotation of a LOD specific device, and Noroozi *et al* devised a method using centrifugal forces and chamber pneumatics [10]. First, centrifugal force is used to push the liquid to be mixed into a sealed chamber. The relaxation of the centrifugal force then releases the pressure in the chamber when the liquids are pushed in and returns them to their original position. A series of these actions are repeated to mix the liquids. Grumann *et al* also proposed a mixing method using magnetic beads [15]. This method is performed by magnetic beads added inside the chamber and a permanent magnet placed outside it. The magnetic beads mix the liquid in the chamber by repeating periodic motions in response to the magnetic field generated by their magnets due to the rotation of the device [15, 17, 22]. On the other hand, a mixing method (Euler-force mixing) using Euler forces generated by the angular acceleration of the rotating coordinate system has also been reported [15, 16, 20–22, 25]. This method is expected to reduce fabrication costs and to be applicable to various applications, as the flow path structure is very simple and can be performed only by rotational control. In previous studies, this mixing method has been investigated for liquids of the same density, but not for liquids of different densities. In the case of different densities, they are supposed to be separated into two layers by centrifugal forces, which would inhibit mixing. However, the densities of samples and reagents required to perform chemical reaction systems usually differ. We have investigated the mixing of liquids of different densities by Euler-force mixing.

Euler-force is an inertial force acting perpendicular to the radius of motion caused by the acceleration and deceleration of angular velocity in a rotating coordinate system and is expressed by equation (5.3) [7–10]. The Euler force is given by the term of the time derivative of the angular velocity vector and the outer product of the position vector, and therefore varies with changes in position and angular velocity in the disk over time. This means that the Euler force varies with position even when the rotational speed is constant and unchanged with time, causing a turning behaviour that contributes to mixing. The vortex created by this difference in

Euler force is used to stir the liquid. If the rotational speed is modulated with time, it is possible to excite even greater swirling behaviour, which allows mixing to proceed quickly.

Since the Euler force is proportional to angular acceleration, it appears during acceleration and deceleration. To use this force to mix liquids, we fabricated a simplified device structure as shown in the figure. The inset of the figure shows the result of trying to mix the liquid by accelerating and decelerating this device with the profile shown in the figure. For details, please refer to the literature. In this way, active use of acceleration and deceleration enables mixing even in regions where laminar flow is dominant within the microchannel.

Here, the flow structure for validating the effectiveness of Euler-force mixing in liquids of different densities is shown in figure 5.9(a). It consists of a reservoir, a mixing chamber and a capillary valve. Initially, the liquid with the lower density is inserted into the reservoir and the liquid with the higher density into the mixing chamber (figure 5.9(c)). At this point, both liquids have the same volume. The rotational speed of the device is then increased to the breaking speed and the liquid is injected into the mixing chamber (figure 5.9(d)). The low-density liquid injected into the mixing chamber stays in the upper layer of the mixing chamber due to the buoyancy generated by centrifugal forces. This prevents agitation of the liquid during injection. With this state as the initial state, Euler-force mixing is performed by repeatedly accelerating and decelerating the device speed (figure 5.9(f)). The LOD with the Euler-force-based mixing mechanism was fabricated by the soft lithography described in the chapter 3.

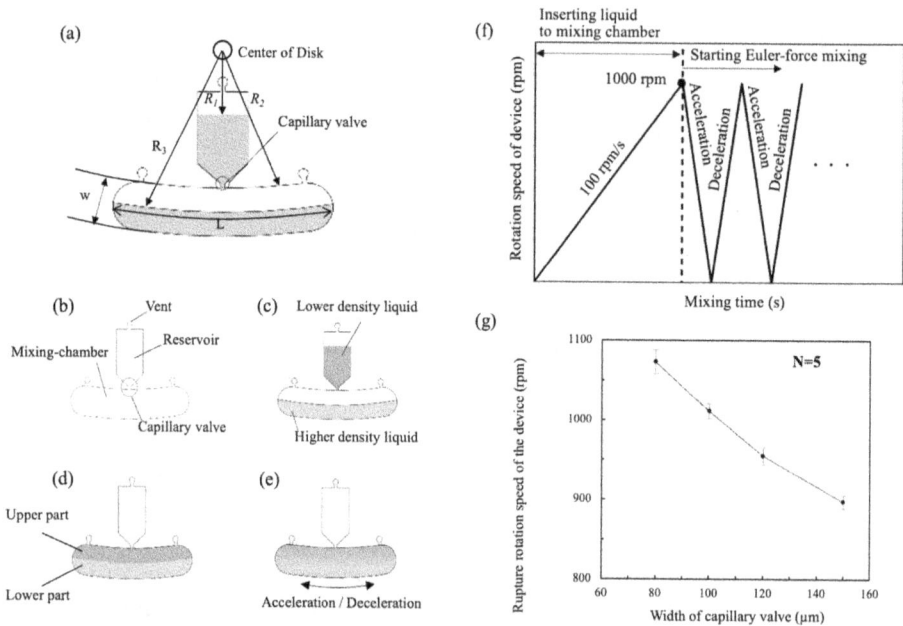

Figure 5.9. (a) Flow path drawing. (b)–(e) Schematic diagram of the operating behaviour according to rotation operation shown in (f). (g) Capillary valve characteristics. Reproduced with permission from [25].

Here is an example of a device that uses Eulerian forces as an example. Readers are encouraged to use the examples to fabricate and evaluate the structure of the device to suit their own requirements. In this example, coloured water with 0.1 wt% safranine (safranine, FUJIFILM Wako Pure Chemical Corp.) added to pure water and a concentration-diluted phosphoric acid solution were used. The device is inserted with the coloured water in the reservoir and the phosphoric acid solution in the mixing chamber. The device then undergoes a rotary operation according to figure 5.9(f). First increase the rotational speed of the device at an angular acceleration of 100 rpm s^{-1}, rupture the capillary valve and insert the coloured water into the mixing chamber. Here, the angular acceleration of the device is set at 100 rpm s^{-1} in order to suppress the agitation caused by the Euler-force mixing that occurs when the liquid in the reservoir is injected into the mixing chamber. This state is the initial state, where the device decelerates and accelerates at a constant angular acceleration, again to a rotational speed of 1000 rpm. This angular acceleration/ deceleration action is repeated to perform mixing of different density liquids by Euler-force mixing. As indicated above, in this experiment, coloured water inserted into the reservoir is injected into the mixing chamber by breaking the capillary valve. In this experiment, $R_1 = 29.1$ mm and $R_2 = 33.6$ mm. From equation (5.20), the rupture turnover can be controlled by varying the width of the capillary valve; figure 5.9(g) shows the rupture turnover against the width of the capillary valve for the coloured water used in this experiment. Here, the depth of the capillary valve is 80 μm. In this experiment, the width of the capillary valve is 120 μm, as it is assumed that the liquid in the reservoir is injected into the mixing chamber when the rotational speed of the device reaches 1000 rpm.

5.4.3 Evaluation of mixing

The following standard deviation methods are often used to assess mixing. Mixing is evaluated from the images taken of the inside of the mixing chamber. Using the ImageJ image processing software, the captured images are converted to 8-bit greyscale and the standard deviation s is calculated from the histogram of luminance values as follows [20]:

$$s = \sqrt{\frac{1}{n}\sum_{i=0}^{255}(m_i - \mu)^2 f_i} , \qquad (5.22)$$

$$\mu = \frac{1}{n}\sum_{i=0}^{255}m_i f_i , \qquad (5.23)$$

$$n = \sum_{i=0}^{255} f_i . \qquad (5.24)$$

where m_i is the gradation value and f_i is the luminance of that gradation value. The smaller the value of this standard deviation, the better the two liquids in the mixing chamber are mixed.

Another method to assess the degree of mixing is to determine the Shannon entropy from images of microchambers where mixing has occurred [21]. Basically, this method is the same as the method for determining the standard deviation described earlier, with the difference that the entropy is calculated. Under the assumption of particle mixing, the system is divided into bins, and the number $n_{j, c}$ articles of species c ($c = 1, 2, 3, 4, ..., C$) is numbered for each bin ($j = 1, 2, 3, 4, ...,$ M). Then, the system entropy S is given by

$$S = -\sum_{j=1}^{M}\sum_{c=1}^{C} p_{j, c} \ln p_{j, c},$$ (5.25)

where $p_{j, c}$ is the join probability of finding a group/complex of P_c particles of species c in bin j:

$$p_{j, c} = \frac{\frac{n_{j, c}}{P_c}}{\sum_{j=1}^{M}\sum_{c=1}^{C} \frac{n_{j, c}}{P_c}}.$$ (5.26)

To simplify and calculate the entropy, we divided the system into pieces and labelled each piece with a number j ($j = 1, 2, 3, ..., n$). Each piece corresponds to a pixel of the obtained image. We assumed that fine particles were distributed in each piece and estimated the distribution probability. This distribution probability corresponds to the pixel gradation, and such logic can be directly applied to the mixing of different liquids. That is, for each pixel, the gradation can be defined as the degree of mixing. Normalizing the gradation for each pixel corresponds to the probability p_j ($j = 1, 2, 3, ...$) of mixing. It can be expressed as the total entropy for the entire evaluation area as the following form:

$$S = -\sum_{j=1}^{M} p_j \ln p_j,$$ (5.27)

where p_j it the probability of a particular gradation value for pixel j.

5.4.4 Example of mixing using Euler force

Here, as an example, the specific gravity difference d (%) of the phosphoric acid solution to coloured water, the angular acceleration α (rpm s^{-1}) of the device and the depth D (mm) of the mixing chamber were varied to verify Euler-force mixing in liquids of different density. Here, R_3, L and w, shown in figure 5.10, were set to 35, 20 and 2.0 mm, respectively. $\alpha = 500$ rpm s^{-1}, $d = 1.0\%$ and D was set to 0.50 and 0.80 mm. Figure 5.10(d) shows the standard deviation versus mixing time in the case of $D = 0.80$ mm, and it was found that the standard deviation is larger for each mixing time. This is due to the smaller Reynolds number due to the smaller volume, which makes turbulence less likely to occur [20, 25].

It can also be seen that the larger the density difference, the shorter the mixing time and the faster the mixing progresses. Conversely, for smaller density

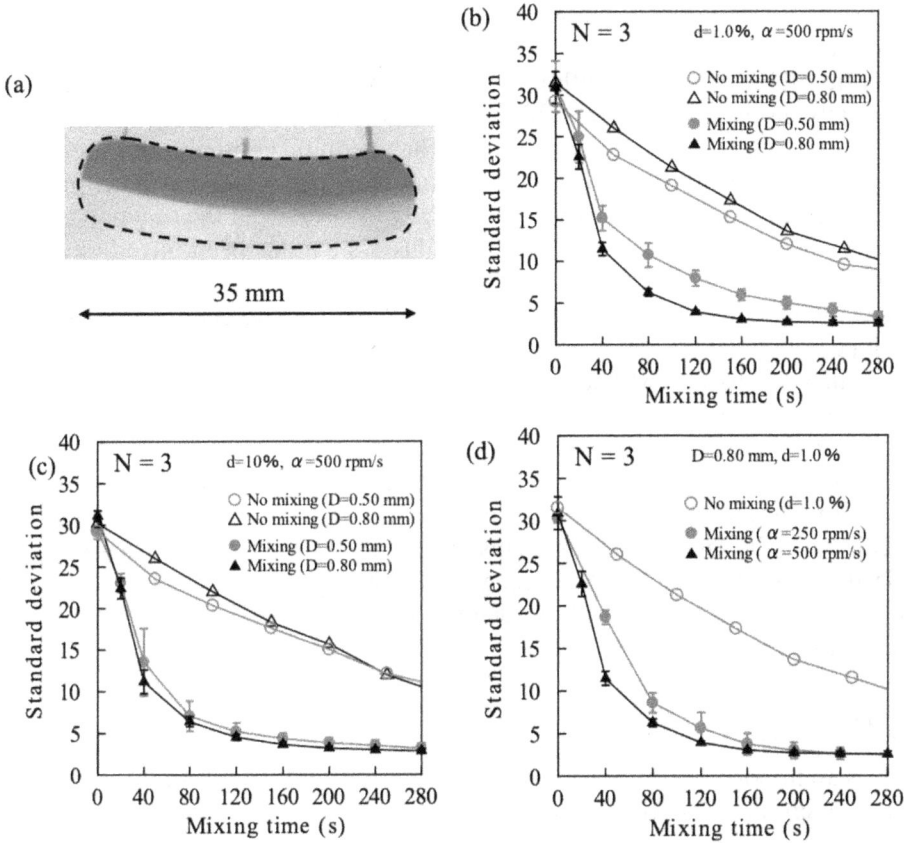

Figure 5.10. (a) Shot image of mixing chamber. Standard deviation dependence of mixing time, where depth of mixing chamber is changed on condition that density difference of liquids is (b) 1.0% and (b) 10%. Standard deviation dependence of mixing time, where angular acceleration is changed [20, 25]. Reproduced with permission from [25].

differences, it can also be seen that mixing is more difficult. However, it was found that mixing, which microfluidic channels are not good at, can proceed quickly by using the Euler force.

5.5 Application of LOD

The field of centrifugal microfluidics has experienced tremendous growth during the past 25 years, especially in applications such as LOD diagnostics. Research and development is taking place all over the world for medical devices. Please refer to the review articles and commentaries from various groups on research trends. Two examples of applications are given here.

In 2009, Lee *et al* demonstrated a portable, disc-based, and fully automated ELISA system enabled to test infectious diseases from whole blood [29]. In their system, a laser-irradiated ferro-wax microvalve was incorporated into a centrifugal microfluidic disk to suppress capillary valve operating variability, allowing the suspension ELISA method with microbeads from whole blood to be constructed completely on the disk. Laser-irradiated valves can actively operate bubbles, thus suppressing operational variability, but they require complex mechanisms. A bead-based suspension ELISA method using magnetic beads has been proposed by the Sysmex Co. Ltd, which has successfully incorporated the functionality of a large testing system into a small system. This system has been introduced to the market.

Because magnetic bead-based assays and controls are so controllable, Wu *et al* developed the liquid-aliquoting-and-syphoning-evacuation (LASE) technique, where all reagents are preloaded into a reservoir prior to testing. Instead, each reagent is loaded only once during the test [14]. Their system simplifies the assay protocol and proposes a low-cost, user-friendly ELISA system.

An unrepresentative number of R&Ds have been carried out here, with LOD excelling the most when it comes to the practical application of microsystems [22, 25–28]. Recently, multi-layered systems have been reported by various groups, and basic and applied research is flourishing [7–18]. Interested readers are encouraged to take the opportunity to contsult academic papers and conference presentations on the latest systems.

5.6 Summary

LOD is an important sub-discipline of microfluidics. It is a system that has proven itself in microfluidics in the field of application or practical application as well as in basic research. The system optimizes processes and enables high-throughput, automated analysis by integrating critical analysis steps onto a single chip. By utilizing rotational mechanics to precisely control fluid dynamics without the need for an external pressure source, centrifugal microfluidics facilitates rapid operation, ideal for emergency medical and field environments. Recently, research, development and social implementation studies have been carried out mainly for POCT. Here, the theoretical principles and practical applications of centrifugal microfluidics for beginners are described with a focus on education. Recent research and development trends are remarkably fast, so it is difficult to give an all-encompassing and comprehensive overview and explanation. However, the basic concepts and principles will never go out of date and are still essential scientific principles for modern systems. Therefore, this chapter introduces the basic operating principles and fluid control mechanisms; presents examples, such as the rudimentary incorporation of ELISA; and describes typical implementations of centrifugal microfluidic platforms in recent immunoassays, nucleic acid testing, antimicrobial susceptibility testing and other tests. If the reader is interested, we also consider the advantages and potential limitations of centrifugal microfluidic platforms and hope to provide suggestions that can be developed into the development of innovative methods for conventional procedures and their deployment in diagnostics.

References

[1] Mark D, Haeberle S, Roth G, von Stetten F and Zengerle R 2010 From microfluidic application to nanofluidic phenomena issue—reviewing the latest advances in microfluidic and nanofluidic research *Chem. Soc. Rev.* **39** 1153

[2] Sin M L, Gao J, Liao J C and Wong P K 2011 System integration—a major step toward lab on a chip *J. Biol. Eng.* **5** 6

[3] Mauk M, Song J, Bau H H, Gross R, Bushman F D, Collman R G and Liu C 2017 Miniaturized devices for point of care molecular detection of HIV *Lab Chip* **17** 382–94

[4] Kimura H, Sakai Y and Fujii T 2018 Organ/body-on-a-chip based on microfluidic technology for drug discovery *Drug Metab. Pharmacokinet.* **33** 43–8

[5] Kim H R, Andrieux K, Delomenie C, Chacun H, Appel M, Desmaële D, Taran F, Georgin D, Couvreur P and Taverna M 2007 Analysis of plasma protein adsorption onto PEGylated nanoparticles by complementary methods: 2-DE, CE and protein Lab-on-chip® system *Electrophoresis* **28** 2252–61

[6] Sheng W, Ogunwobi O O, Chen T, Zhang J, George T J, Liu C and Fan Z H 2014 Capture, release and culture of circulating tumor cells from pancreatic cancer patients using an enhanced mixing chip *Lab Chip* **14** 89–98

[7] Madou M, Zoval J, Jia G, Kido H, Kim J and Kim N 2006 Lab on a CD *Annu. Rev. Biomed. Eng.* **8** 601–28

[8] Kong L X, Perebikovsky A, Moebius J, Kulinsky L and Madou M 2016 Lab-on-a-CD: a fully integrated molecular diagnostic system *J. Lab. Autom.* **21** 323–55

[9] Strohmeier O, Keller M, Schwemmer F, Zehnle S, Mark D, von Stetten F, Zengerle R and Paust N 2015 Centrifugal microfluidic platforms: advanced unit operations and applications *Chem. Sov. Rev.* **44** 6187

[10] Noroozi Z, Kido H, Micic M, Pan H, Bartolome C, Princevac M, Zoval J and Madou M 2009 Reciprocating flow-based centrifugal microfluidics mixer *Rev. Sci. Instrum.* **80** 075102-1–8

[11] Siegrist J, Gorkin R, Clime L, Roy E, Peytavi R, Kido H, Bergeron M, Veres T and Madou M 2010 Serial siphon valving for centrifugal microfluidic platforms *Microfluid. Nanofluid.* **9** 55–63

[12] Thio T H G, Soroori S, Ibrahim F, AI-Faqheri W, Soin N, Kulinsky L and Madou M 2013 Theoretical development and critical analysis of burst frequency equations for passive valves on centrifugal microfluidic platforms *Med. Biol. Eng. Comput.* **51** 525–35

[13] Shih C, Lu C, Wu J, Lin C, Wang J and Lin C 2012 Prothrombin time tests on a microfluidic disc analyzer *Sens. Actators* B **161** 1184–90

[14] Wu H -C, Chen Y -H and Shih C -H 2018 Disk-based enzyme-linked immunosorbent assays using the liquid-aliquoting and siphoning-evacuation technique *Biomicrofluidics* **12** 054101

[15] Grumann M, Geipel A, Riegger L, Zengerle R and Ducrée J 2005 Batch-mode mixing on centrifugal microfluidic platforms *Lab Chip* **5** 560–5

[16] Khorrami Jahromi A, Saadatmand M, Eghbal M and Parsa Yeganeh L 2020 Development of simple and efficient lab-on-a-disc platforms for automated chemical cell lysis *Sci. Rep.* **10** 11039

[17] Uddin R, Burger R, Donolato M, Fock J, Creagh M, Hansen M F and Boisen A 2016 Lab-on-a-disc agglutination assay for protein detection by optomagnetic readout and optical imaging using nano- and micro-sized magnetic beads *Biosens. Bioelectron.* **85** 351–7

[18] Romero-Soto F O, Weber L, Mager D, Aeinehvand M M and Martinez-Chapa S O 2023 Characterization of the flow rate on lab-on-a-disc by a low-powered electrolysis pump for wireless-controlled automation of bioanalytical assays *Sens. Actuators* B **377** 133025

[19] Abe T, Okamoto S, Taniguchi A, Fukui M, Yamaguchi A, Utsumi Y and Ukita Y 2020 A lab in a bento box: an autonomous centrifugal microfluidic systems for an enzyme-linked immunosorbent assay *Anal. Methods* **12** 4858

[20] Takeuchi M, Fujitani K, Ishimoto A, Yamaguchi A and Utsumi Y 2020 Mixing of different density liquids by euler-force on lab-on-a-disc *IEEJ Trans.* C **140** 465

[21] Yamaguchi A, Ishimoto A, Saegusa S, Sugiyama M, Amano S and Utsumi Y 2021 Liquid mixing evaluation using entropy in alab-on-a-disc platform *Sensors Mater.* **33** 4371–82

[22] Hori K and Kakuta M 2021 Lab-on-a-chip and lab-on-a-CD ed A Yamaguchi, A Hirohata and B Stadler *Nanomagnetic Materials, Fabrication, Characterization and Application* (Amsterdam: Elsevier) ch 9 section 9.4

[23] Kim E and Whitesides G M 1997 Imbibition and flow of wetting liquids in noncircular capillaries *J. Phys. Chem.* B **101** 855–63

[24] Zeng J, Deshpande M, Greiner K B and Gilbert J R Fluidic capacitance model of capillary-driven stop valves *IMECE2000-1149* 581–7

[25] Takauchi M 2020 Construction and application of a synchrotron radiation processing platform for the creation of high-precision, high-aspect-ratio microstructures *PhD Thesis* (University of Hyogo)

[26] *Roche* https://roche.com/

[27] *Abaxis* https://abaxis.com/

[28] *Samsug Healthcare* https://Samsunghealthcare.com/en/

[29] Lee B S, Lee J-N, Park J-M, Lee J-G, Kim S, Cho Y-K and Ko C 2009 A fully automated immunoassay from whole blood on a disc *Lab on a Chip* **9** 1548–55

IOP Publishing

Nano–Microsystems
Science and applications
Akinobu Yamaguchi

Chapter 6

Actuator systems based on surface acoustic wave devices

Recently, an active actuator for the transportation, mixing, and separation of liquid and powder is strongly required because of its need for physical operations in various miniaturized systems such as Lab-on-a-Chip (LOC) and micro-total analysis systems. However, it is difficult to efficiently and accurately handle both powder and liquid because they are treated not only as a solid but also as a liquid or gas in the miniaturized systems. A surface acoustic wave (SAW) device allows us to treat both the powder and liquid in the systems. Here, we outline the principle of SAW devices, and introduce the manufacturing method and procedures. Finally, we show some examples of incorporating them into miniaturized systems.

6.1 Introduction

Mixing small amounts of fluids and transport for small amounts of powder are usually delicate tasks. For liquids confined to small volumes, interaction between the fluid and the walls of the container become stronger as the surface-to-volume ratio becomes larger. As described many times, Reynolds's number parameterizes the strength of the interaction for the system, indicating the inertial forces are more significant than the viscous forces as the system becomes smaller. In general, Reynold's number, usually being a small quantity in microfluidic systems and to some extent equivalent to an increase of the apparent viscosity, significantly influences the hydrodynamic behaviour of a liquid. Consequently, only laminar flow processes are possible. Most of the time, almost the only way for small fluid volumes to mix effectively is by diffusion if one does not do anything in particular.

Methods such as an external pump and centrifugal liquid transport have been proposed as a liquid transport mechanism in a microfluidic system, and each has its own characteristics. Both liquid transport mechanisms are very useful and available for the creation of microfluidic systems. However, for many applications, such as

doi:10.1088/978-0-7503-3111-1ch6
6-1

biochips and combinatorial chemical reaction chips, a deliberate and controlled agitation of the fluid under study is very important. Herein, another actuator mechanism using SAWs, which is completely different from these methods, will be described. It is the most important feature of the drive mechanism based on SAW, and the difference from the above two mechanics is that a SAW actuator can transport not only liquid but also powder. This feature can provide a novel idea and microchemical system composed of the SAW actuator [1]. SAWs can also be used for sensor applications, integrated into Nano–Microsystems [2]. Microchemical systems often manipulate fluids, but few systems can manipulate powders. In order to realize actual chemical synthesis and automatic analysis mechanism, not only fluid manipulation but also powder manipulation is indispensable. Therefore, a SAW mechanism as an actuator that can be miniaturized and can be mounted on a microsystem is attracting attention.

This chapter begins with some historical background on acoustics, microfluidics, and piezoelectric materials. The section that follows is on the theoretical outline for SAWs and SAW-induced phenomena such as liquid transport, atomization, and particle displacement. Until now, many applications have been reported in microfluidics for the use of acoustic waves and SAWs. Here, we follow some previous studies and summarize them for the sake of developing the state-of-the-art new devices or systems by using the SAW.

6.2 Background

Focusing on acoustics, it is related to musical instruments and music. Faraday, Helmholtz, Lord Rayleigh, etc were the first to conduct academically systematic research on various phenomena that are well known as acoustics. Readers who want to know more about the detailed historical background are encouraged to read the references. Here, we will immediately describe various devices that use SAWs. First of all, when taking about SAWs, we must not forget the SAW filter, which is often used in the communication field. White and Voltmer developed a desire to generate acoustic waves on the planar surface of a piezoelectric substrate [3]. Their achievement in part resulted in a broad development of ultrasonic devices in communication applications that continues today. Next, research and development (R&D) are very advanced and support today's wireless communication technology. In this process, the development of materials with excellent properties such as lithium niobate ($LiNbO_3$) has also been also promoted. Product development has progressed due to improvements in manufacturing technology and production efficiency for single-crystal materials. Furthermore, the propagation characteristics of SAWs change depending on the crystal orientation, which led to the development of SAW filters suitable for the application.

To date, there have been many researches and developments on acoustics and SAW, and it is difficult to cover them all. This textbook is also an educational textbook focusing on basic R&D and recent examples of Nano–Microsystems. Therefore, this chapter describes the role of SAW in Nano–Microsystems or vice versa, and examples of the creation of Nano–Microsystems that use SAW as a basic technology. Therefore, we will not explain in detail the physical mechanism related to SAW, physical property research, and application examples to electronics. Readers who are interested in these physical mechanisms and their applications to electronics and wireless communications should read the references. In the following, we will explain the development of Nano–Microsystems, especially droplet transport and powder transport systems using SAW [1, 4–27].

The discipline of microfluidics relies on the ubiquity of fluids in performing tasks in chemistry, biology, and materials science. In LOC, working at a scale where the fluid physics is dominated by surface tension and viscosity, the flow are nearly laminar. In the systems, the flow is laminar and thus the mixing predominately relies on diffusion; chemical reactions on which most practical applications rely proceed very slowly even though the diffusion distances are smaller than for conventional laboratory reaction vessels. To solve the problem, the methods should be provided as passive and active mixers that can mix fluids chaotically at the microscale. For the passive mixers, intricate designs and nano/microscale precise fabrications are usually required to achieve high effective performance while avoiding unacceptably high loss. The passive mixers are typically described in chapter 2. On the other hand, active mixer usually moves the requisite complexity from the structure to the materials. Effective mixing devices and systems that make use of piezoelectric materials can generate acoustic waves and act to improve the mixing in the microscale microchannels. In particular, the devices that incorporates SAWs will be the main devices for achieving the effective mixing in the LOC. Friend and Yeo provided a comprehensive review that describes the physical mechanism of SAW and its applications to microfluidic devices [1]. In the following, we will introduce what is necessary and interesting for beginners, referring to their review and recent works.

Since a SAW is a physical vibration wave, it can also be used for powder transportation and its control. Currently, commonly used powder operation methods are intermittent air transportation, conveyors and feeders using spiral screws that rotate inside the tube, and so on. These operating methods are often used in large systems for mass production. On the other hand, when conducting drug discovery or searching for new chemical substances, it is often the case that synthesis is performed at the laboratory level and the physical properties of the substance are evaluated, and it is required to manipulate trace powders. At this time, the precise and quantitative manipulation of the trace powder is a factor that determines the composition and physical properties of the substance to be synthesized. At present, there is almost no method for manipulating trace powders of several mg or less with powders having a diameter of several tens of metres or less, and they are often operated by human hands. Therefore, the substances that can be created are limited, and it becomes the rate-determining factor in material exploration. The applicant is currently conducting R&D on a combinatorial substance synthesis system using a microchemical system, and is proceeding with research aiming at the realization of a system in which liquid and powder can be operated in units at the same time. By using SAW as the basic unit operation actuator, the applicant has realized the transportation of trace liquids and powders to date [1, 4–27]. Based on the results so far, we have succeeded in transporting and weighing powders of mg or less in combination with a micro feeder [17–19]. If a combinatorial chemical synthesis system can be realized by combining these elemental technologies, it is expected that a system that can perform automatically combinatorial synthesis substances using data from AI and material informatics can be realized.

However, in reality, it was found that there are many unsolved problems regarding the transport characteristics of the powder material. Furthermore, we realized again that there is a fundamental problem caused by using SAWs. It is the contribution to the powder transport characteristics due to the contribution of electrostatic force and intermolecular force, which can be ignored on the macro scale, and the stoppage of powder transport due to the standing wave excitation of SAWs. Since the substrate that excites SAWs is a ferroelectric substance, the contribution of electrostatic force is particularly large. Furthermore, the generation of standing waves can be used to stop powder transportation in controlling powder transportation, but it stops transportation, so it is a problem that must be solved in order to realize unit operations. In this chapter, we will outline a mechanism to suppress standing waves and conduct R&D with the aim of establishing basic unit operations to realize a combinatorial chemical synthesis system. Furthermore, the infrastructure of the combinatorial chemical synthesis system will be improved by combining unit operations.

The following outlines the basic principles of SAW. We will briefly introduce how to make various devices that incorporate SAW generation. Then, some empirical examples will be described. As an example, devices involved in liquid transportation and liquid manipulation will be described. For example, Sritharan et al demonstrated acoustic mixing at low Reynold's numbers [13]. They showed that the interaction between SAWs and a fluid confined to a microfluidic device induces pronounced streaming effects which in turn act as an internal stirrer to the fluidic. Vuskasinovic et al investigated the onset and frequency distribution of the consequent capillary wave and followed the phenomena that a sessile drop purely vibrated purely in a direction transverse to the solid surface through to bursting and atomization of the drop as the amplitude of the vibration was increased [8]. Girado et al combined SAW in $LiNbO_3$ substrate with polydimethylsiloxane (PDMS) to form a sort of micropump with a fluid interface [9]. Langelier et al were also able to pump fluids at about 100 Pa through their microfluidics deice as desired among the four input channels [10]. There are so many R&D cases that it is not possible to introduce and describe them all, but we will introduce typical cases that are easy to understand.

6.3 Basics of surface acoustic waves

Direct generation of SAWs on a piezoelectric substrate using interdigital electrodes was first reported by White and Voltmer [3]. The devices are fabricated by micro/nanofabrication techniques combined with lithography and film deposition to pattern a metallic coating, typically Al or Cr/Au, into the interdigital pattern.

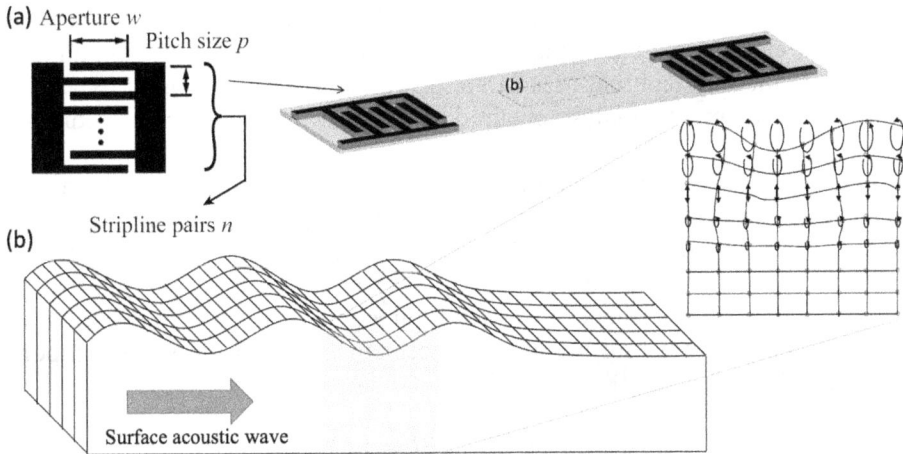

Figure 6.1. SAW devices: (a) design of the comb electrode structure and its placement on the substrate, and (b) schematic diagram of acoustic wave propagation on the piezoelectric substrate surface.

As shown in figure 6.1, the SAW is generated along the surface of a piezoelectric substrate by applying a high-frequency electromagnetic signal to the interdigital transducer (IDT). SAW-powered microfluidic systems are capable of powder transport as well as liquid transport. Furthermore, in liquid transport in the microchannel, the Reynolds' number itself is small and a laminar flow state occurs, so that it may be difficult to mix the solutions required for the chemical reaction or the antigen–antibody reaction. The mixing process using a SAW is extremely efficient and enables the melting process. In addition, the SAW enables the atomization, transportation, and vibration of liquids as shown below. The transportation of powder can also be achieved. SAW excitation requires semiconductor microfabrication on a pezoelastic substrate, but depending on how it is used and combined, it becomes an elemental member responsible for very important unit chemical operations.

6.4 Principle of SAW excitation

A SAW is a wave that propagates along the surface of a medium in the same way as a wave that propagates on the water surface, and is also called a surface acoustic wave because 90% or more energy is concentrated within one wavelength from the surface. There are several wave types of SAW. Most of the waves in devices using SAW are Rayleigh waves. Rayleigh wave propagation velocity is characterized by being constant regardless of frequency. However, the propagation speed depends on the crystal cut surface and the propagation direction of the crystal of the substrate. A Rayleigh wave is a wave in which a longitudinal wave having a variation component in the propagation direction and a transverse wave having a displacement component perpendicular to the propagation plane are combined. Since these two waves have a phase difference of π from each other, the mass points on the propagation plane move in an elliptical orbit with respect to the travelling direction of the waves. Piezoelectric elements that are generally used include $LiNbO_3$ and lithium tantalate. Here, we outline a system that uses $LiNbO_3$ as a substrate. Although $LiNbO_3$ is an anisotropic medium, its analysis method is outlined as an isotropic medium for simplicity.

Figure 6.2 illustrates the fluid–solid half-space model of the first-order acoustic field. The elastic solid-medium moves along an elliptical locus in the counter-clockwise direction due to the SAW when coupled with air. Here, we define the Rayleigh wave propagation directly along x_1-axis, as schematically illustrated as figure 6.2. As the displacement of the depth direction x_3-component decreases with becoming deeper from the surface, the displacement u_i of the x_i-direction component can be described by [1, 3, 4]

$$u_i = A_i e^{-\frac{\alpha\omega}{vx_3}} e^{j\left(\omega t - \frac{\alpha x_i}{v}\right)} (i = 1, 2, 3). \tag{6.1}$$

Here, v is the velocity of SAW and α is the damping constant along the x_3-direction. Both parameters are determined by medium constants and boundary conditions.

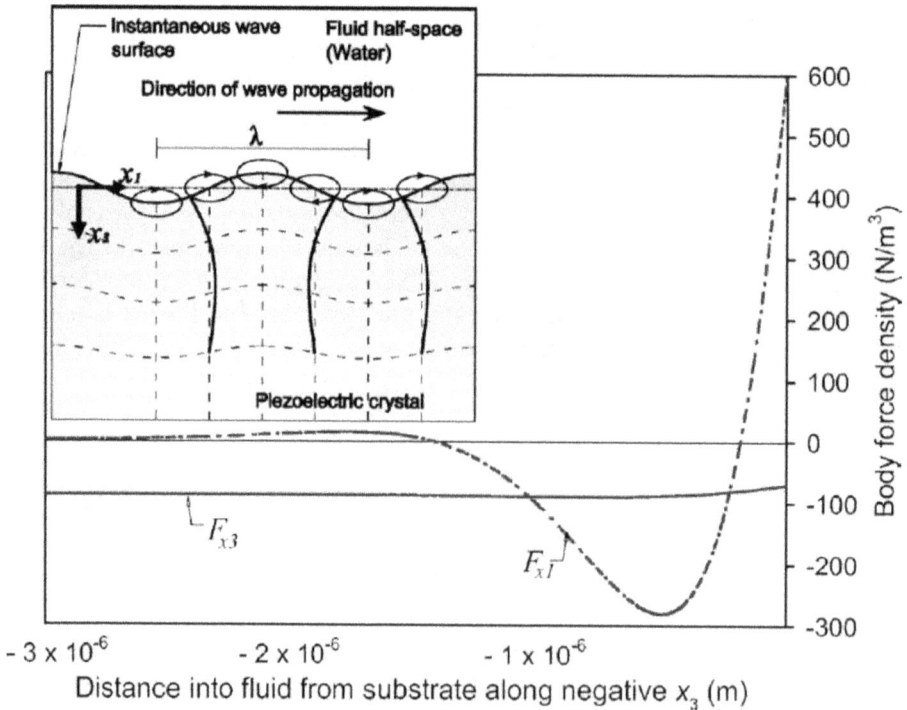

Figure 6.2. SAW on the semi-infinite $LiNbO_3$ (LN) substrate coupled with the half-space air. The x_1 and x_3 components of streaming body force density are plotted along the direction of the waves propagation in the fluid and at $x_1 = 0$. The inset shows the motion of the solid particle elements of the LN substrate. Reprinted from [4] with the permission of AIP Publishing.

Solving by considering medium constants and boundary conditions, the displacements $u_i (i = 1, 2, 3)$ are obtained as follows:

$$u_1 = F_1 \left(e^{-\frac{\alpha_3 \omega}{v x_3}} - p e^{-\frac{\alpha_p \omega}{v x_3}} \right) e^{j\left(\omega t - \frac{\alpha x_i}{v}\right)}$$ (6.2)

$$u_2 = 0$$ (6.3)

$$u_3 = -j \frac{F_1}{\alpha_3} \left(e^{-\frac{\alpha_3 \omega}{v x_3}} - \frac{1}{p} e^{-\frac{\alpha_p \omega}{v x_3}} \right) e^{j\left(\omega t - \frac{\alpha x_i}{v}\right)},$$ (6.4)

where $p = -\frac{F_2}{F_1}$, F_1, and F_2 are amplitude coefficients. From these equations, it can be seen that the displacement component has only propagating direction and depth direction of the wave, and u_3 is advanced in phase $\pi/2$ with respect to u_1. This indicates that the wave is elliptical backward in the propagating direction near the surface of the medium. The acoustic streaming body force density can be given by solving

$$-F_{\mathrm{dc}} = \frac{1}{c^2} \left\langle P_1 \frac{\partial u_1}{\partial t} \right\rangle + \rho_0 \langle (u_1 \cdot \nabla) u_1 \rangle,$$ (6.5)

where c and ρ_0 are the speed of sound in air and the density of air, respectively. The symbol $\langle \rangle$ in equation (6.5) refers to time averaging, the subscript 'dc' refers to second-order steady-state terms, P_1 is the first-order fluidic pressure, and u_1 is the first-order fluidic velocity. The SAW is generated on the substrate and interacts with a fluid drop, causing the drop to deform into an asymmetric conic leaning approximately at the Rayleigh angle. The SAW propagation can induce the streaming and atomization of the droplet, as shown in figure 6.3 [5]. Tan *et al* numerically calculated the acoustic streaming body force density for 30 MHz SAW propagating on a 128°-rotated Y-cut X-propagating LiNbO$_3$ substrate coupled with air, resulting in the plot in figure 6.4 [6]. Their numerical results are plotted in figure 6.4 for the x_1 and x_3 components of the body force density at the solid–liquid interface. The results are indicated as follows. The body force density tangent to the substrate surface (F_{x_1}) is 100 times higher in the viscous boundary layer than in the bulk fluid region. It is directed in the positive x_1 direction at the solid–fluid interface because of the counterclockwise surface motion that gives rise to a large inertial force in the fluid at the interface. Due to the low viscosity of air, the weak attenuation of this longitudinal surface acceleration suggests that only the fluid adjacent to the surface is accelerated in the same direction as that of the propagating SAW. The atomization of droplets is also easier with SAW, as demonstrated in figure 6.5 [8].

Thus, the mass points of the propagation plane draw an elliptical direction as shown in figure 6.6 with respect to the propagating direction of the SAW. On the surface of the medium, these points have a rear elliptical rotation with respect to the propagating direction of the SAW, and the displacement is only in the depth

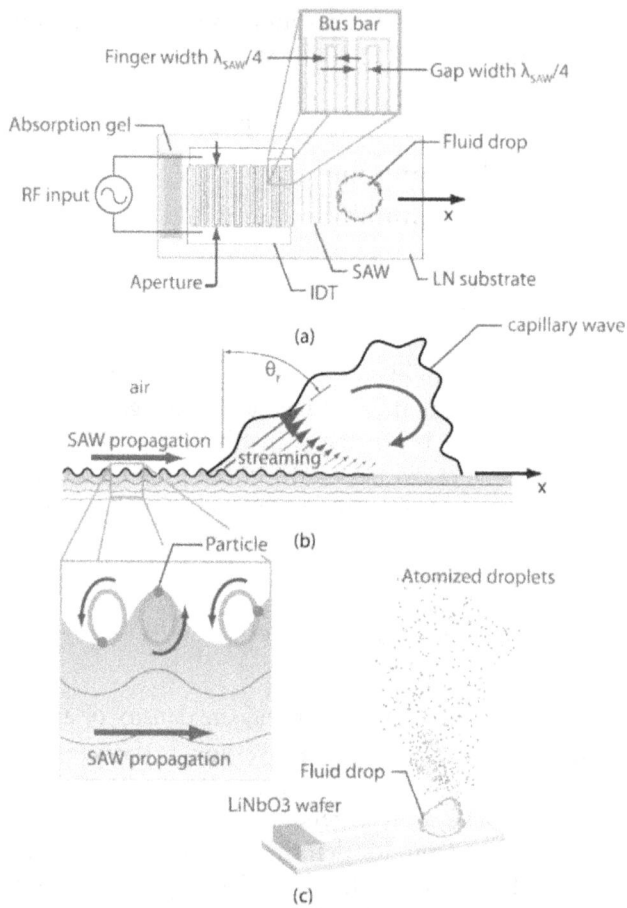

Figure 6.3. Schematic of droplet transport by SAW propagation excited on a piezoelectric substrate: (a) from the top, (b) from the side; the SAW itself is a point on the surface moving anti-clockwise in an ellipse. It has the highest amplitude at the surface of the substrate and decays exponentially to almost negligible levels within four to five wavelengths. Interaction of the SAW with a droplet causes the droplet to deform into an asymmetric conical shape leaning roughly at an angle corresponding to the Rayleigh angle θ_r. The acoustic irradiation causes drop deformation through first-order effect on the time scale of the acoustic wave and bulk fluid recirculation on a hydrodynamic time scale. It is acoustic streaming. (c) Schematic of atomization of a fluid drop. It occurs from the free surface of the irradiated drop. Reprinted from [5] with the permission of AIP Publishing.

direction at the position of about 1/5 wavelength. They are forward elliptical rotating underneath the position of it.

As will be described later, when a Rayleigh wave is excited by this SAW propagation characteristic: (1) In the case of droplets illustrated in figure 6.6(a), the droplet travels in the same direction as the SAW propagation direction. (2) In the case of powders illustrated in figure 6.6(b), the powder travels in the direction opposite to the SAW propagation direction [18, 25, 27]. However, in the higher order wave of Rayleigh wave, powder transport does not always occur in the

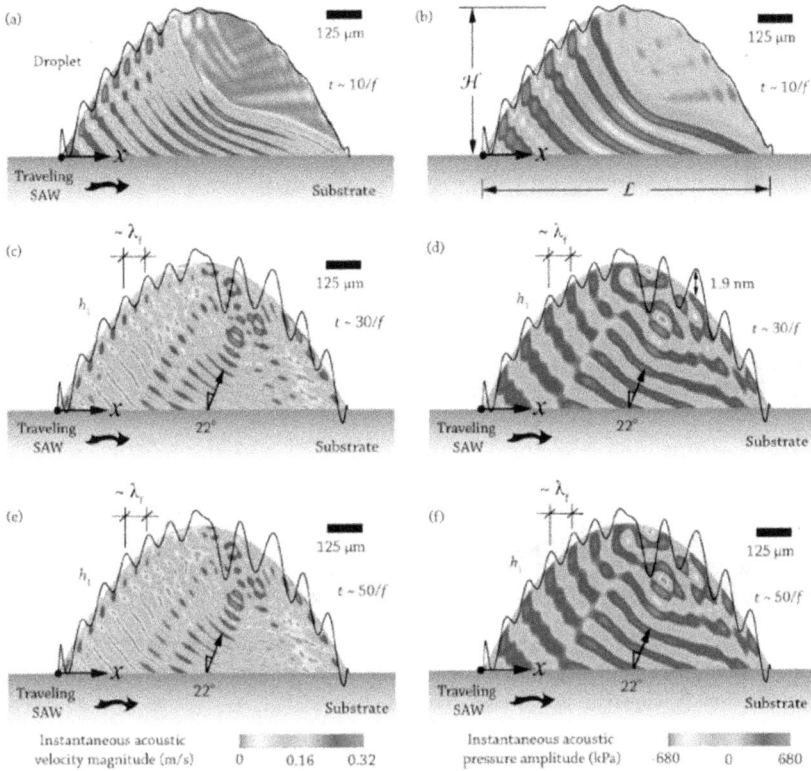

Figure 6.4. Calculation results of the transmission of acoustic waves into a sessile drop after (a), (b) 10, (c), (d) 30, and (e), (f) 50 periods. Instantaneous acoustic velocity magnitude is represented in (a), (c), and (e). Instantaneous acoustic pressure amplitude is shown in (b), (d), and (f). The simulation was performed in the condition of 128YX LiNbO$_3$ substrate with a standard interdigital transducer (IDT) at a resonant frequency of 20 MHz. Reprinted from [39] with the permission of AIP Publishing.

direction opposite to the propagation direction of SAW, as described below. The above points will be described in detail in actual LOCs.

6.5 How to generate a SAW

To generate the SAW, in general, piezoelectric substrates are used. In particular, LiNbO$_3$ is one of the most famous and useful materials. SAW can be easily excited by forming regularly intersecting comb-tooth electrodes or an interdigital transducer (IDT) on a piezoelectric material and applying a high-frequency signal. The applied high-frequency signal is converted into a SAW, and the centre frequency can determine a specific frequency band by the processing period of IDT and the physical properties of the piezoelectric body. For that reason, it is often used as a filter for communication and broadcasting equipment. A device using a SAW in this way is generally called a SAW device.

If the substrate consists of 127.68° rotated Y–X cut, X-propagating LiNbO$_3$, Rayleigh waves can be generated in which two mechanical acoustic wave

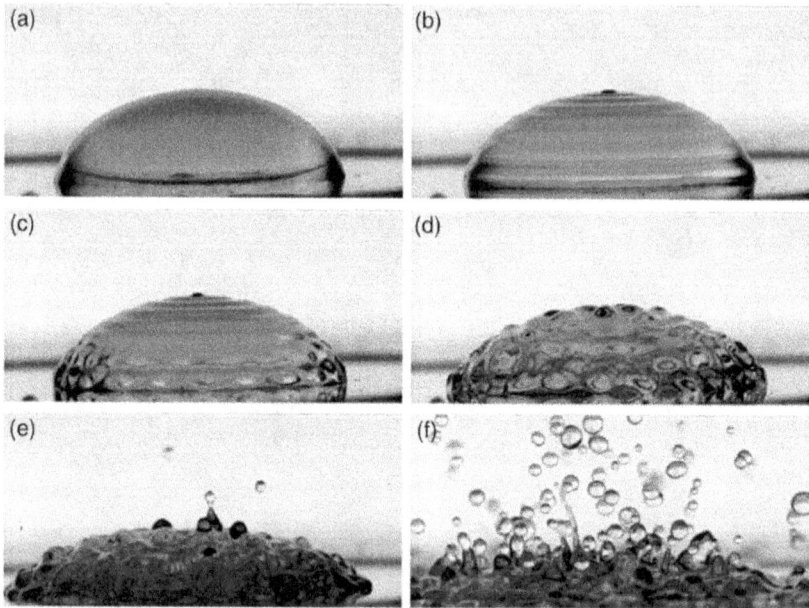

Figure 6.5. (a) A 100 μl droplet paced on a LiNbO₃ substrate, SAW applied, asymmetrically distorted droplet as shown in (b), continued application of SAW from (c) to (e) and eventual destruction of the droplet as shown in (f). Reprinted from [8] copyright Cambridge University Press 2007.

components are generated, as described in the above equations, a compressional component along the direction of propagation and a transverse component perpendicular to the surface. Here, a typical IDT electrode structure is shown in figures 6.1, 6.3, 6.6 and 6.7. A commonly used IDT electrode is a structure in which two combs are engaged. The comb-tooth pair number N is the number of electrodes for one comb, and the number of electrodes in the entire IDT is twice the comb-tooth pair number N. The width between the comb-tooth electrode L and the comb-tooth distance is equal, and the comb-tooth pitch is $p = 4L$ as shown in figure 6.1. Assuming that the SAW is a Rayleigh wave, which is the fundamental wave, and the propagation velocity is v, the following relationship holds between the pitch, the propagation velocity, and the frequency f_0.

$$f_0 = \frac{v}{p}. \tag{6.6}$$

When the electric field with frequency of f_0 is applied, the SAW is the most strongly excited.

Piezoelectric substrates are required to have smoothness on the order of micron or submicron metres, which is the wavelength of SAWs, and reproducibility of frequencies of several hundred ppm or less. Therefore, film formation is performed on piezoelectric single crystals or single crystals with small variations in material

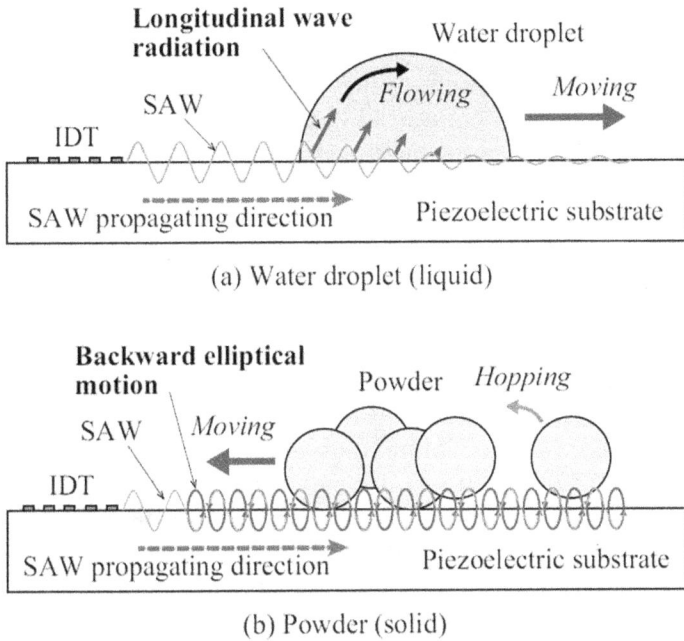

(a) Water droplet (liquid)

(b) Powder (solid)

Figure 6.6. Schematic of object transport mechanism by SAW for (a) droplet (liquid) and (b) powder. Reprinted from [18], copyright (2017) with permission from Elsevier.

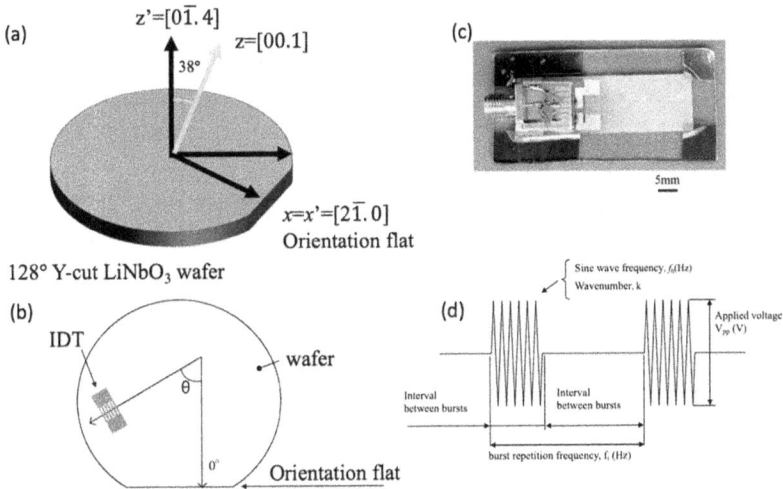

Figure 6.7. (a) 128° Y-cut LiNbO$_3$ substrate surface orientation and cur surface alignment. Orientation flat (OF) direction is parallel to $x = x' = [2\bar{1}.0]$. (b) IDT alignment settings, angle θ defined with the OF direction as $\theta = 0°$. (c) Optical micrograph of the fabricated SAW excitation chip, which was fabricated with $\theta = 0°$. The scale bar is 5 mm. (d) Waveform of applied AC voltage. Continuously generating large amplitudes may cause distortion and heat, which may cause destruction and the input is a sine wave with an interval in burst mode.

constants. Piezoelectric thin film is often used. In order to use a piezoelectric thin film, it is necessary to align the crystal orientation of the thin film like a single crystal, which makes the manufacturing process difficult. When using it as an actuator in a nano/micro-system instead of using a filter in communication equipment, it is a good idea to use a piezoelectric single crystal with a large electro-mechanical coupling coefficient. Therefore, the most commonly used is 128°-Y-cut-LiNbO$_3$ substrate.

The IDT electrodes for excitation of the SAW are generally fabricated by using a conventional semiconductor process combined with lithography and film deposition. For example, Al, Au, and Cu films are often used as electrode. The details of fabrication procedure are described in chapter 3 and section 6.6. Upon excitation, the displacement of the substrate normal to the substrate surface is larger than the displacement along the direction of SAW propagation since the elastic solid is more free to vibrate transverse to the surface. Manipulation of a liquid droplet on SAW substrates depends strongly on the diffraction of the compressional wave into the droplet. Within the substrate, the compressional and transverse wave displacements decay exponentially with increasing distance from the substrate surface, as described in equations (6.1)–(6.4). As a result, a water droplet is placed on the surface of the substrate to interact with the SAW. The in-plane axially polarized compressional displacement component is diffracted at the Rayleigh angle into the droplet. The Rayleigh angle is defined as $\theta_R = \sin^{-1}(c_w/c_s)$, where c_w and c_s are the Rayleigh wave velocity in the liquid and that in the solid, respectively. The difference between the velocities generates the radiation leakage, especially, the leaky component of the SAW induces an acoustic pressure gradient inside the droplet, giving rise to acoustic streaming. This streaming force is the result of acoustic energy flux dissipation within the fluid; it is balanced with forces due to fluid viscosity and inertia. When the high-power ultrasonic excitation is applied, the induced streaming motion imports momentum to the fluid, giving acoustic-streaming jets or even atomization [1, 4–27].

6.6 Fabrication of a SAW electrode and system with a SAW actuator

Here, as an example, the simplest and most reliable device manufacturing method and procedure will be outlined. Basically, by following the microfabrication method and procedure described in the chapter 3, the reader can easily and surely create a device. First, the IDT manufacturing process will be outlined below. A schematic diagram of the fabrication process is shown in figure 3.1. Photolithography technology, which is one of the microfabrication technologies, was used to fabricate the IDT. For example, first, 10 nm of Cr and 1000 nm of Al are formed on the piezoelectric substrate by radio frequency (RF) sputtering. A positive resist (OFPR-800 manufactured by Tokyo Ohka Kogyo Co., Ltd) is applied on it, UV exposure is performed, and the IDT pattern drawn on the glass mask is transferred. Then, IDT can be produced on the substrate by etching Al and Cr based on this pattern. Figure 3.22 shows a typical manufacturing process of the glass mask used for UV exposure, as described in chapter 3. If a projection drawing device using a digital mirror device (DMD) is used, a glass mask can be fabricated or patterning can be done directly on the resist. The process can be selected according to the application and requirements. If the same pattern is needed for a large quantity, fabrication of a glass mask and normal UV lithography is better. On the other hand, if the pattern is to be adjusted or changed for research applications, direct drawing is used.

6.7 Fluid manipulation

This section shows the several examples and applications appearing in recent years and the ways the technology is being used to address them. Here, we will describe oscillation and transport of droplets, atomization, pumping, jetting, colloidal case, and applications to the mixer.

6.7.1 Oscillation and transport

There are many phenomena induced by the application of SAWs. For example, vibration, translation, levitation (jetting and drops), pumping, internal flow, patterning, and atomization, etc have been reported by many research groups. Here, we begin with an area that predates most activities in acoustic microfluidics, yet systems were researched and developed for both basic and application investigations. Since the Rayleigh wave propagates on the piezoelectric substrate, as shown in figure 6.6, if there is an object in the propagation direction, the SAW exerts a force or torque onto the object. In liquids, longitudinal waves are radiated and attenuated in the Rayleigh angle θ_R direction. Rayleigh waves generated from IDT electrodes radiate longitudinal waves into the liquid while propagating through the interface between liquid and substrate. Due to the difference in radiation pressure before and after the longitudinal wave propagation direction, the liquid causes a flow along the SAW propagation direction, as shown in figure 6.8. Generally, when the liquid is water, the Rayleigh angle is $\theta_R = 23°$. By increasing or decreasing the

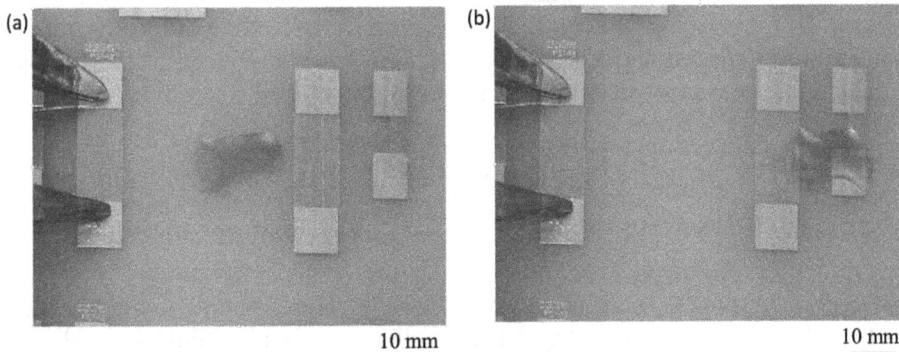

(a)

(b)

10 mm

10 mm

Figure 6.8. Observation of a droplet transported by SAW, moving from (a) to (b).

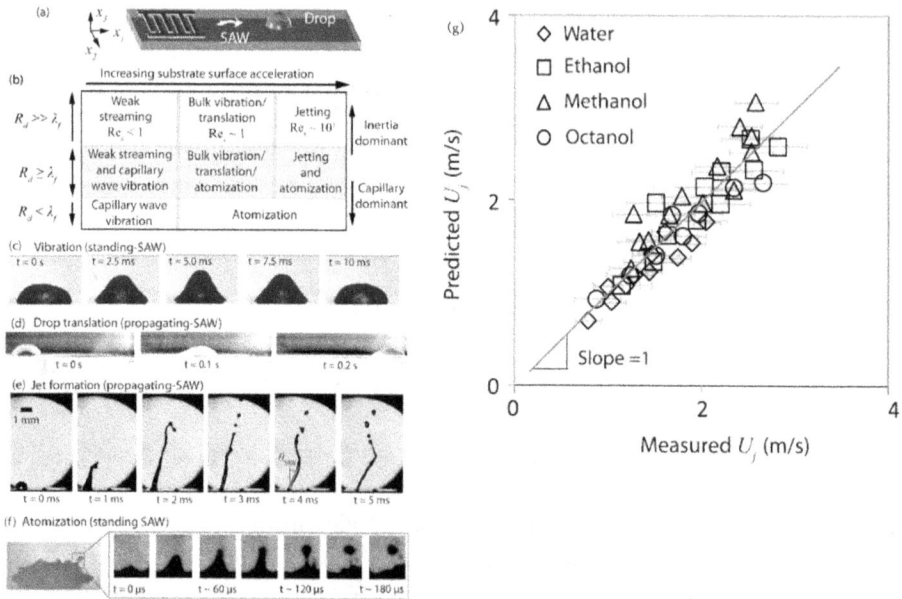

Figure 6.9. (a) Schematic diagram of a SAW device, with 3 ml deionized water droplets paced in the SAW propagation path. Radiation absorption gel is placed at the substrate edge to prevent reflected waves from being generated from the substrate edge. (b) Summary of droplet behaviour, where R_d is droplet size, λ_f is the SAW propagation wavelength at 20 MHz, Re_s is the Reynold number. (c) Vibration of the droplet due to SAW. (d) Translational behaviour of a droplet due to a propagating SAW. (e) Jetting at the Rayleigh angle SAW as a result of the propagating SAW irradiation. (f) Atomization of droplets by stationary SAW. (g) Comparison between the experimentally measured jet velocity with the prediction given by equation (6.7). Reproduced with permission from [11], copyright (2009) by the American Physical Society.

power and frequency supplied to the IDT electrodes, the radiation pressure of the longitudinal wave fluctuates, and the liquid is transported, vibrated, and atomized, as summarized in figure 6.9 [11]. The figure shows that even when the liquids are changed to water, ethanol, methanol, or octanol, the jet velocities by SAWs are universal. By introducing an acoustic forcing term to the leading order axisymmetric jet momentum balance derived by Eggers [28], Tan *et al* arrived at the following relationship that permits prediction of the axial jet velocity U_j [11]:

$$U_j \cong \left[2L_j\left(F_s^y - g\right)\right]^{\frac{1}{2}}, \tag{6.7}$$

where $F_s^y \approx \alpha_0 \beta \xi_{x_3}^2 \, Re_A$ is the force associated with the acoustic streaming in the jet and g is the gravitational acceleration. α_0 is the acoustic attenuation coefficient, given by $\alpha_0 = \pi b f/(\rho c_l^3)$, where $b = 4\mu/3 + \mu_B$. Here, ρ and μ are the liquid density and viscosity, respectively. μ_B is the bulk viscosity of the fluid. c_l is the sound velocity in liquid. ξ and x_3 are the substrate's surface displacement and direction perpendicular to substrate, respectively. $\beta = 1 + B/2A$ is the coefficient of nonlinearity. B/A is experimental value dependent on the liquid [11, 29]. The coordinate system is shown in figure 6.2 [4]. $\dot{\xi}_{x_3}$ represents the first-order time derivative of the ξ along the x_3 direction. Here, Re_A is the acoustic Reynolds number, given by $Re_A = \rho \dot{\xi}_{x_3} \lambda_f / (2\pi b)$.

Using these characteristics, an example of using SAWs as actuators in a microsystem is shown below. We give an overview of the SAW-operated pump for continuous fluidic feed, mixer, and flow channel. We also explain the characteristics of mixing of two liquids with different concentrations on the LOC and demonstrate its effectiveness as a reactor of a microsystem with only a SAW as the drive source.

6.7.2 Continuous fluidic feed

Figures 6.10–6.15 show an example of continuous liquid transport in a microfluidic channel using SAW as a pump or mixer, an example showing its role as a mixing mechanism during laminar flow transport, and an example of transporting droplets by manipulating the timing of transport by changing the frequency, and implementing mixing, etc at the desired timing, respectively. Figure 6.10 shows schematics of a microfluidic channel with the SAW pump [8]. Schematic diagrams incorporating the PDMS microchannel and after integration are shown in figures 6.10(a) and (b), respectively. If the microchannel is formed in direct contact with the SAW substrate, it is considered that the SAW is attenuated at the channel wall and the mechanical

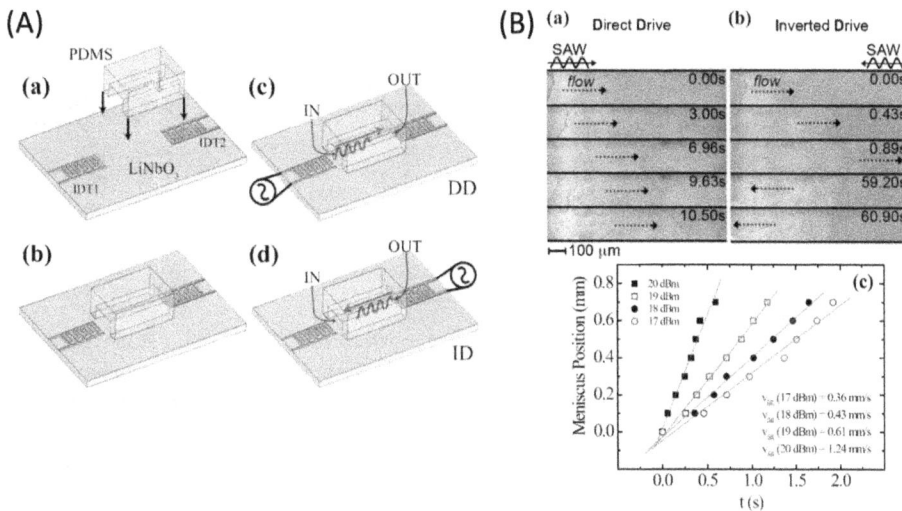

Figure 6.10. (A) Schematic for the assembling of microfluidic devices and activation of the liquid motion into microchannels as shown in (a). Final device structure is shown in (b). SAWs were excited from one IDT to the channel entrance along the channel toward its outlet (direct drive (DD)) in (c). The SAWs were launched in the opposite direction from OUT to IN (inverted drive (ID)) in (d). (B) Snapshots of the time evolution of water filling process for DD configuration in (a). Filling process for ID in (b). (c) Meniscus position as a function time t for ID configuration at different input power P_{SAW}. Reprinted from [12] with the permission of AIP Publishing.

(a)

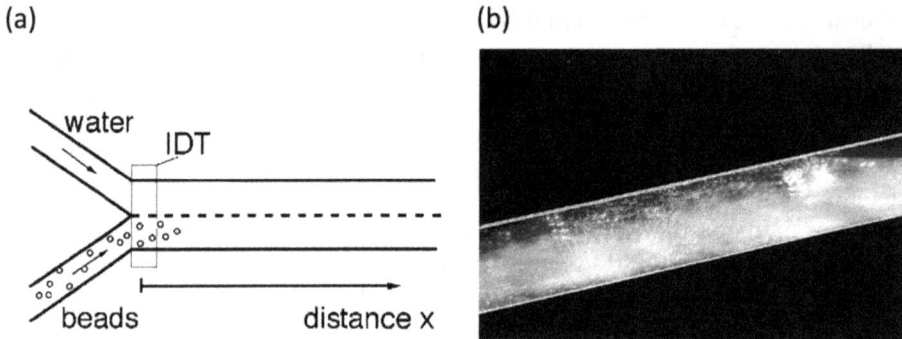

(b)

Figure 6.11. (a) Schematic of Y-shaped microfluidic channel for investigating the SAW-induced mixing. The two inlets were flushed with pure water, to one of them fluorescent beads were added. (b) Snapshot of the SAW-induced mixing in the microfluidic channel. Vortices are formed in the channel from the laminar flow due to the SAW-induced mixing. Reprinted from [13] with the permission of AIP Publishing.

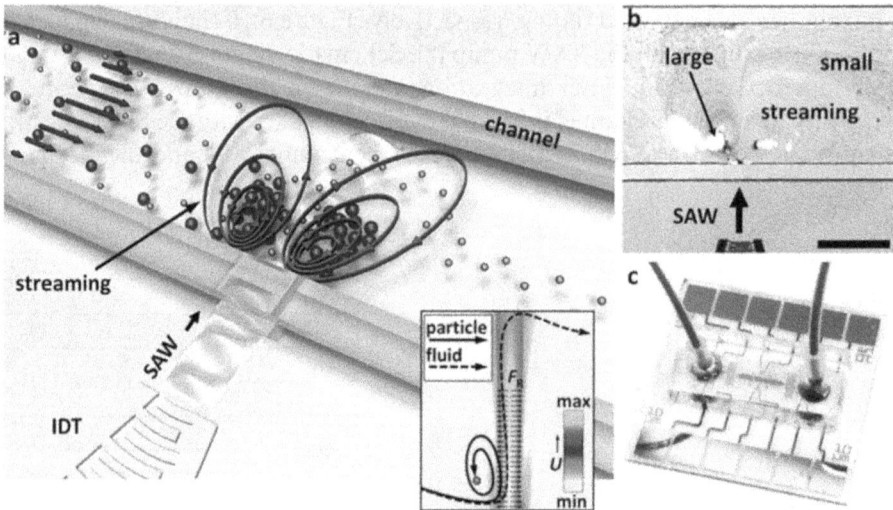

Figure 6.12. (a) A continuous flow of a mixed particle/cell population is subjected to strong, actively acoustic streaming generated by an IDT on a piezoelectric substrate. SAW-induce acoustic streaming serves to align incoming fluid streamlines net the beam and selectively capture particles from incoming flow according to their physical properties. (b) Example image shows the selective capture of larger 2 μm from a mixed suspension of these and smaller 1 μm particles. Scale bar is 200 μm. (c) Photograph of the microfluidic device and image of system, comprising a microfluidic channel on top of a SAW device. Reprinted from [14] with permission from the Royal Society of Chemistry.

vibration required for liquid transport does not propagate to the liquid. Therefore, to make the SAW propagate across the substrate surface, creating gaps in the lower side wall is necessary, resulting that the SAW reaches the bottom of the flow channel without attenuating [8, 15, 16]. The liquid in the flow channel shifts in the direction of propagation of the SAW due to the horizontal component radiating

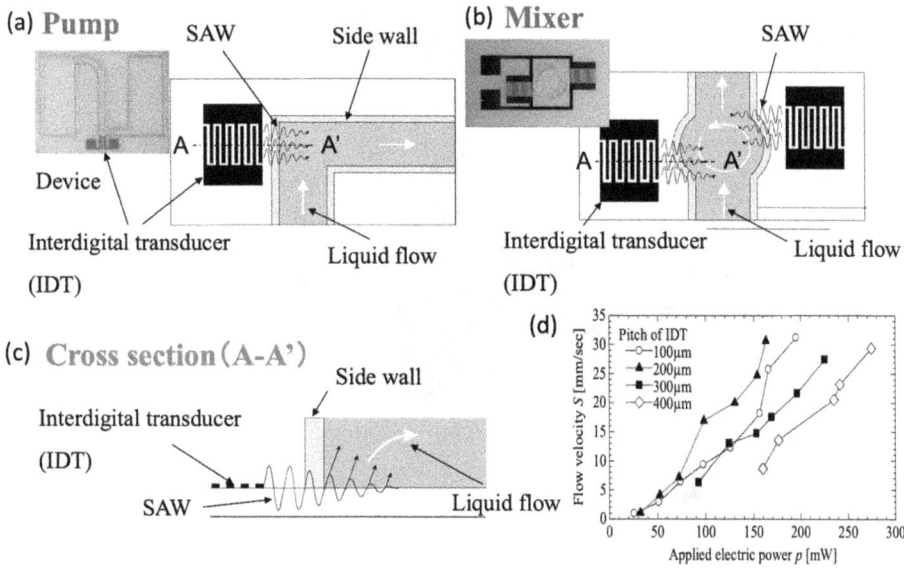

Figure 6.13. Creation of functions by IDT arrangement: (a) Pump and (b) mixer. (c) Cross-sectional view of these devices. (d) Dependence of the liquid transport velocity on the input power and on the pitch of the IDT.

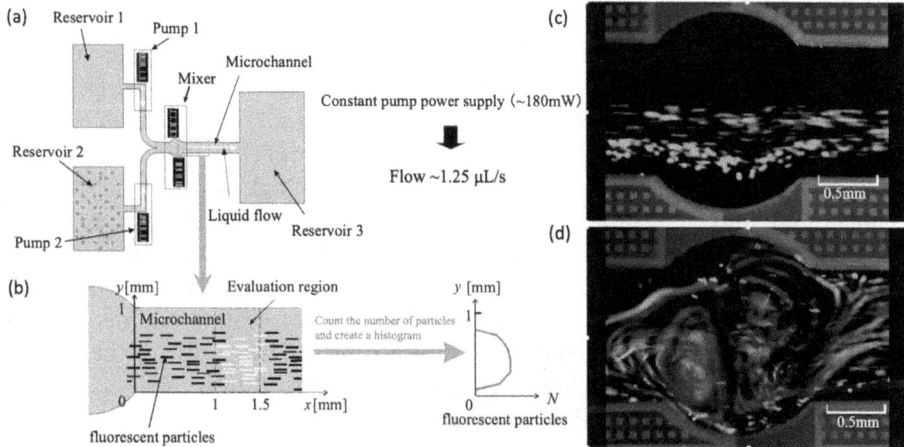

Figure 6.14. (a) Example of a microsystem: overall view of a system in which liquids are transported from two reservoirs 1 and 2, mixed in a SAW mixer in the middle of a Y-shaped flow channel and transported to reservoir 3. Pure water flows from reservoir 1 and fluorescent particle dispersed water is pumped from reservoir 2. (b) Part of the flow channel structure after the mixer and the evaluation area. Observed images of mixing (c) without mixing, (d) with mixing; in (c), the flow is laminar in the absence of SAW excitation. On the other hand, in (d), with mixing, it can be seen that vortices are generated and the solution is stirred.

longitudinally from the SAW. Also, the perpendicular component of the longitudinal radiation has a liquid surface near the IDT up to the position where it is balanced with the force gravity. As a result, for a pump, the liquid flows from the inlet to the outlet.

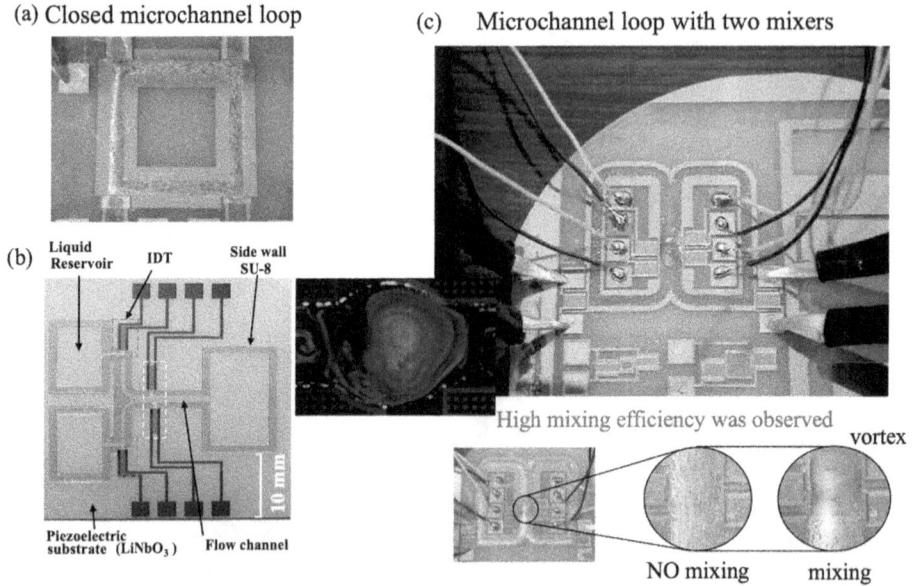

(a) Closed microchannel loop

(b)

(c) Microchannel loop with two mixers

High mixing efficiency was observed

vortex

NO mixing   mixing

Figure 6.15. Examples of various continuous fluid systems using SAW: (a) closed microchannel loop, (b) system for transporting and mixing two liquids. (c) microchannel loop with two mixers.

Sritharan *et al* demonstrated the mixing experiment in a simple Y-shaped microfluidic channel structure, as schematically illustrated in figure 6.11 [13], where two different fluids are injected into a common channel, using the application of SAWs in 3D microfluidic devices containing a set of IDTs for SAW generation, glass, plastic, or silicon. The channel system was 75 μm high and 100 μm wide. The two inlets were filled with water and water with fluorescent bead (diameter is 1 μm) to visualize the streaming patterns under a fluorescent microscope. A steady flow velocity of approximately $v = 250$ μm s^{-1} was achieved by application of a constant back pressure to the inlet. The bead distributions at the injection and near laminar flow downstream channel are shown in figure 6.11(a). By applying the SAW at 146 MHz, the mixing was achieved, as shown in figure 6.11(b). The acoustically induced mixing in the microfluidic channel with the fluid flowing from left to right was achieved, as shown in figure 6.11(b). As seen in this snapshot, the occurrence of vortices along the channel accompanying with downstream is clearly observed. They revealed the acoustic streaming induced complex material folding lines which significantly enhance the mixing efficiency in such microfluidic devices.

Langelier *et al* demonstrated programable operation of continuous streams of fluid by generating customizable acoustically switchable concentration gradients [10]. In their device, dynamically reconfigurable gradients are generated by connecting the device to four reservoirs of coloured water and altering the strength and tonal composition of the acoustic output. When an acoustic signal is introduced, a gradient rapidly forms with a composition representative of the tones and amplitudes present. The output pressure of each cavity was set to the same value (100 Pa) in this demonstration. Adjustment of the relative amplitudes of the input tones allows one to tune and control the component flow ratios.

Acoustic streaming has emerged as a promising technique for refined microscale manipulation, where strong rotational flow can give rise to particle and cell capture. Recently, Collins *et al* demonstrated acoustic streaming vortices to selectively capture 2 μm from a mixed suspension with 1 μm particles and human breast adenocarcianoma cells from red blood cells in a continuous flow microsystem with SAW [14]. Figure 6.12 shows their acoustic streaming based capture demonstration. SAW irradiation is used to excite streaming and separate particle size and type while continuous fluid flows through a microfluidic channel.

6.7.3 Microfluidic systems based on SAW actuators

The ability to pump liquids with SAWs has been shown with droplets; continuous pumping of liquids is also possible with SAWs. Here are three examples. In all three examples, an open channel system must be used when pumping liquid with SAWs. The reason is that if the liquid is confined in the channel by a lid, the SAW will be damped and the liquid will not be driven by friction or other effects. Based on this point, examples of R&D using microfluidic channels in open systems are given. Figure 6.13 shows a channel that selectively pumps reagents by choosing the frequency of SAW excitation. The second example is a system that combines pumping and mixing. As shown in figure 6.13, a pump is placed at a location where the flow path is refracted, and a system that excites flow in the liquid and pumps the liquid is possible using SAW. Furthermore, by shifting the centre line of the IDTs and placing them opposite each other, a mixer can be created by generating a rotational force. The SAW is emitted in a position shifted parallel to the central axis of the cylindrical reservoir, and thus the liquid in the flow channel flows while rotating. Combining these two functions, as shown in figure 6.14, liquid 1 and liquid 2 are pumped from reservoir 1 and reservoir 2, and liquid 1 and liquid 2 are mixed in the mixer and transported to reservoir 3. Fluorescent particles are mixed in reservoir 2, and the system is able to check if the mixer is functioning. The figure shows the results without and with SAW mixing. Without the mixer, the flow distribution of the glowing particles clearly shows that the system is in a laminar flow state. On the other hand, when a mixer is used, vortices are generated and it is clear that mixing is occurring. Figure 6.14 shows the evaluation of this mixing state, and it was found that the mixing was almost uniform when the supplied power was 250 mW or higher. The third example shows that placing a SAW in a bent flow path enables pumping, and looping, and it can also create a liquid circulation system. An example of a system combining these two is shown in figure 6.15(a), with an example of a closed flow path structure created and operated by combining four SAW pumps. Figure 6.15(b) shows a system that combines a pump and a mixer to transport and mix two liquids. Figure 6.15(c) is an integrated system that combines all liquid transport, circulation, and mixing. As described above, it can be seen that the use of SAWs makes it easy to pump and mix liquids electrically. Although it has the disadvantage of operating in an open system, the liquid pumping and mixing mechanism using SAW is a very effective means of solving the dead volume problem caused by pump wiring and the mixing problem in microfluidic systems.

6.8 Sensors for chemical reaction

SAW technology can be also available for chemical sensing applications. Wohltjen and Dessy first demonstrated chemical sensing by using the SAW [30, 31]. SAW sensors are indirect probes of various physical and chemical quantities [30–38]. There are two types of device configurations, namely the delay lines and the resonators which are commonly used. In the case of the delay line configuration, sensors can be performed by the measurement of delay time and phase shift when the substance to be measured is present or absent between the IDTs. It is also possible to use a single IDT for both purposes, excitation and detection, by providing a reflector. This case is the one-port or reflective delay line configurations, as shown in figure 6.16(a). At this time, sufficient impedance matching between the IDTs of the device is a necessary condition for increasing the sensitivity of the device. It has a simple device structure and is easy to use.

On the other hand, in the resonator type device, two IDTs for emission and detection of the acoustic waves and grating reflectors are necessary to be placed outside of each IDT so that a resonating cavity is formed. Both of these configurations have the same mechanism response. The output responses from these devices are similar. Experimentally, the acoustic wave velocity change is evaluated by measuring the resonance frequency or phase of the wave. The following relations are given by the measured changes in the resonance frequency and the phase with and without substance exposure such as gas and chemical liquid [30–32]:

$$\frac{\Delta v}{v_0} = \frac{\Delta f}{f_0} = -\frac{\Delta \phi}{\phi_0}, \tag{6.8}$$

where v_0, f_0, and ϕ_0 are the velocity, resonance frequency, and phase of the SAW device without any perturbation. When perturbed, these parameters change to v, f, and ϕ. Devices that measure phase or frequency changes are commonly used, since velocity changes are small changes.

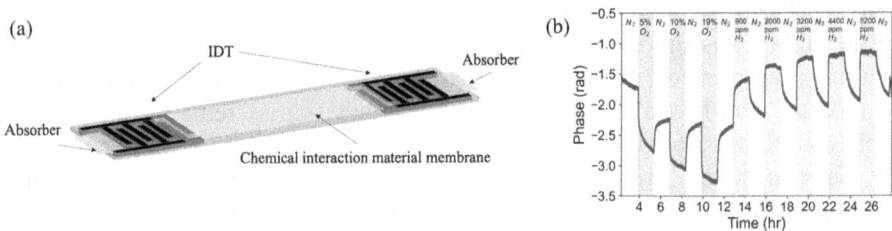

Figure 6.16. (a) The schematic structure of the SAW relative humidity sensor. (b) Phase responses of In$_2$O$_3$-coated SAW sensor in real time under 800–5200 ppm H$_2$ and 5%–19% O$_2$ at 350 °C. Reprinted from [35], copyright (2022) with permission from Elsevier.

Devkota *et al* have investigated the application of indium oxide (IO) and indium tin oxide (ITO) films on langasite (La$_3$G$_5$SiO$_{14}$, LGS)-based SAW reflective delay line sensor devices for monitoring hydrogen at 350 °C [35]. They modelled the effect of the IO and ITO sensing layer thickness on the wave velocity, attenuation, and effective electromechanical coefficient. Figure 6.16(b) shows the response of an LGS/IO sensor to various concentrations of H$_2$ and O$_2$ balanced with N$_2$ after compensating for the temperature effect. The result of figure 6.16(b) shows that the phase changes with the type of gas introduced: H$_2$ increases the phase of the reflected SAW wave, while O$_2$ decreases the phase of the reflection. This trend is due to the following mechanisms. Reducing gases such as H$_2$ increases the conductivity of the membrane and decreases the wave velocity. On the other hand, oxidizing gases such as O$_2$ decrease the conductivity of the IO film through oxidation. The defects such as oxygen vacancies and interstitial atoms, which affect the electrical properties of metal oxides, at high temperatures, strongly influence the electrical properties of metal oxides. Specifically, the concentration of free charge carriers depends on the concentration of oxygen vacancies, which affects conductivity. As a result, the circuit parameters of the equivalent circuit associated with SAW propagation change, so that phase changes occur and gas adsorption and desorption can be sensed. Interested readers should refer to the references, as there are many examples of research. A representative example is introduced below. SAW sensors are capable of detecting chemicals at very low concentrations ($\sim$ppb levels). Venema *et al* fabricated LiNbO$_3$ SAW sensors with different operating frequencies, coated with three different thicknesses of metal-free PC films [32]. It was shown that the fabricated device can detect NO$_2$ gas with very high sensitivity.

6.9 Powder transport

As shown in figure 6.6(b), when the powder is placed on the Rayleigh wave propagation surface, the powder comes into contact only at the wave front of the SAW, and the powder moves so that it is repelled by the frictional force at the contact surface. At this time, the SAW makes a rearward spheroidal rotation, and the force acts in the direction opposite to the SAW propagation direction at the wave front, so the powder moves in the opposite direction to the SAW propagation direction.

The propagation velocity and propagation characteristics of SAWs are strongly dependent on the substrate crystal and crystal orientation. One of the most user-friendly and suitable substrates for SAW excitation is 128° Y-cut LiNbO$_3$ wafer. The profile of its 128° Y-cut LiNbO$_3$ wafer is shown in figure 6.7(a), where the IDTs can be freely arranged on the wafer by using the semiconductor lithography technique. Here, the arrangement of the IDTs is considered as the angle θ from the orientation flat (OF) direction. Below, fixed at $\theta = 0°$, which has the highest SAW excitation amplitude, is used. Figure 6.7(c) shows an optical photograph of a typical powder transport device by SAW. Basically, the structure is the same as that of a device for droplet transport. It is wired with SubMiniature Version A (SMA) connectors for ease of use. When SAWs are excited in the device, the piezoelectric effect causes a large distortion, and the device is destroyed by the distortion and heat. Therefore, the electric voltage for SAW excitation is input in burst mode, as shown in figure 6.7(d). When SAW devices are incorporated into the experimental setup shown in figure 6.17(a), powder transport by SAW excitation can be evaluated in real time and in real space. First, this SAW device is used to investigate powder transport by drive frequency. One IDT was prepared on the wafer and cut out in a strip shape using a dicing saw, so that the influence of unexpected reflected waves from other IDTs and the cut surface of the wafer was eliminated. Furthermore, in order to minimize the influence of reflection, the power supply is supplied by the SMA terminal. Furthermore, although not shown here, a vibration-absorbing structure consisting of rubber on the cut end surface can significantly suppress reflected waves. Reflected waves interfere with travelling waves to form standing waves, which can affect powder and droplet transport. It is important to configure the system to cancel out or take advantage of standing waves, depending on the device application. Electric leads were joined by soldering to the SMA terminal and the printed circuit board, and ultrasonic wire bonding was used to connect the printed circuit board and IDT

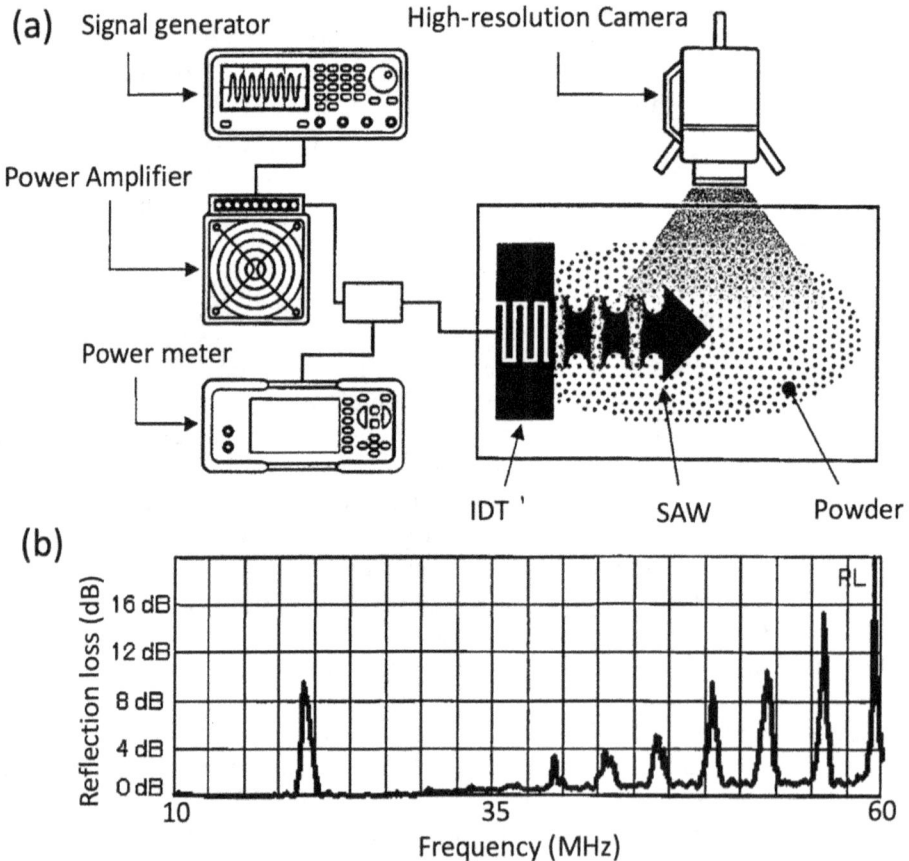

Figure 6.17. (a) Schematic diagram of the setup of the powder transport experiment. (b) Harmonic characteristics of the SAW device used.

using pressure and ultrasonic vibration. The OF of the wafer was set to 0°, and the clockwise direction was set to θ, as shown in figure 6.7(b). In this experiment, the IDT angle was set to $\theta = 0°$, which is the most commonly used angle. This is the direction in which the propagation velocity of a Rayleigh wave (3960 m s^{-1}) is the highest. The IDT parameters were set to a pitch of 200 μm, an intersection width of 5000 μm, and a logarithm of 20, with reference to those that have been investigated in the powder transport experiment of this study.

First, a high-frequency voltage is generated by a signal generator (AFG3252 manufactured by Tektronix) and amplified by an amplifier (ALM00110-2840FM manufactured by R & K). Then, a SAW is generated by applying the amplified voltage to the IDT. The actual resonance frequency actually is measured for each IDT using a network analyzer (AA-230PRO manufactured by RigExpert). The thickness of the LiNbO$_3$ substrate used was 500 μm ($\pm$30 μm), and the surface roughness was 0.3 nm or less. As shown in figure 6.17(b), there are multiple resonance modes between the IDTs. The multiple peaks are strongly dependent on the crystalline direction of the LiNbO$_3$ substrate and pitch distance of the IDT electrode. As the high-frequency voltage applied to the IDT, a burst voltage waveform, as shown in figure 6.7(b), was used to prevent damage to the piezo-electric substrate due to heat generation. This waveform consists of a burst period and a rest period in which a sine wave continues for k cycles. By using such a burst wave, it is possible to reduce the energy supplied to the IDT from the continuous wave to $k \cdot f_i/f_b$. Here, f_b is the frequency of the sine wave during the burst period, and is determined by the IDT shape and the like. In addition, f_i is the reciprocal of the burst interval from one sine wave to the next sine wave, and is set to 1 kHz here. In the experiment, k is 2000. In the experiment, the power P is measured by the RF power meter (Rohde & Schwarz NRP-291) instead of measuring the amplitude V_{pp} of the voltage applied to the IDT. The resonance frequency of Rayleigh wave excitation was 19.2 MHz in this SAW device. Figure 6.18 shows the results of comparing the powder transport characteristics when SAWs were excited at 19.2 MHz, the fundamental resonance frequency, and at 42 MHz, the higher harmonic frequency. Figures 6.18(a) and (b) show the respective powder distributions at $t = 0.2$ and 4 s after time $t = 0$ s at the respective frequencies. Powders that were uniformly distributed at time $t = 0$ in both cases, gathered at the IDT in the case of fundamental frequency excitation in figure 6.18(a) and moved away from the IDT in the case of harmonic frequency excitation in figure 6.18(b) as time elapses. At this time, only the width of the IDT shows SAW propagation, indicating that the fundamental wave is faster and transported more powders. On the other hand, in the case of harmonic excitation, a periodic powder distribution appears, indicating that standing waves are being excited. Repeating similar experiments with varying resonance frequency, the powder transport characteristics by SAWs were investigated. Figure 6.19 summarizes the powder transport characteristics. It was found that changing the excitation frequency significantly changed the powder transport properties. This can be attributed to the change in the propagation characteristics of the SAW excited by the substrate. Paradoxically, it can be considered that frequency control is sufficient to achieve

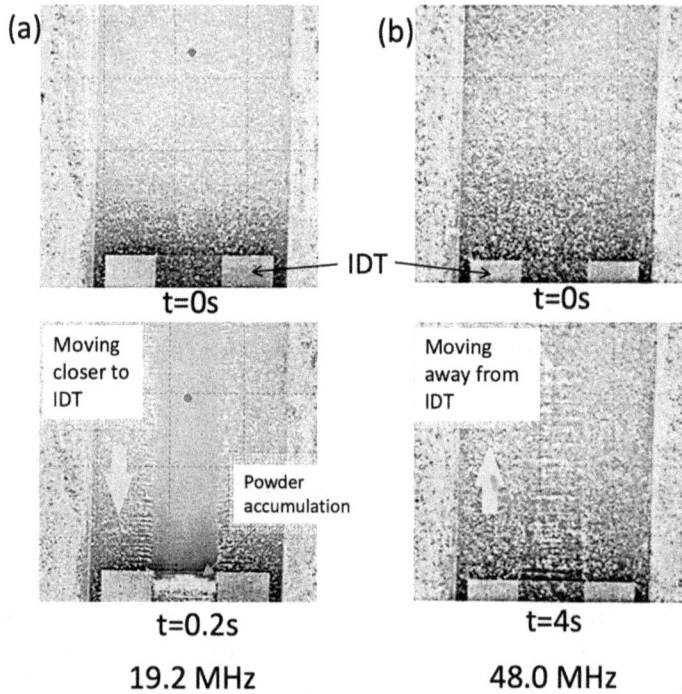

Figure 6.18. Results of powder transport experiments (powder directly placed on the LiNbO$_3$ substrate): Excitation frequencies of (a) 19.2 and (b) 48.0 MHz, respectively. The direction of powder transport is dependent on the frequency. Frequency control enables the powder transport.

Frequency (MHz)	19.2	34.4	36.9	40.5	44.2	48.0	51.8	55.9	59.4
Transport speed (mm/s)	25	-	-	-	0.1	0.8	0.3	0.3	1.0
Input power (mW)	2903	201	478	709	1060	1376	1338	1095	738
Incidence rate (%)	78	24	49	56	63	82	85	91	85
Transport direction									

Figure 6.19. Summary of powder transport properties: the Rayleigh frequency in the device used was 19.2 MHz. Summary of powder transport speed, input power, incidence rate and transport characteristics when excited at harmonics.

the desired powder transport. This is one of the features that is very important and needs to be taken into account when specifically implementing and functioning SAW actuators in microchemical systems.

Other features of powder transport by SAWs, which are not explained here, are that SAW propagation is characterized by reflection and refraction, etc. Therefore, by arranging multiple IDTs so that SAWs satisfy Bragg's diffraction conditions, the operation of reflection and refraction can also be performed for powder transport. By taking these features into account, it is possible to develop applications for various microchemical systems.

6.10 Powder feeder

The powder transporting behaviour can be controlled by SAWs as shown above. Here, as one example, the powder transporting on the miniature feeder is demonstrated, as schematically illustrated in figure 6.20 [18, 19]. A typical photograph of fabricated miniature feeder driven by SAWs is displayed in figure 6.21. The miniature feeder was prepared with monitoring one powder

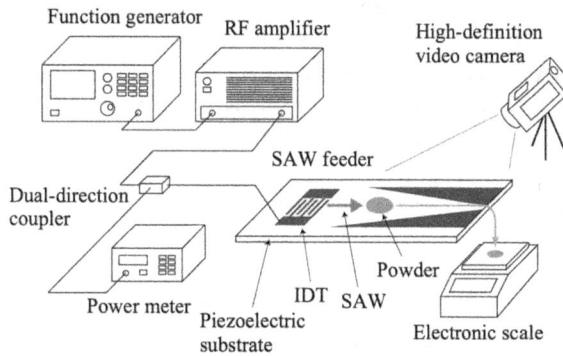

Figure 6.20. Conceptual diagram of the experimental setup for controlled experiments on trace powder transport [18, 19].

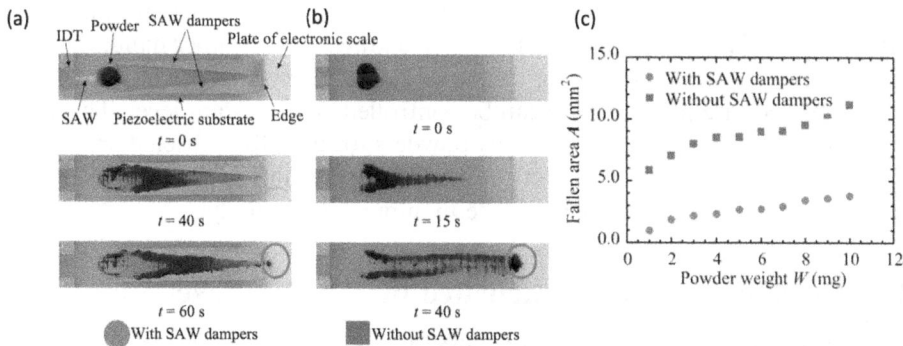

Figure 6.21. Observations from powder transport experiments (a) with and (b) without damper structures. (c) Relationship between powder fall area and powder volume.

storage hopper and tow guide walls on the SAW actuator to adjust the powder amount control precisely. Both guide walls had heights of 2 mm and lengths of 35 mm. The guide walls were located at a convergent angle of 20° against an outlet with a 2 mm width. The hopper and the guide walls were made of hard transparent resin (Fullcure720) and fabricated using a 3D printer (Connex500, Stratasys Inc.).

A temporal change in the weight of the powder fallen is plotted as a function of elapsed time. The powder was supplied from the miniature feeder driven by SAWs. It was found that the miniature feeder began to supply the powder at $t = 3$ s. This is the time it takes for the powder supplied from the hopper to be transported by SAW, dropped onto an electronic balance, and measured. This time can be manipulated by changing the SAW transport length and SAW power associated with the powder transporting velocity.

To select drive frequencies for transporting powder toward the SAW downstream side, 30 mg of dry copper powder was placed in front of the IDT. The average particle size of the powder was about 97 μm and the standard deviation was 26 μm. Under the application of 1 kHz burst waveforms consisting of 2000 cycles of 19.2 MHz sine waves by using a function generator, the observation was performed, as shown in figure 6.20. Figure 6.21 shows the results of an actual powder transport experiment with SAW. Here, two devices were prepared. One is a device with a damper, a metallic Al structure that suppresses SAW propagation and spatially limits powder transport. When a metallic structure is patterned on the SAW substrate, SAWs do not propagate where the metallic structure exists. This is due to the fact that the SAW is damped by the metal structure. As can be seen from figures 6.21(a) and (b), in a device with a damper, the powder is transported only in the transport path restricted by the damper structure. Therefore, it was found that the amount of powder transported to the limited range can be precisely controlled compared to devices without dampers. Figure 6.21(c) shows the relationship between the powder feed rate and the area of powder spilled from the device. It shows that devices with dampers can reduce the area where powder falls and spreads more minutely than devices without dampers. Furthermore, the relationship between transport elapsed time and powder supply weight for these two devices is shown in figure 6.22. In the device without a damper, the powder falls all at once, resulting in a sudden increase in the powder feed rate. On the other hand, the device with a damper can limit the powder feed rate and keep it constant, which means that the powder feed rate per unit time can be controlled at a constant level. This can be used in microchemical systems to supply powders to spatially restricted areas, and it can also be used in more precise and automated dispensing systems for dispensing in pharmacies, etc, because powders can be manipulated precisely.

6.11 Centrifugation devices actuated by the SAW excitation

The SAW micro-centrifugation and superstrate concepts have led to the development of an on-chip centrifugal micromotor, with azimuthal streaming in the fluid coupling layer, as shown in figure 6.23 [26]. The SAW devices were also fabricated

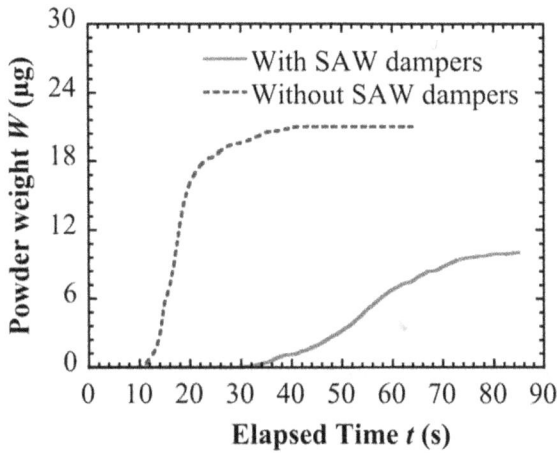

Figure 6.22. Powder feeder characteristics with SAW: time dependence of powder fall rate with and without damper structure.

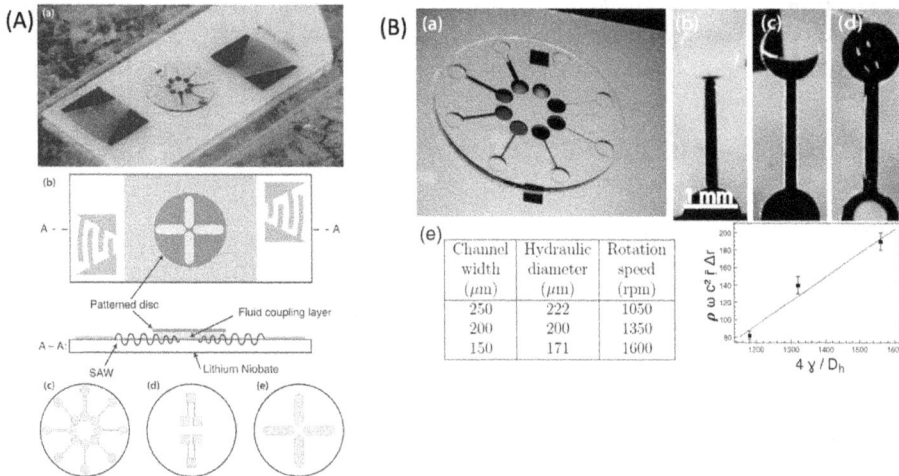

Figure 6.23. (A) miniaturized Lab-on-a-Disc (miniLOAD) device (a) image and (b) schematic of the device, which comprises a pair of offset elliptically focussing single-phase unidirectional transducers fabricated onto a teflon AF (Dupont, Wilmington, DE, USA)-coated LiNbO$_3$ substrate. Schematics of discs for demonstrations of (c) capillary valving, (d) mixing, (e) particle concentration. (B) Demonstration of capillary valving in the miniLOAD device. Reprinted from [26] John Wiley & Sons. Copyright 2012 WILEY-VCH Verlag GmbH & Co. KGaA, Weinheim.

using standard microfabrication techniques. The 10 mm diameter discs, fabricated out of SU-8 photoresist using two-step photolithography technique, are achieved. The size is significantly smaller than the current-state-of-the-art Lab-on-a-Disc. Their disc rotation, at speeds up to 1400 rpm, was used to demonstrate valving and mixing as two fluids. Glass *et al* demonstrated the mixing two fluids using the miniaturized Lab-on-a-Disc (miniLOAD) device [26]. Opening the capillary valves formed by two channels connects separate inlet reservoirs into an outlet reservoir on rotation, as shown in figure 6.23, two fluid species housed in the inlet reservoirs can be driven into the outlet reservoir and mixed. They reported a mixing time of just 4 s after disc rotation commenced that was achieved by evaluating a pixel intensity analysis based on a standard deviation evaluation technique. They successfully developed a SAW actuation device and disc system including a camera-battery-powered driver circuit and being small enough to sit in the palm of one's hand. In their demonstration, the miniature Lab-on-a-Disc device with diameter of 10 mm does not require moving parts to drive rotation of the disc, unlike a macroscopic Lab-on-a-Disc. The disc is driven to rotate using SAW irradiation incident upon a fluidic coupling layer from a pair of offset, opposing single-phase unidirectional IDT patterns on a $LiNbO_3$ substrate, as shown in figure 6.23. The SAW irradiation causes azimuthally oriented acoustic streaming with sufficient intensity to rotate the disc at several thousand revolutions per minute. Simple capillary valve operations were demonstrated on their platform. The feasibilities of mixing two fluidics and particle concentration using their platform have been also studied. They revealed that their platform had an adequate ability to demonstrate valving and mixing, and to concentrate particle suspensions. Their study exhibits that the Lab-on-a-CD functionality can be reproduced at these small scales for the development of more miniaturized and potable devices for real-time filed-use diagnostics and sensing [24].

6.12 Application of disposable microchip platform with removable SAW actuators

A SAW device capable of separating the microchip from the piezoelectric substrate, serving as the drive source, is described below. The miniLOAD device is one of the SAW devices capable of separating the microchip from the substrate [24]. These systems can be repeated reused because contamination of the piezoelectric substrate can be prevented by simple replacing the chip [20–27]. In addition, the dead volume can be significantly reduced, and pump-less and valve-less microsystems can be fabricated. Here, three example applications of the SAW-driven microfluidic device are reported: acceleration of the antigen alternation reaction by introducing a mixing process, droplet transport, and powder transport are simultaneous transport and mixing of liquids and powders [27].

6.12.1 Operating principle

Figure 6.24 shows the principle of the microdisposable stirrer. SAWs are generated by supplying high-frequency signals to the comb electrodes (IDTs) on the piezo-electric substrate, which propagate on the piezoelectric substrate surface and radiate longitudinal waves to the coupling solution, which propagate to the coupling solution, the tip, and the solution on the tip. The longitudinal waves propagate to the coupling liquid, the chip, and the solution on the chip, and the solution is agitated by shifting the IDTs in parallel and patterning them. The solution in the

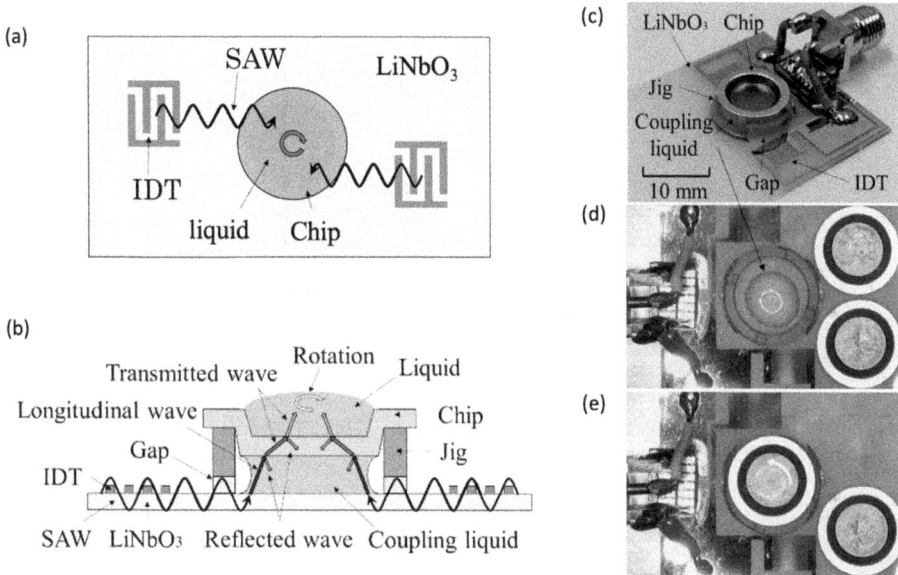

Figure 6.24. Conceptual diagram of a removable micro-mixing system. (a) Top view: the IDTs are opposing each other with their central axes shifted. (b) Conceptual cross-sectional structure of the system; the coupling solution is placed on the LiNbO$_3$ substrate where the SAW propagation takes place and the aluminium (Al) container is loosely coupled with a jig to prevent its displacement. (c) The actual fabricated system. (d) Photograph of two Al containers and SAW-induced mixing device. The coupling solution is placed in the centre of the swirl. (e) Observed image of SAW-induced mixing with one Al container placed on the coupling solution.

aluminum container can be mixed, as shown in figures 6.24(c)–(e), and the aluminum cup can be freely replaced at any time.

Let us consider now the principle of operation. Here, the liquid medium is assumed to be considered as an area surrounded by boundaries and filled with microparticles. Sound waves can be thought of as propagating as the vibration of these microparticles. As shown in figure 6.25, different media are in contact with each other. When a sound wave with a pressure amplitude of P_i is vertically incident on this boundary, sound waves with pressure amplitudes of P_r and P_t are reflected and refracted. Let the densities of medium 1 and medium 2 be ρ_1 and ρ_2, respectively, and the sound velocities be c_1 and c_2, respectively. The relationship between these amplitudes is determined by the boundary conditions. If the sound pressure amplitudes are different on the left and right sides of the boundary surface, the sound pressure amplitude will be continuous on the left and right sides of the boundary surface because the force will be applied to the boundary of zero mass. The relationship between these amplitudes is obtained as follows:

$$P_i + P_r = P_t. \tag{6.9}$$

If the particle velocity differs on the left and right sides of the boundary, the particle velocity will be continuous because the medium on the left and right sides will either

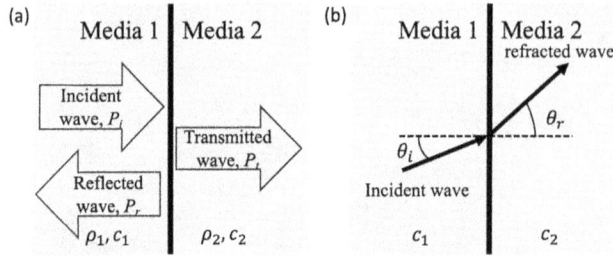

Figure 6.25. Schematic diagrams of (a) reflection and transmission of incident sound waves from medium 1 at the interface between medium 1 and medium 2 (b) the relationship between refraction of sound waves.

overlap or there will be a gap at the boundary. Then, the following relationship is obtained:

$$\frac{P_i}{\rho_1 c_1} - \frac{P_r}{\rho_1 c_1} = \frac{P_t}{\rho_2 c_2}, \tag{6.10}$$

Solving equations (6.9) and (6.10), we obtain

$$\frac{P_r}{P_i} = \frac{\rho_2 c_2 - \rho_1 c_1}{\rho_1 c_1 + \rho_2 c_2}, \tag{6.11}$$

This equation (6.11) is the reflectance of the sound pressure amplitude. Since the sound intensity (energy per unit area of ultrasonic wave (W m^{-2})) is proportional to the square of the amplitude, the transmission coefficient T of the sound intensity is expressed using the acoustic impedance ($Z = \rho c$) for each medium,

$$T = 1 - \left(\frac{P_r}{P_i}\right)^2 = \frac{4 Z_1 Z_2}{(Z_1 + Z_2)^2}. \tag{6.12}$$

From equation (6.12), it can be seen that the transmissivity of the sound intensity of the longitudinal wave depends on the density and sound velocity of the medium. It can be said that optimizing the materials of the coupling liquid and the tip in the device proposed in this study can drive the fluid with higher efficiency. As shown in figure 6.25(b), a wave incident from medium 1 at an angle of incidence θ_i is refracted into medium 2 at a refraction angle θ_r. This is obtained from Snell's law using the speed of sound c_1 and c_2 for each medium,

$$\frac{\sin \theta_i}{c_1} = \frac{\sin \theta_r}{c_2}. \tag{6.13}$$

During the propagation process, ultrasonic waves are attenuated depending on the medium, distance, and frequency. There are two main mechanisms for ultrasonic attenuation: absorption and scattering. In absorption attenuation, ultrasonic energy is absorbed into the medium and converted mainly into thermal energy. In scattering attenuation, ultrasonic waves hit a sufficiently small object and are scattered.

In general, ultrasonic attenuation is characterized by the following exponential decrease of the sound pressure amplitude P and the sound intensity I with the distance x.

$$P = P_0\, e^{-\alpha x} \qquad (6.14)$$

and

$$I = I_0\, e^{-2\alpha x}, \qquad (6.15)$$

where P_0 and I_0 are the sound pressure amplitude and sound intensity at $x = 0$, respectively. α is the attenuation coefficient of the medium. Coefficient 2 in the exponential term of the sound intensity equation is the result of converting the sound pressure amplitude into sound intensity, since sound intensity is proportional to the square of the pressure.

6.12.2 Fabrication procedure

The fabrication method of the micro hot stirrer is shown in figure 3.22. The device was fabricated using lithography and etching techniques used in semiconductor manufacturing processes. A piezoelectric substrate ($LiNbO_3$: 128° rotated Y-plate X propagation) is used as the starting substrate. The thickness of the piezoelectric substrate used is 500 μm and its surface roughness is less than 0.3 nm. On this piezoelectric substrate, films of Cr and Au are deposited in this order using an RF sputter (CFS-4EP-LL, Shibaura Mechatronics). The thicknesses of the Cr and Au films are 10 and 100 nm, respectively. Next, UV exposure is carried out using positive resist (Tokyo Ohka Kogyo, OFPR800LB-20cP) and the IDT pattern on the glass mask is transferred using a mask aligner (SUSS Micro Tec, MA6). Then, based on this pattern, etching of Au and Cr is carried out and the resist is peeled off to fabricate the IDT and heater as the source of the SAW. The device is then completed by attaching a liquid reservoir wall (Stratasys, Vero White) produced by a 3D printer (Stratasys, Connex500) to this piezoelectric substrate using adhesive (Toagosei, Aron Alpha) [23–25, 27].

6.12.3 Device characterization

Supply power dependence and response time:
1 kHz burst voltage consisting of 2000 cycles of a 19.12 MHz sine wave generated by a signal generator (Tektronix, AFG3252) is supplied to the IDT and amplified by an RF amplifier (ALM00110-2840FM, R&K) is used to amplify the burst voltage. The power supplied to the IDT was adjusted by the voltage of the signal generator and checked with an RF power meter (Rohde & Schwarz, NRP-Z91 and AR, DC3001M1). When amplified in-phase RF voltage is supplied to the IDT, SAWs are generated and propagate across the substrate surface, reaching the coupling liquid and radiating longitudinal waves. The longitudinal waves propagate to the coupling liquid, the chip, and the liquid on the chip. In the experiment, 60 μl of pure water was rotated on an Al chip and a zirconia ball (0.9 mm diameter, 4.0 g cm^{-3} density) was floated as a tracer.

Figures 6.26(a) and (b) show images of tracers when 0.10 and 0.50 W of power was supplied to the IDT, and the dependence on the supplied power was calculated.

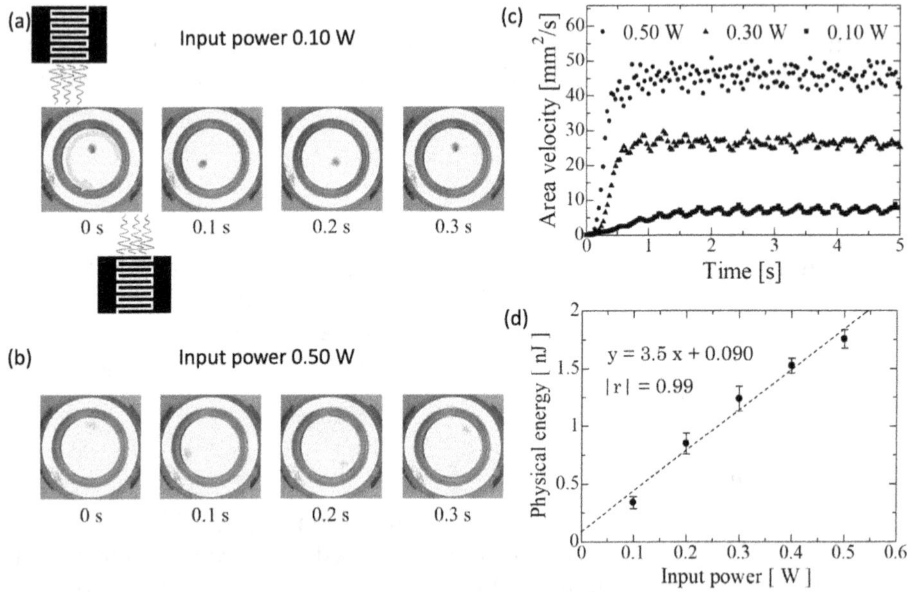

Figure 6.26. Evaluation of the mixing mechanism using the SAW mixing system shown in figure 6.24: results of monitoring the swirling behaviour of floating zirconia balls (0.9 mm diameter, 4.0 g cm^{-3} density) in pure water and exciting SAW at input powers of (a) 0.1 W and (b) 0.5 W in order to assess their rotational characteristics. (c) Input power and time dependence of area velocity. (d) Relationship between estimated zirconia ball kinetic energy and input power.

The IDTs are located at the upper left and lower right outside the images, so the tracer rotates counterclockwise. Although the angular velocity of the tracer appears to remain almost the same with respect to the supplied power, the radius of the rotating orbit makes a difference: when 0.50 W of power is supplied to the IDT, the tracer rotates at 188 rpm. Figure 6.26(c) shows area velocity of tracer as a function of time under the power application of 0.1, 0.3, and 0.5 W. When it is accelerated from rest to reach swivelling behaviour, it is found to be moving in a circular motion at a nearly constant angular velocity. The reason for the oscillating angular velocity is that the power used for SAW excitation is supplied in burst mode. Figure 6.26(d) shows the kinetic energy of the tracer when 0.10, 0.20, 0.30, 0.40, and 0.50 W of power was supplied to the IDT. A linear approximation with the power supplied on the horizontal axis as x and the kinetic energy on the vertical axis as y, $y = 3.5x + 0.090$, yielding a correlation coefficient of 0.99 and a high degree of proportionality. This indicates that the device has a very high dependence on the power supply and can easily control the rotation speed of the solution on the chip.

Here, kinetic energy due to swirling (rotational motion) is evaluated as physical energy. The evaluation will be performed as follows. The law of conservation of angular momentum is a law of constant area velocity. The time difference of the area dS is given by [25]

$$dS = \frac{L}{2m}dt, \tag{6.16}$$

where L is the magnitude of the angular momentum and m is the mass of the ball. Then, the area velocity is given by

$$\frac{dS}{dt} = \frac{L}{2m}.$$ (6.17)

Here, assuming that the ball is moving in nearly circular motion, with a radius of rotation, r, due to the SAW agitation the kinetic energy of ball, K, is given by the following equation:

$$K = \frac{1}{2}m\left[\left(\frac{dr}{dt}\right)^2 + \frac{L^2}{m^2r^2}\right].$$ (6.18)

Considering the rotational radius is constant with no change in time, that is $dr/dt = 0$ when enough time has passed, the kinetic energy, K_{eq}, of the rotational behaviour is given by

$$K_{eq} = \frac{1}{2}m \cdot \frac{L^2}{m^2r^2} = \frac{2m}{r^2}\left(\frac{dS}{dt}\right)^2.$$ (6.19)

Dependence of longitudinal wave propagation distance:
Figure 6.27(a) shows the kinetic energy of the tracer when 0.50 W of power is supplied to the IDT and the propagation distance of the longitudinal wave propagating through the coupling liquid is 1.1, 1.6, 2.2, 2.7, and 3.3 mm. The horizontal axis is the propagation distance of the longitudinal wave through the coupling liquid, and the vertical axis is the kinetic energy of the tracer. The result of figure 6.27(a) shows that as the propagation distance increases, the kinetic energy of the pure water on the tip decreases.

Dependence of coupling solution on NaCl concentration:
Figure 6.27(b) shows the kinetic energy of the tracer when 0.30 W of power is supplied to the IDT and NaCl solutions (mass percent concentrations of 2.5%, 10%,

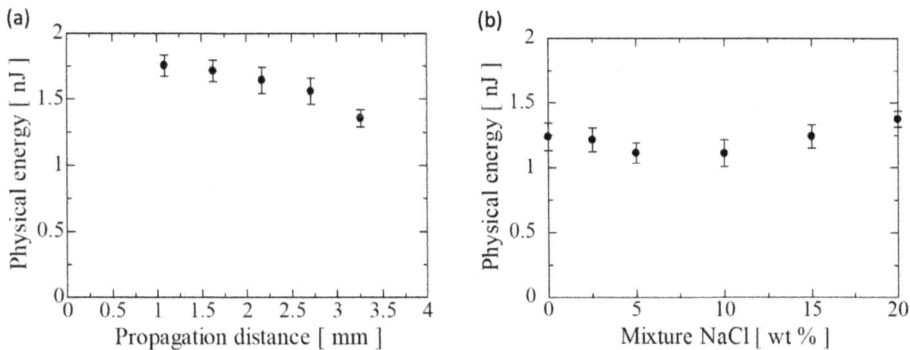

Figure 6.27. (a) Dependence of physical (kinetic) energy on propagation distance. Summary of results obtained from the evaluation of the ball swirl velocity by varying the height of the Jig. (b) Variation of the physical energy at different NaCl concentrations in the coupling liquid solution.

15%, and 20%) are used as coupling solutions. The horizontal axis is the concentration of sodium chloride solution and the vertical axis is the kinetic energy of the tracer. When NaCl solution of 5% or less is used as the coupling solution, the energy efficiency becomes worse with the amount of NaCl added. In contrast, the kinetic energy of NaCl solutions of 10% or more increased with the amount of NaCl added, and when 20% NaCl was added, the energy efficiency was higher than that of pure water. This is considered to be the effect of the increase in the absorption coefficient of longitudinal waves and the increase in the transmission of sound intensity due to the addition of NaCl. First, the addition of NaCl to the coupling solution increases the intermolecular force in pure water, resulting in greater absorption of ultrasonic waves during propagation. In contrast, increasing the amount of NaCl added increases the density of the coupling liquid and the strength of the sound transmitted to the Al chip. In other words, the energy transmitted to the liquid on the tip depends on both the absorption coefficient and density of the coupling liquid.

Dependence of coupling solution on glyerince concentration:

Figure 6.28 shows the kinetic energy of the tracer when 0.30 W of power is supplied to the IDT and aqueous glycerine solutions (10%, 20%, 30%, 40%, 50%, 60%, 70%, 80% mass concentration) are used as coupling solutions. The horizontal axis is the viscosity coefficient of the aqueous glycerol solution and the vertical axis is the kinetic energy of the tracer. A logarithmic approximation using the viscosity coefficient on the horizontal axis as x and the kinetic energy on the vertical axis as y, $y = -1.4 \ln x + 1.2$, yielding a correlation coefficient of 0.98, indicating a high correlation. As the concentration of glycerine is increased, the density increases and the intensity of sound incident on the Al tip increases. However, the viscosity coefficient also increases, resulting in a larger attenuation of the longitudinal wave in

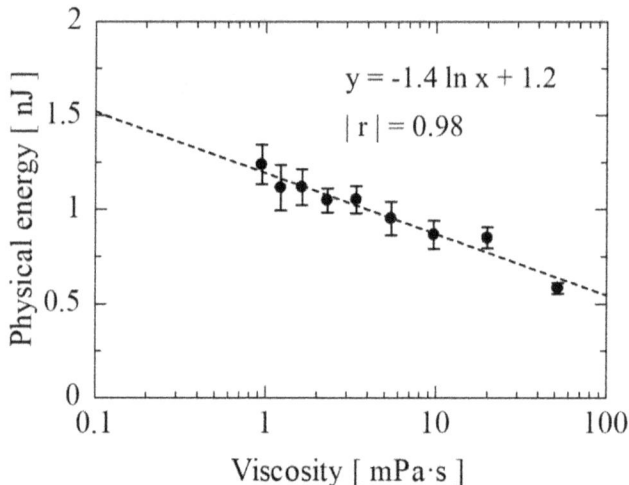

Figure 6.28. The physical energy of the tracer ball when 0.30 W of power is supplied to the IDT and aqueous glycerine solutions with various concentrations are used as coupling solutions. The relationship between physical energy and viscosity of coupling liquid solution.

the coupling liquid. In other words, the energy transmitted to the liquid on the chip strongly depends on the viscosity coefficient rather than the density of the coupling liquid. Therefore, to further improve the energy efficiency of this device, it is necessary to select a liquid with a low viscosity coefficient.

6.12.4 Applications

This mechanism can be used for a variety of applications, one example being the mixing of two droplets. A cover glass is placed over the device with the coupling liquid as shown in figure 6.29. The edge of the cover glass is circled by a dotted line in the figure. A blue-coloured droplet and a transparent droplet are placed on the cover glass at a distance from each other; when the SAW mixing device is operated, the two droplets collide while rotating to form a single droplet, as shown in the figure. It can be seen that vortices are also formed inside the droplet. Thus, different droplets can be mixed without contamination.

Next, a microdisposable stirrer driven by SAW is used to react the substrate with the enzyme-labelled antibody used in Enzyme-Linked Immunosorbent Assay (ELISA). In this demonstration [27], Anti-IgG(H+L), Mouse, Goat-Poly, horseradish pyruvate oxidase (HRP) (1 mg ml^{-1}) was used as enzyme-labelled antibody and the tetramethylbenzidine (TMB) Microwell Peroxidase Substrate System (2-Component System) as substrate solution. By mixing these two solutions, HRP (horseradish pyruvate oxidase), an enzyme labelled on the antibody, decomposes hydrogen peroxide in the substrate solution to generate active enzyme. The active enzyme generated is highly unstable and acts as an oxidant. This active enzyme oxidizes TMB to generate an oxidized dye. The dye also absorbs light at wavelengths of 370 and 650 nm. In this experiment, the reaction volume of the enzyme-substrate reaction was measured by measuring the absorbance of the reaction reagent at a wavelength of 650 nm. In the experiment, 55 μl of substrate solution and 5 μl of enzyme-labelled antibody solution diluted to 100, 30, 3, 1, 0.3, and 0.1 ng ml^{-1} in DPBS (phosphate buffered saline) were mixed on an Al tip for a total of 60 μl. The room temperature was 20 °C.

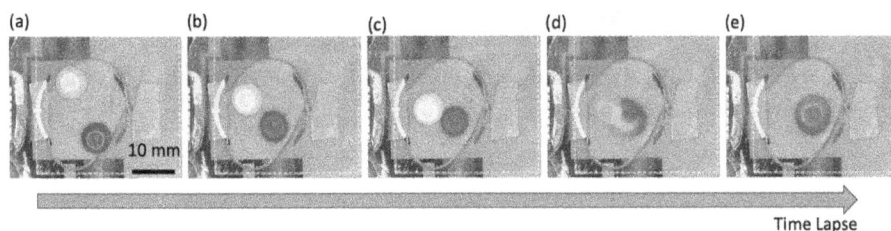

Figure 6.29. A removable/disposal SAW mixer system is used to create a system in which the LiNbO$_3$ substrate is coupled to a teflon-coated cover glass via a coupling solution. The yellow dotted line shows the cover glass edge. When blue-coloured water and pure water were placed as shown in (a) and SAW excitation was performed, the two droplets approached each other with a swirling motion over time and finally fused and mixed, as shown in (b) to (e).

The RF voltage supplied to the IDT is a 1 kHz burst voltage consisting of 2000 cycles of a 19.12 MHz sine wave generated by a signal generator (Tektronix, AFG3252) and amplified by an RF amplifier (R&K, ALM00110-2840FM). The power supplied to the IDT was adjusted by the voltage of the signal generator and checked with an RF power meter (Rohde & Schwarz, NRP-Z91 and AR, DC3001M1). Here, an RF voltage of 1.0 W is supplied to the IDT. When the amplified in-phase RF voltage is supplied to the IDT, SAWs are generated and propagate across the substrate surface, reaching the coupling liquid and radiating longitudinal waves. The longitudinal waves propagate to the coupling liquid, the chip, and the liquid on the chip to agitate the solution. The distance from the substrate to the tip is 1.0 mm, and 40 μl of pure water is used as the coupling liquid. Here, when 1.0 W of power is supplied to the IDT, the solution temperature is heated to 30 °C. Therefore, in the experiment without stirring for comparison, the experiment was conducted while heating the solution to 30 °C on a hot plate.

The absorbance of each concentration when 0 and 1.0 W of power is supplied to the IDT is shown in figure 6.30. This result shows the concentration of the enzyme-labelled antibody solution on the horizontal axis and the absorbance of the reaction solution on the vertical axis. Here, the absorbance at 9 min after the start of solution mixing is shown. From this result, it can be seen that the amount of reaction increases with agitation by SAW at each concentration. This is because agitation by SAW increases the number of physical collisions between the enzyme-labelled antibody and the substrate. Therefore, the use of SAW-driven microdisposable agitators for enzyme-substrate reactions in ELISA is expected to increase the reaction efficiency and improve sensitivity.

Images of copper powder (40 μm particle size) transported on a cover glass are shown in figure 6.31. This figure shows images at 0, 10, 20, 30, 60, 120, and 180 s

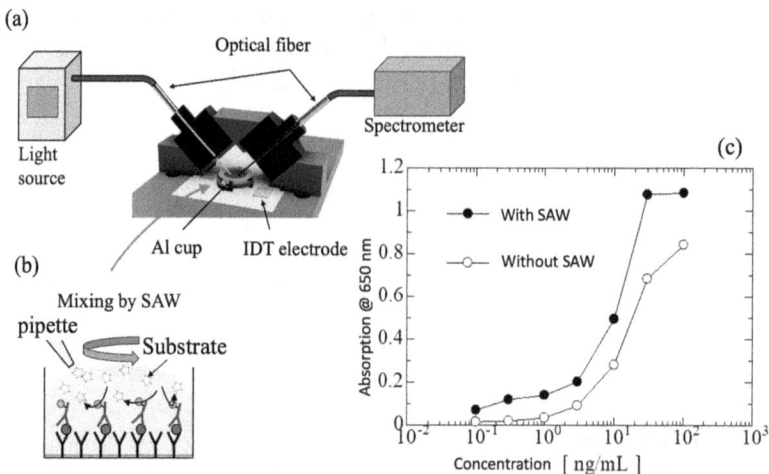

Figure 6.30. Proposed mixing system for ELISA using a removable and disposable SAW-induced mixing system. (a) Schematic diagram of the entire system. Light is irradiated from the light source to the aluminium container by an optical fibre and the reflected light is spectroscopically measured. (b) Schematic diagram of substrate mixing. (c) Substrate concentration dependence of absorbance with and without mixing by SAW.

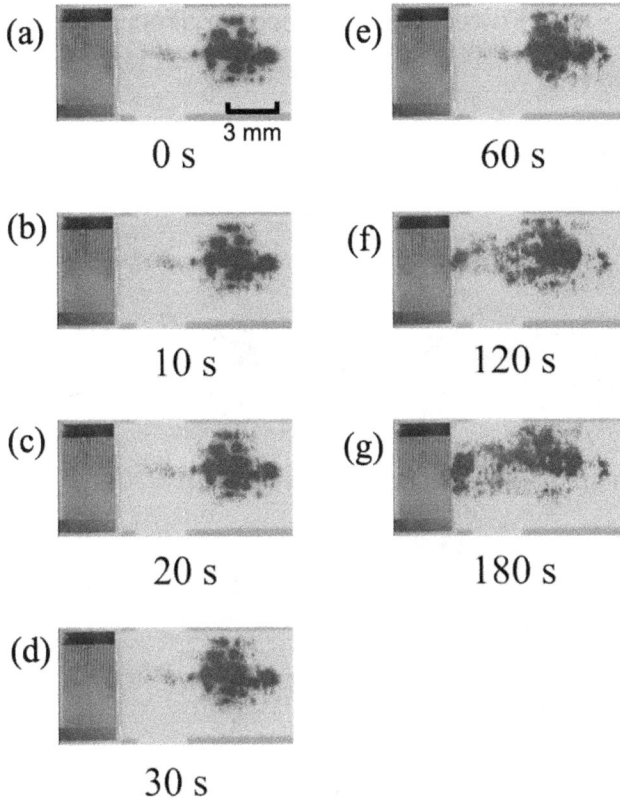

Figure 6.31. Snapshots of SAW-induced transport experiments on powders placed on a cover glass bonded with a coupling liquid at (a) 0, (b) 10, (c) 20, (d) 30, (e) 60, (f) 120 and (g) 180 s.

after the start of power supply to the IDT. From this result, it can be seen that the powder is transported in the opposite direction of the SAW generated by the IDT. This suggests that Rayleigh waves are also generated on the cover glass surface. However, this is the case when cover glass is used, and we believe that it depends on the material properties of the chip used. Many variables are possible, such as the atomic structure and interatomic distance of that material, and the atomic interactions of the powder being transported.

Finaly, we have shown that it is possible to transport droplets and powders using the disposable microchips. A droplet and a powder were placed on the cover glass chip of the chip, which was coupled through the coupling liquid on the piezoelectric substrate, as shown in figure 6.32(a). The directions of the propagation of the liquid and powder were opposite, which means that they collided and the powder was encapsulated in the droplet. After 25 s SAW application, a large vortex structure formed within the droplet, as show in figure 6.32(b). These results show that the disposable system is capable of simultaneously performing unit operations on powders and droplets. Combining this SAW-based micropump enables the microfluidic device to be fully disposable.

(a) $t = 0$ s   Powder is in the opposite direction of travel as SAW

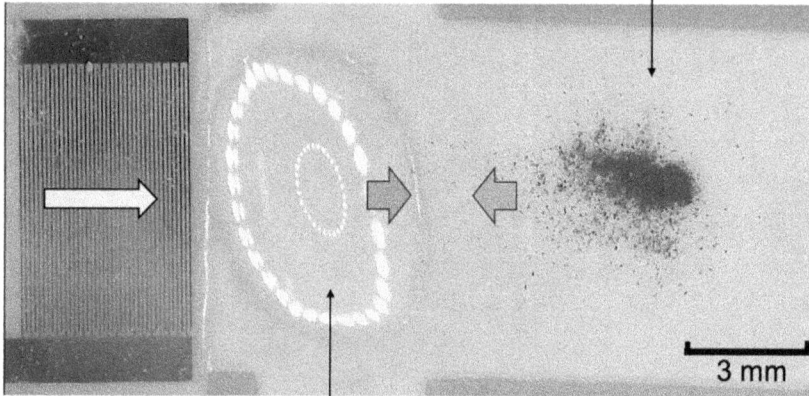

Liquid is in the same direction of travel as SAW

(b) $t = 25$ s

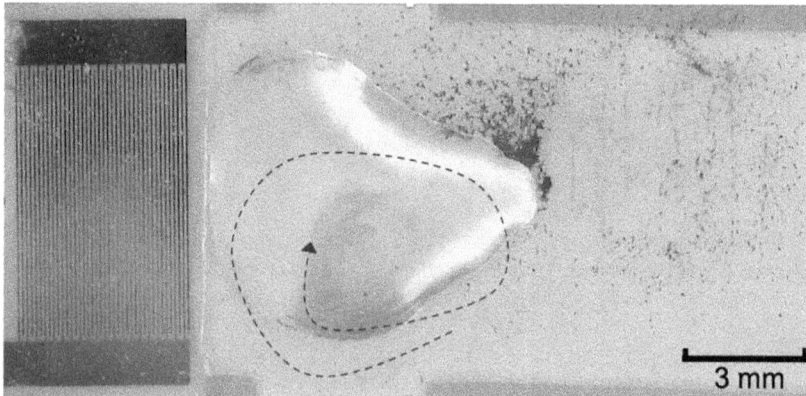

Figure 6.32. (a) Droplet and powder on cover glass placed via coupling liquid on LiNbO$_3$ substrate. (b) Photograph observed 25 s after application of SAW. The droplet and powder travel in opposite directions, allowing them to mix. The powder mixed with the droplet is in a swirling motion, indicating the formation of a vortex structure in the liquid.

6.13 Summary

In this chapter, Nano–Microsystems using SAWs as actuators are presented; after an overview of the SAW generation mechanism, examples of droplet vibration and transport using SAWs are presented. It was shown that not only droplets but also continuous fluids can be driven by combining microfluidic channels, and that SAWs can be used to transport powders as well as liquid phases. With Rayleigh mode, the direction of SAW propagation and the direction of droplet transport are the same in droplet transport. On the other hand, when powder transport is performed in Rayleigh mode, the direction is opposite to that of droplet transport, and the powder moves in the opposite direction to the direction of SAW propagation. In powder transport, the harmonic mode can be used to align and control the SAW propagation direction and the powder transport direction in the same direction. The above is an example of a case where the sample is placed directly on the piezoelectric substrate. If the sample is placed directly on the substrate, the transport characteristics are efficient, but problems such as sample contamination arise. Especially when handling biological samples in microfluidics, sample contamination cannot be tolerated. Therefore, by inserting a coupling liquid between the piezo-electric substrate and the microchip, it was shown that the SAW substrate as an actuator and the microchip can function as a removable system without contamination. Using this system, we showed that both liquid and powder samples can be driven in the system simultaneously without contamination.

References

[1] Friend J and Yeo L Y 2011 Microscale acoustofluidics: microfluidics driven via acoustics and ultrasonics *Rev. Mod. Phys.* **83** 647
[2] Ballantine D S, White R M, Martin S J, Ricco A J, Zellers E T, Frye G C and Wohltjen H 1997 Acoustic Wave Sensors *Theory, Design, and Physico-Chemical Applications* (New York: Academic)
[3] White R M and Voltmer F W 1965 Direct piezoelectric coupling to surface elastic waves *Appl. Phys. Lett.* **7** 314
[4] Tan M K, Friend J R and Yeo L Y 2007 Direct visualization of surface acoustic waves along substrates using smoke particles *Appl. Phys. Lett.* **91** 224101
[5] Qi A, Yeo L and Friend J 2008 Interfacial destabilization and atomization driven by surface acoustic waves *Phys. Fluids* **20** 074103
[6] Tan M K, Tjeung R, Ervin H, Yeo L Y and Friend J 2009 Double aperture focusing transducer for controlling microparticle motions in trapezoidal microchannels with surface acoustic waves *Appl. Phys. Lett.* **95** 134101
[7] Tan M K, Yeo L Y and Friend J 2010 Unique flow transitions and particle collection switching phenomena in a microchannel induced by surface acoustic waves *Appl. Phys. Lett.* **97** 234106
[8] Vukasinovic B M, Smith M K and Glezer A 2007 Dynamics of a sessile drop in forced vibration *J. Fluid Mech.* **587** 395
[9] Girardo S, Cecchini M, Beltram F, Cingolani R and Pisignano D 2008 Polydimethylsiloxane–LiNbO$_3$ surface acoustic wave micropump devices for fluid control into microchannels *Lab Chip* **8** 1557–63
[10] Langelier S M, Chang D S, Zeitoun R I and Burns M A 2009 Acoustically driven programmable liquid motion using resonance cavities *Proc. Natl Acad. Sci.* **106** 12617–22
[11] Tan M K, Friend J R and Yeo L Y 2009 Interfacial jetting phenomena induced by focused surface vibrations *Phys. Rev. Lett.* **103** 024501
[12] Cecchini M, Girardo S, Pisignano D, Cingolani R and Beltram F 2008 Acoustic-counterflow microfluidics by surface acoustic waves *Appl. Phys. Lett.* **92** 104103
[13] Sritharan K, Strobl C J, Schneider M F, Wixforth A and Guttenberg Z 2006 Acoustic mixing at low Reynold's numbers *Appl. Phys. Lett.* **88** 054102

[14] Collins D J, Khoo B L, Ma Z, Winkler A, Winkler R, Weser R, Schmidt H, Han J and Ai Y 2017 Selective particle and cell capture in a continuous flow using micro-vortex acoustic streaming *Lab Chip* **17** 1769–77

[15] Saiki T, Okada K and Utsumi Y 2010 Micro liquid rotor operated by surface-acoustic-wave *Microsyst. Technol.* **16** 1589–94

[16] Saiki T, Okada K and Utsumi Y 2011 Highly efficient liquid flow actuator operated by surface acoustic waves *Electron. Commun. Jpn.* **94** 10–6

[17] Saiki T, Matsui Y, Arisue Y, Utsumi Y and Yamaguchi A 2014 Powder transport by surface acoustic wave actuator using Bragg reflection *IEEJ Trans. Electron. Inf. Syst.* **134** 1934–5

[18] Saiki T, Tsubosaka A, Yamaguchi A, Suzuki M and Utsumi Y 2017 Interdigital transducer generated surface acoustic waves suitable for powder transport *Adv. Powder Technol.* **28** 491–8

[19] Saiki T, Takizawa Y, Kaneyoshi T, Iimura K, Suzuki M, Yamaguchi A and Utsumi Y 2021 Transporting powder with surface acoustic waves propagating on tilted substrate *Sens. Mater.* **33** 4409

[20] Kondoh J, Muramatsu T, Nakanishi T, Matsui Y and Shiokawa S 2003 Development of practical acoustic wave liquid sensing system and its application for measurement of Japanese tea *Sens. Actuators* B **92** 191

[21] Shiokawa S and Kondoh J 2004 Surface acoustic wave sensors *Jpn. J. Appl. Phys.* **43** 2799

[22] Terakawa Y and Kondoh J 2020 Numerical and experimental study of acoustic wave propagation in glass plate/water/128YX-LiNbO$_3$ structure *Jpn. J. Appl. Phys.* **59** SKKC08

[23] Takahashi M, Yamaguchi A, Utsumi Y, Takeo M, Amaya S, Sakamoto H and Saiki T 2022 Micro stirrer with heater mounted on SAW actuator for high-speed chemical reaction *J. Photopolym. Sci. Technol.* **35** 147

[24] Saegusa S, Saiki T, Amaya S, Fukuoka T, Takizawa Y, Amano S, Utsumi Y and Yamaguchi A 2023 Aggregation of Au colloids using surface acoustic waves *J. Photopolym. Sci. Technol.* **36** 127

[25] Yamaguchi A, Takahashi M, Saegusa S, Utsumi Y and Saiki T 2024 Removable and replaceable micro-mixing system with surface acoustic wave actuators *Jpn. J. Appl. Phys.* **63** 030902

[26] Glass N R, Shilton R J, Chan P P Y, Friend J R and Yeo L Y 2012 Miniaturized lab-on-a-disc (miniLOAD) *Small* **8** 1881

[27] Yamaguchi A, Takahashi M, Amaya S and Saiki T Disposable microchip platform with removable actuators using SAW excitation *ACS Meas. Sci. Au* https://doi.org/10.1021/acsmeasuresciau.5c00027

[28] Eggers J 1997 Nonlinear dynamics and breakup of free-surface flows *Rev. Mod. Phys.* **69** 865

[29] Beyer R T 1998 *Nonlinear Acoustics* ed M F Hamilton and D T Blackstock (New York: Academic)

[30] Wohltjen H and Dessy R 1979 Surface acoustic-wave probe for chemical-analysis. 1. Introduction and instrument description *Anal. Chem.* **51** 1458–64

[31] Wohltjen H and Dessy R 1979 Surface acoustic-wave probes for chemical-analysis. 2. Gas-chromatography detector *Anal. Chem.* **51** 1465–70

[32] Venema A, Nieuwkoop E, Vellekoop M J, Nieuwenhuizen M S and Barendsz A W 1986 Design aspects of SAW gas sensors *Sens. Actuators* **10** 47–64

[33] Khlebarov Z P, Stoyanova A I and Topalova D I 1992 Surface acoustic wave gas sensors *Sens. Actuators B Chem.* **8** 33–40

[34] Caliendo C, Verona E and Anisimkin V I 1997 Surface acoustic wave humidity sensors: a comparison between different types of sensitive membrane *Smart Mater. Struct.* **6** 707–15

[35] Devkota J, Mao E, Greve D W, Ohodnicki P R and Baltrus J 2022 A surface acoustic wave hydrogen sensor with tin doped indium oxide layers for intermediate temperatures *Sens. Actuators:* B **354** 131229

[36] Liu B, Chen X, Cai H, Mohammad A M, Tian X, Tao L, Yang Y and Ren T 2016 Surface acoustic wave devices for sensor applications *J. Semicond.* **37** 021001

[37] Ricco A J, Martin S J and Zipperian T E 1985 Surface acoustic wave gas sensor based on film conductivity changes *Sens. Actuators* **8** 319–33

[38] Andle C and Vetelino J F 1994 Acoustic wave biosensors *Sens. Actuators* A **44** 167

[39] Tan M K, Friend J R, Matar O K and Yeo L Y 2010 Capillary wave motion excited by high frequency surface acoustic waves *Phys. Fluids* **22** 1121125

IOP Publishing

Nano–Microsystems
Science and applications
Akinobu Yamaguchi

Chapter 7

On-chip synthesis of organic and chemical materials

In this chapter, we will outline on-chip synthesis of chemical materials. The Lab-on-a-Chip can provide precise unit chemical operation, resulting in controlling the chemical reactions by combination of various operations. Frequently, a wide variety of chemical syntheses are carried out in systems combined with droplet transport and analyzed as a screening. Microsystems in which this droplet manipulation is developed are also presented as examples. Droplet manipulation is expected to be a very good method in terms of realizing a wide variety of chemical syntheses in small quantities. In chemical reactions, temperature and concentration control are important chemical operations, which can be incorporated into the system relatively easily, as seen in previous microsystem examples. Heating using micro-heaters or lasers, or photo-excited chemical reactions using lasers have also been proposed as reaction mechanisms. Although there are many examples, only examples of heating and microwave irradiation mechanisms are presented here. It is also well known that temperature and concentration control alone have limitations in improving the efficiency of reactions. Therefore, in order to promote reactions efficiently, the configuration of a microchemical system equipped with a microwave irradiation mechanism is introduced. The example of a microchemical system equipped with a microwave irradiation mechanism is a fluid, but it can also be combined with droplet manipulation.

7.1 Introduction

Microfluidic systems, as has been described, allow chemical laboratories of the size of the palm of your hand to freely combine unit chemical operations. This makes it possible to combine chemical operations performed by humans or by robots in a very compact manner and to carry out various combinations of chemical operations

doi:10.1088/978-0-7503-3111-1ch7

at high speed and in small quantities. Such a system is very suitable for drug discovery and the search for new materials. For example, if the system can be developed into a system that can synthesize infusions and crystal samples, changing their composition ratios combinatorially as designed, and automatically evaluate whether the expected products are produced, it can be developed for the creation of new drugs. From the above perspective, the microchemical system, which has been used mainly for analysis, is now considered to be configured as a system for chemical synthesis.

Several chemical operations are required to carry out chemical synthesis. Examples include mixing, heating and extraction. As a simple example, consider the ester reaction. Esters are materials that are found in many everyday products and are involved in everyday life. For example, they are found in perfumes and soaps. They are also found in many medicines for medical use. For example, betamethasone butyrate propionate is a synthetic corticosteroid that exhibits anti-inflammatory effects by stimulating glucocorticoid receptors. Aspirin (acetylsalicylic acid), a commonly used drug for the treatment of headache, toothache and neuralgia, is another type of salicylic acid ester compound. Ester compounds are thus used in various places and function as active ingredients in medicines. So, how can ester compounds be synthesized? Ester compounds are synthesized by condensation reactions between carboxyl groups and alcohols [1]. For example, ethyl acetate, which is the easiest to synthesize, can be efficiently synthesized by heating acetic acid and ethanol using sulfuric acid as a catalyst for dehydration-condensation and extracting the ethyl acetate produced by continuous distillation. In esterification reactions, a microreactor system connected by tubes is often used for the purpose of using sulfuric acid and extracting large quantities. The difference between a microreactor and a lab-on-a-chip is the size, whereas a lab-on-a-chip contains the chemical reaction system on a single chip, a microreactor is achieved by connecting several elemental devices. Microreactors are advantageous for mass production.

Now, we have digressed a little, but in order to realize heating and non-equilibrium reactions, it is necessary to provide the system with a heating mechanism. As a heating mechanism, a heater can be incorporated, but simply heating the system does not increase the reaction efficiency. Droplet reaction chips that implement a heating mechanism using electrical control are attracting attention. With droplets, it is possible to produce large quantities of droplets while exciting reactions with a small and quantitatively flexible change in the mixing concentration, so that statistical data can also be obtained. It is well known to work very well as a screening method in drug discovery. There are very many research reports on this technique. Two typical examples are briefly presented here.

On the other hand, microwave chemical reactions have recently been attracting attention. In the case of microreactors, the individual sizes are not as small as in the case of microchips, so microwave irradiation structures, such as cavity structures to confine microwaves, can be implemented more easily than in microchips. As microchips are the subject of this section, microwave irradiation structures mounted on microchips are described below.

7.2 Microfluidic platform for combinatorial synthesis based on droplet control

Early-stage screening for drug discovery has an inherently low success rate. Therefore, there is a need for methods and capabilities that enable efficient screening of large quantities of compounds, especially in the early stages of drug discovery, where the success rate is low. Moreover, as reagents are expensive, there is a need to establish screening methods with minimal reagent consumption and with an emphasis on speed of development. Combinatorial chemistry is therefore often used as a method to create structurally diverse libraries. Droplet-based microfluidics can strongly enhance the combinatorial chemistry platform for early-stage drug discovery described above, as it can generate a large number of droplets with small reagent consumption and comprehensively assess them, compared to conventional microtiter plate technology. Using droplet-based microfluidics allows for individual experiments using six to eight orders of magnitude less starting material than microtiter plate approaches, greatly reducing reagent consumption and improving efficiency through automation and online analysis.

Theberge *et al* presented a droplet-based microfluidic platform for miniaturized combinatorial synthesis. They prepared the platform shown in figure 7.1 to produce a library of small molecules for early-stage drug screening [2]. By combining droplet

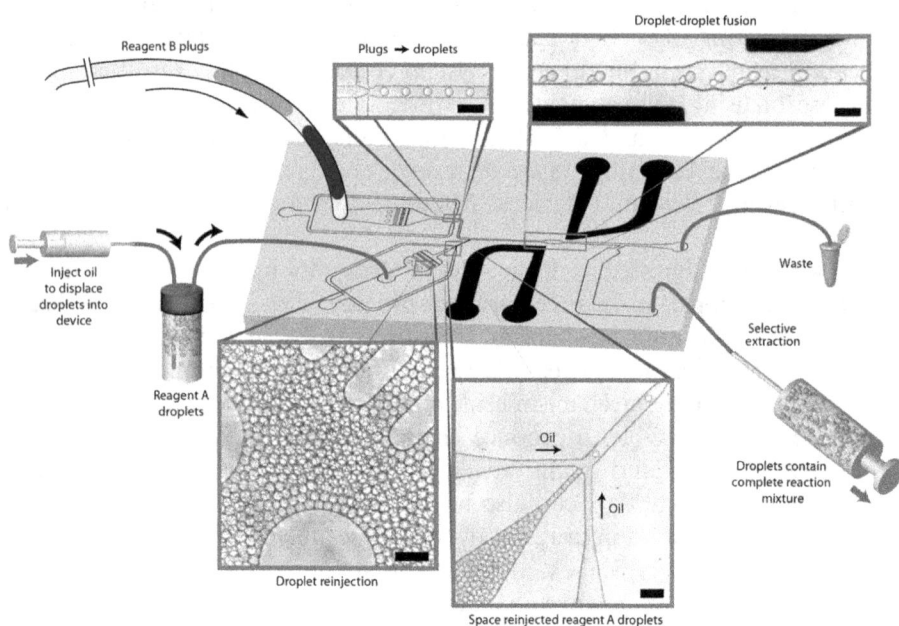

Figure 7.1. Schematic diagram of combinatorial synthesis in microdroplets with optical micrographs for key droplet manipulations. Droplets of the two reagents are prepared and combinatorially fused at different concentrations. Fusion was achieved using electrocoalescence by applying 200–300 V with frequency of 35 kHz sine wave. The scale bars are 50 μm. Reprinted from [2] with permission from the Royal Society of Chemistry.

formation, mixing, reinjection, fusion, and storage modules, they demonstrated combinatorial synthesis, performing a total of 20 Ugi-type reactions in droplets. Here, the Ugi reaction is a multi-component reaction involving a ketone or aldehyde, an amine, an isocyanide and a carboxylic acid to form a bis-amide. The reactions are called Ugi-type reactions after the name of the discover who reported them first in 1959 [3–5]. In their platform [2], sufficient replicate droplets of each reagent were produced to ensure that all reaction combinations arise statistically by droplet fusion. Droplets containing the combinatorial reactions were produced at over 2 kHz. Their method drastically reduces reagent consumption, allowing reactions to be conducted in volumes six orders of magnitude lower than in microtiter plates.

Keng *et al* have reported microchemical synthesis of molecular probes on an electronic microfluidic device [6]. They have developed an all-electronic digital micro-fluidic device for microscale chemical synthesis in organic solvents, operated by electrowetting-on-dielectric (EWOD). They have demonstrated the multistep synthesis of volatile organic solvents such as acetonitrile to produce 2-[^{18}F] fluoro-2-deoxy-D-glucose ([^{18}F] FDG), the most common radiotracer for imaging of living subjects with positron emission tomography (PET). Their EWOD-based micro-reaction technology was optimized for performing unit operations (transporting, heating, mixing, and solvent exchange) on organic or aqueous droplets, which can be combined to perform multistep synthesis at the microscale. We succeeded in demonstrating the synthesis of [^{18}F] FDG with high and reliable fluorination efficiency (88% ± 7%, $n = 11$) and quantitative hydrolysis with 22% ± 8% ($n = 11$) radiochemical yield of the purified product, [^{18}F] FDG. They have showed that [^{18}F] FDG can be prepared on a chip in sufficient quantities to administrered to multiple animals. In other words, they have shown that such a method and chip can be used to synthesize and use the required reagents on the spot, as required. This is considered to be a very effective means of using reagents with short half-lives, for example.

The microfluidic synthesis and characterization of micrometre-sized actuating Janus particles containing a liquid crystalline elastomer (LCE) was demonstrated by Hessberger *et al* in a microfluidic droplet channel combined with UV light irradiation [7]. The setup of the microreactor is illustrated in figure 7.2. They prepared the microreactor by combination of two PEEK T-junctions connected to each other by a PEEK tube. In addition, two glass capillaries (ID: 100 μm, OD:

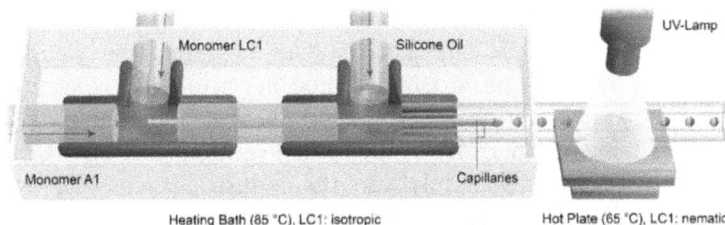

Figure 7.2. Schematic illustration of the capillary based microfluidic system for production of liquid crystalline Jenus particles in a continuous flow of silicone oil. Reprinted from [7] with permission from the Royal Society of Chemistry.

165 µm) were situated inside the T-junctions. One is to provide the monomer mixture LC1 from the inside of the T-junction on the left. The other one is directly connected to a polytetrafluoroethylene (PTFE) tube to provide the monomer mixtures. The polymerization tube was irradiated over a distance of 1 cm on the hot plate using an UV lamp with 323–385 nm wavelength. The synthesis is based on the dispersion of two immiscible monomer mixtures in a continuously flowing silicone oil, using two glass capillaries side by side to form Janus microdroplets of different morphologies. Thus, it is possible to construct microchemical systems without the microfluidic channels being formed on glass or silicon substrates as semiconductor devices. In this case, the system is called a microreactor. Since scalability is possible in fluid mechanics, microsystems can be constructed in combination with unit chemical operations such as droplet manipulation, even if slightly larger systems are constructed in this way.

To realize chemical reactions on microchips, it is common to implement heating, microwave irradiation and photo-induced reactions [8–18]. In the case of heating, the flow path can be placed on a heater or a heater can be introduced into the microfluidic channel. For non-contact heating, methods such as laser irradiation [10], induction heating [8, 9] or microwave irradiation have been proposed [12–18]. Recently, graphene can now be laser-patterned, and systems using this as an electrode and electrochemical sensor with a heating mechanism have also been proposed [11]. In the following, the description will concentrate on microwave chemical reactions with high reaction efficiency and good integration with microfluidic channels.

7.3 Microwave-induced reaction

A promising approach for performing sequential and automatic chemical reaction systems on scaled-down systems is microfluidic devices, including microreactor technology [2, 6–16]. Microreactors have a high specific surface area and small diffusion lengths of reacting molecules, which enable precise temperature control and high reaction rates and yields. They are used not only for chemical reactions but also in biochemical fields such as enzyme catalysis, immune reactions and cell culture.

On the other hand, the field of research on chemical synthesis using microwave energy as a heat source has also attracted attention [12–26]. Microwave heating is characterized by 'selective heating', in which only specific substances can be heated, and 'internal heating', in which the solution can be heated from the inside without heating the container [13, 25, 26]. This enables high reaction yields and rapid chemical synthesis, and has been demonstrated in various fields such as organic/inorganic chemistry, polymer chemistry, metal chemistry and catalytic chemistry, and is recognized as an innovative chemical synthesis method. Commercially available microwave sources mainly use the industrial, scientific and medical (ISM) band of 2.45 GHz, so square waveguide systems and multimode cavities are often employed, and the dimensions of the equipment are relatively large.

In addition, the inside of the waveguide is isolated and closed from the outside. The microwave oven, which is widespread, is a perfect example. Microwave heating using microwave ovens is already on a commercial basis and is used for various chemical syntheses. The microsystems described in this book have also been proposed to combine microwave ovens and another microwave antenna structure to make microchips, as shown in figure 7.3 [12]. In the example in this diagram of figure 7.3, a microwave antenna arranged on a flat surface is combined with microfluidic channels to induce droplet transport and microwave chemical reactions. However, the reaction efficiency is not good because the microwaves do not irradiate the entire droplet and the irradiated area is biased. Thus, due to their large size and low response efficiency, the structure described below, in which the side walls of a square waveguide are made up of a series of multiple metal posts, i.e. a post-wall waveguide as illustrated in figure 7.4, is considered suitable for making a microsystem [14–17]. As will be

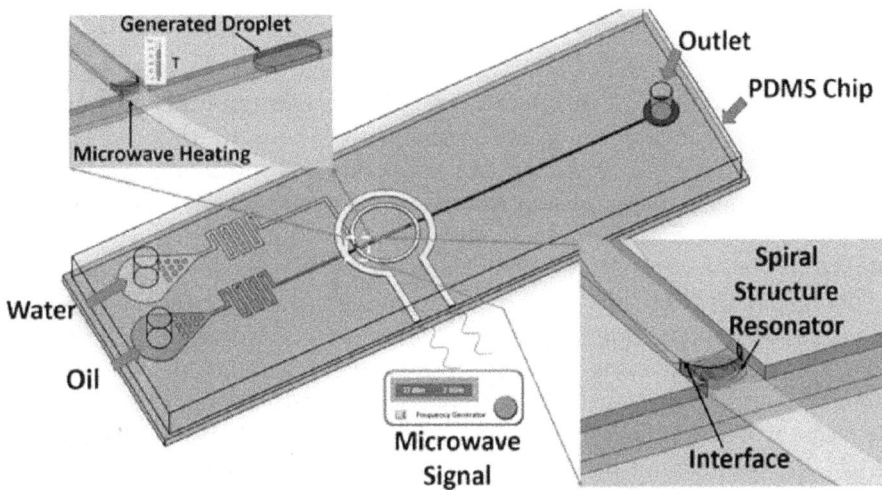

Figure 7.3. Schematic of the microfluidic chip integrated with microwave heater. Reprinted from reference [18], copyright (2020) with permission from the American Chemical Society.

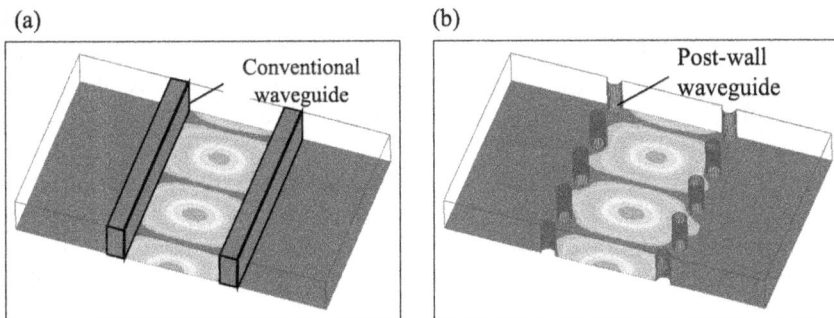

Figure 7.4. Electric field propagation distribution for (a) conventional metal-walled and (b) post-wall waveguides simulated by the COMSOL Multiphysics [27].

Nano–Microsystems

discussed below, if the microwave chemical reaction area is surrounded by a metal wall to confine the microwaves, the flow path cannot be freely arranged. On the other hand, the use of post-wall channel breakage provides freedom in the arrangement of microfluidic channels. This post-wall waveguide enables high integration of microwave components at low cost and has been applied to slot array antennas, leaky wave antennas and cross-shaped directional couplers [14–17]. Here, a microwave chemical system consisting of these post-wall waveguides [28] and microfluidic channels that pass between their metal posts is presented as an example. Instead of the 2.45 GHz ISM band, which is usually used for commercial applications to reduce the size of the waveguide, the shorter 24.125 GHz ISM band is used in this system (ITU RadioRegulation 5.150—ISM Bands) [29].

7.4 Principle of microwave heating

Microwaves are electromagnetic waves with a frequency of 300 MHz to 300 GHz (wavelength of 1 m to 1 mm) [21–26]. Microwaves are used not only for communications, but also for astronomical observations using radio telescopes, radar-based moving-object monitoring systems and GPS positioning systems, which we all know from car navigation systems. Another application is heating. The microwave oven in the home is exactly microwave heating itself. Microwave heating is defined as the heating of dielectric materials by the action of electromagnetic waves between 300 MHz and 300 GHz, generating heat mainly through molecular motion and ionic conduction. It is explained that when, for example, the permanent dipole follows the oscillations of the microwave field with a slight delay, i.e. when the permanent dipole changes with a phase delay relative to the change in the microwave field, this delay acts as a resistance force against the change in the microwave field and the permanent dipole is heated. The permanent dipole is heated by the microwave electric field. Here, it is the water molecules in the microfluidic channel that form the permanent dipole. A water molecule is formed from one oxygen and two hydrogen atoms. The water molecule consists of two hydrogen atoms bonded to an oxygen atom at an angle of approximately 104.5 degrees, forming a dipole. In the absence of an external field, the dipoles of the water molecule are oriented in a random direction. On the other hand, when an electric field is applied, the dipoles of the water molecules align in the direction of the electric field. Here, if an AC electric field is applied, the dipoles of water molecules will change their direction in time along the AC electric field. In the case of a household microwave oven, this would result in 2.45 billion oscillations per second, in which the positive and negative dipoles are swapped 2.45 billion times. If a water dipole is irradiated with an alternating electric field at different frequencies, how would the dipole behave? If the frequency is too high, the dipole cannot follow the electric field change and does not change. At this time, no heating occurs. On the other hand, if a low frequency is irradiated, the dipole can instantly follow the changing electric field and change direction, so no heat is generated. If the frequency of the AC electric field is set near the response frequency of the dipole, the dipole will follow the changes in the AC electric field with a time delay. During this time delay, water absorbs energy from the electromagnetic field and generates heat. In the case of water, resonance frequencies exist around 2.45 and 24.15 GHz.

When an object is to be heated, the method normally used is to apply heat from outside the object by means of flames, hot air, steam, electric heaters, etc. The heat absorbed by the object is transferred to the inside of the object by heat conduction. The heat absorbed by the object to be heated is transferred to the inside of the object by thermal conduction. This heating method is called external heating. In contrast, high-frequency dielectric heating using microwaves is called internal heating. This is because while the radio waves are propagating through the heated object, energy is given to the heated object by the above-mentioned phenomenon called dielectric heating, and the heated object is almost uniformly heated both externally and internally. In other words, the heated object itself is considered to have become a heating element.

When mathematically describing the behaviour of dipoles with respect to changes in the electric field, the relative permittivity is described by a complex number. In order to express the phenomenon mathematically, if a microwave-irradiated dielectric is represented by an equivalent circuit as shown in figure 7.5, the current flowing from the power supply can be regarded as a composite of I_1, which is in phase with the electric field, and I_2, which is 90° out of phase, so the power P consumed in the equivalent circuit is described by the following formula [21–26].

$$P = j\omega\, C_0\varepsilon_r^* V = j\omega\, C_0\left(\varepsilon_r - j\varepsilon_r'\right)V = (I_2 + I_1)V = j\omega C\left(1 - j\frac{G}{\omega C}\right)V. \quad (7.1)$$

If the power consumed in the dielectric is P_0, the P_0 is obtained by the following equation:

$$P_0 = VI_1 = GV^2. \quad (7.2)$$

Here, the phase delay δ of I_2 with respect to I_1, the $\tan\delta$ is given by $\tan\delta = \frac{I_1}{I_2} = \frac{G}{\omega C} = \frac{\varepsilon_r'}{\varepsilon_r}$.

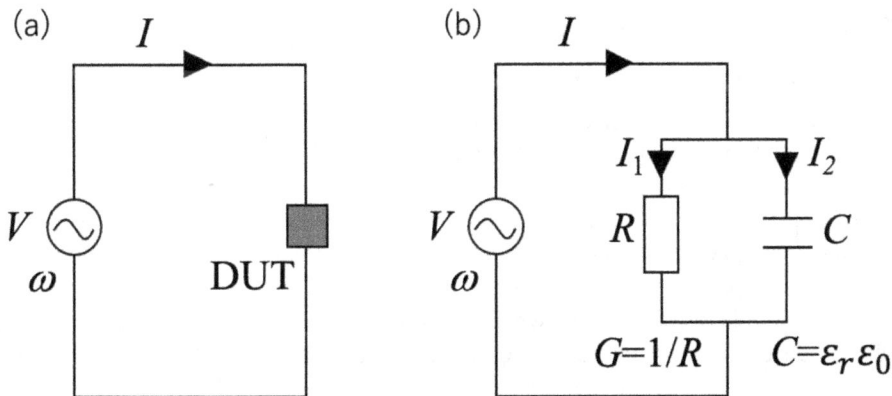

Figure 7.5. (a) Schematic circuit diagram of microwave irradiation of a dielectric sample (device under test: DUT). (b) Equivalent circuit of a dielectric sample.

Then, as the conductance is $G = \omega C \tan \delta$, the power consumption can be rewritten by

$$P_0 = \omega C V^2 \tan \delta. \tag{7.3}$$

If the area of the parallel-plate capacitance in the equivalent circuit is A and the electrode spacing is d, equation (7.3) can be rewritten by

$$P_0 = \omega \frac{\varepsilon_0 \varepsilon_r A}{d} (Ed)^2 \tan \delta \tag{7.4a}$$

or

$$P = \omega \varepsilon_0 E^2 \varepsilon_r \tan \delta = 2\pi f \cdot \varepsilon_0 E^2 \varepsilon_r \tan \delta. \tag{7.4b}$$

In practice, some dielectrics as heated objects actually conduct some electricity, in which case, in addition to dielectric absorption due to dispersion of orientation polarization, energy absorption by electrical conduction occurs. If equation (7.4) is generalized and rewritten to include this energy absorption by electrical conduction, the following equation is obtained:

$$P = E^2(\sigma + 2\pi f \cdot \varepsilon_0 \varepsilon_r \tan \delta), \tag{7.5}$$

where σ is the conductivity. As can be seen from the above, the larger the $\varepsilon_r \tan \delta$ (loss factor) of a material, the more easily it is heated, and the higher the frequency and field strength used, the more energy is absorbed per unit volume and converted into heat. When a material with high energy absorption is irradiated with radio waves, the waves are sometimes absorbed near the surface and do not penetrate into the dielectric material. The penetration depth D is used to express the depth of penetration of radio waves; D is the dimension at which the power density of the radio wave becomes half that of the surface and is called the power half-depth. Since the power density is proportional to E^2, it can be said that, as a rough guide, an object with a thickness of $2D$ can generate heat almost uniformly all the way to the centre.

7.5 Design and fabrication of post-wall waveguides and microfluidic channels

If a microwave waveguide and a microfluidic channel allowing access to the inside of the waveguide from the outside can be integrated, a sequential and automatic reaction form can be constructed. An evaluation system can also be constructed in conjunction. Figure 7.6 shows the basic structure of a system combining post-wall waveguides and microfluidic channels. The system consists of a post-wall waveguide with densely arranged metal posts and microchannels inserted between the posts. The post-wall waveguide is made of metal posts, which allow liquid to be inserted into the waveguide from the outside, and a structure that easily integrates a microreactor and a microwave heating device while retaining the function of the waveguide [14–17]. This allows a series of operations to be carried out in a flow system: inserting a solution into the waveguide, microwave irradiating the solution, and removing the solution from the waveguide.

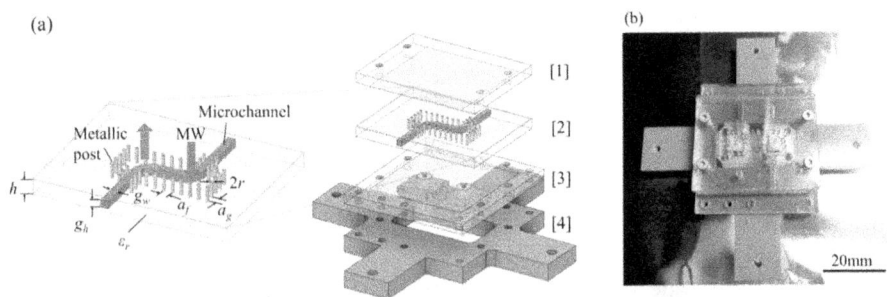

Figure 7.6. Schematic of microfluidic system for the microwave-induced chemical reaction in a microchannel embedded post-wall waveguide. (a) [1], [3] Glass with ITO film, [2] PDMS channel with integrated post-wall waveguide, [4] device fixing platform. (b) Assembled whole system. Reproduced from [15], copyright 2023 The Japan Society of Applied Physics. All rights reserved.

7.6 Structural materials

For the structural material of the microwave waveguide, it is necessary to select a material with a low microwave loss coefficient ($\varepsilon_r \cdot \tan \delta$) among dielectric materials. The power-halving depth D can be expressed by the following equation [25, 26].

$$D = \frac{3.31 \times 10^7}{f \sqrt{\epsilon_r} \cdot \tan \delta}[\text{m}] \qquad (7.6)$$

This formula indicates that PTFE, quartz and polystyrene are suitable as structural materials for microwave waveguides. However, in order to incorporate microwave waveguides into microfluidic devices, not only excellent dielectric properties but also heat resistance, chemical stability and ease of processing are required. PTFE, which combines excellent dielectric properties, heat resistance and chemical resistance, is commonly used as a component material for microwave applications. Utsumi *et al* proposed a microwave chemical chip in which post-walls and channels are formed in PTFE and brass plates are sandwiched between them from above and below, enabling selective heating in the channels [14]. Using this chip, the highly efficient synthesis of metal complexes (Ru complexes) and the particle size control of Au nanoparticles have been successfully achieved [14–17].

In the following, the principle and examples of this microwave waveguide chip are presented. The initial chip structure was a brass plate for microwave shielding, but the low light transmissivity of the brass and PTFE made *in situ* observation through the structural material difficult when observing reactions during microwave irradiation. This disadvantage is particularly significant in microfluidic devices where detection by absorbance analysis or fluorescence measurement is required. To solve this disadvantage, Fujitani *et al* have proposed a microwave chemistry chip that enables reaction monitoring using a low dielectric loss fluorinated resin FEP and a transparent conducting film indium tin oxide (ITO) [16]. This chip can efficiently raise the temperature of solvents such as water while monitoring reactions using a small microwave power supply device, and has also successfully synthesized metal complexes.

However, these materials have two challenges: first, they are not suitable for multistep reactions for combinatorial chemistry such as drug screening on microwave microfluidic devices due to the difficulty and high cost of high-precision processing of waveguides and microfluidic channels. Secondly, the low transmittance of visible light makes it difficult to observe and analyze reactions during microwave irradiation. On the other hand, one of the most typical materials used in microfluidic devices is polydimethylsiloxane (PDMS), an elastomer with excellent microchip fabrication properties PDMS is a silicone resin used as a material for microfluidic devices due to its transparency, biocompatibility and durability [30]. The precursor of PDMS is gel-like and can penetrate into gaps of a few microns, making it possible to transfer patterns such as micro-sized flow paths. They are also highly transparent after curing and can be easily observed, making them suitable for microfluidic devices.

In the following, in order to solve the above two problems, a chip capable of continuous microwave irradiation using ITO and PDMS as structural materials is attempted. It is also a material with excellent light transmittance in the visible light range. Similarly, PDMS also exhibits excellent transparency, but with chemical stability and heat resistance.

7.7 Waveguide design, fabrication and evaluation

This section describes the basic design for integrating post-wall waveguides and microfluidic channels. Figures 7.6(a) and (b) show a schematic diagram and a photograph of a fabricated chip to enable continuous microwave irradiation consisting of a post-wall waveguide and microfluidic channels. The dielectric material was PDMS ($\varepsilon_r = 2.6$) and the dimensions of the post-wall waveguide were determined for the 24.125 GHz band with a cut-off frequency of 16 GHz for the TE10-like mode and an attenuation of 0.1 Np m^{-1}. A 50 Ω coaxial transmission line was then incorporated into the post-wall waveguide so that microwaves were input and propagated through the post-wall waveguide. The metal columns r are arranged as 0.365 mm with a horizontal width spacing a_f of 1.8 mm and a vertical width a_g of 1.7 mm. The height h of the chip is 3 mm and the flow paths incorporated between the post-walls have a square structure with cross-sections g_h and g_w of 1.6 mm and are arranged using the spacing between the metal pillars. In incorporating the channels into the post-wall, the spacing between the metal columns at the locations where the channels are inserted is adjusted to 1.4 mm, as shown in figure 7.7(a). Then, two metal pillars are added at each channel insertion point at an interval a_g 1.7 mm to prevent leakage due to the wavelength of the microwaves in the liquid shrinking and advancing. These ensure that microwave leakage at the sides is kept extremely low (figure 7.7).

The height of the post-wall waveguide is generally used to be sufficiently smaller than 1/2 wavelength and has a constant structure in the height direction, so that the propagation mode is a TE10-like mode, similar to the TE10 mode [14–17, 28]. The centre conductor length l_c of the coaxial line is 1.5 mm and the distance l_d from the waveguide end is 3.2 mm, and the SubMiniature version A (SMA) connector is

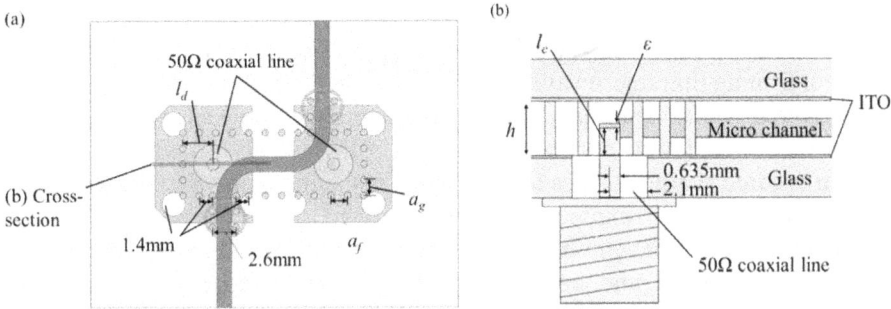

Figure 7.7. (a) Top view and (b) cross-section, being at red line shown in (a), of excitation structure using coaxial transmission line to post-wall waveguide transformer. Reproduced from [15], copyright 2023 The Japan Society of Applied Physics. All rights reserved.

attached. Furthermore, an air gap was created by setting the hole ε machined on it to 0.5 mm. The outline of the chip fabricated using the mould is shown in figure 7.6. The device fabricated in this study consists of two PDMS sheets of different thicknesses: one sheet with a thickness h_a of 2.3 mm, which includes the channel structure and post-wall waveguide, and the other sheet with a thickness h_b of 0.7 mm, which only includes the post-wall waveguide. A PDMS (SILPOT 184, Dow Corning Toray Co. Ltd, Japan) precursor (liquid mixture of PDMS monomer and cross-linker at 10:1) is poured into the machined mould and heated on a hot plate at 50 °C for 6 h [30]. The PDMS was then pulled off the mould and treated with oxygen plasma for 5 min. After the oxygen plasma treatment, the two sheets of PDMS were laminated together and a metal pin with a radius r of 0.365 mm and a height h_s of 3 mm was inserted to fabricate the device. As a result, the device combines the following features:

    (i) injection of sample into the microchannel;
    (ii) microwave irradiation of the sample in the post-wall waveguide;
    (iii) removal of the sample during microwave irradiation.

In previous studies [14–17], metal plates such as brass were used to prevent microwave leakage above and below, making it impossible to observe the reaction process. However, here, a glass plate with a thin film of ITO, a transparent conductive film, was used to visualize the reaction process. The PDMS sheet was sandwiched between the glass plates after thin film formation to prevent microwave leakage from the top and bottom. The thin film formation on the glass plate was fabricated by RF sputtering: sputtering was carried out for 3.5 h in an Ar atmosphere at 0.6 Pa pressure, using In_2O_3 10 wt% SnO_2 as the target while rotating at an RF output of 200 W [31, 32]. The visible light transmittance of the thin-film-formed glass plates was about 80% in the visible light region from 400 to 700 nm.

The temperature and electric field profiles of water simulated using COMSOL Multiphysics are shown. The complex permittivity of water was assumed to be 32.5 $-$j34.98, thermal conductivity 0.615 W (m K)$^{-1}$ and heat capacity 4.2 kJ (kg K)$^{-1}$. Microwaves are input at 24.125 GHz and 4 W output at an ambient temperature of 25 °C. The simulation results show that the temperature decreases with distance from the microwave input port. This indicates that the microwaves are mostly absorbed by the solvent near the microwave input. The maximum temperature was 383 °C, confirming that microwaves can be heated with almost no leakage, even with PDMS as the dielectric material. Electric field concentration was also observed near the microwave irradiation port of the channel. Ethylene glycol, which is used as a solvent in organic synthesis, rose to 100 °C in 70 s, confirming that it can be heated with an output of approximately 4 W [15].

There are two possible causes for the lower temperature than in the simulation: first, heat vaporization and heat conduction to the dielectric due to the phase transition, which was not considered in the simulation environment; second, lack of device accuracy, such as gaps between the PDMS and sputtered glass plates and PDMS sheets, as well as heat and microwave leakage from the ITO film formed on the glass plates. Heat and microwave leakage from the glass plate on which the ITO film is formed may be considered. The frequency characteristics of the fabricated devices were also measured by a vector network analyzer (VNA). The frequency characteristics of S11 with and without water filling were compared. It is found that the characteristics deteriorated with and without filling with water, but a frequency of around −10 dB is ensured in the 24.125 GHz band. This value is close to the simulation value, confirming that the prototype device functions as designed.

Here, a microwave microfluidic device made of PDMS is proposed, which is a silicone resin commonly used in microfluidic devices due to its high transparency and easy fabrication from a mould. The post-wall waveguide is used as the microwave irradiation structure, which enables the connection of the channels from between the post-walls. ITO-sputtered glass plates are used on the top and bottom of the waveguide, so that the inside of the channel can be observed during microwave irradiation. The fabricated devices were evaluated by electromagnetic field and temperature rise simulations, S-parameter measurements using a VNA and ethylene glycol temperature rise experiments. It was confirmed that the microwaves did not escape from the channel structure and were directed to the terminator. In addition, the results of a temperature rise experiment of ethylene glycol using a fibre-optic thermometer at a microwave output of approximately 4 W showed that a temperature rise to 100 °C was recorded in 70 s, confirming heating of the solvent in a short period of time. This suggests that the fabricated device is suitable for chemical reactions using ethylene glycol as a solvent. In addition, as the device uses PDMS as the structural material, various flow patterns can be fabricated using a mould. Therefore, the device could be applied to sequential chemical reactions such as combinatorial chemistry by fabricating a device combining a microwave heating mechanism and units such as mixing, separation and detection.

Previous examples include the successful synthesis of ruthenium complexes, iridium complexes and gold nanoparticles [14, 16, 17]. As an example, figure 7.8 shows an example of the synthesis of a ruthenium complex, where the transparency of PDMS is utilized to show the progress of the synthesis and fluorescence emission upon microwave irradiation. By using a transparent structure to construct a chip in this way, it is possible to directly observe chemical reaction monitoring, etc during the reaction *in situ*.

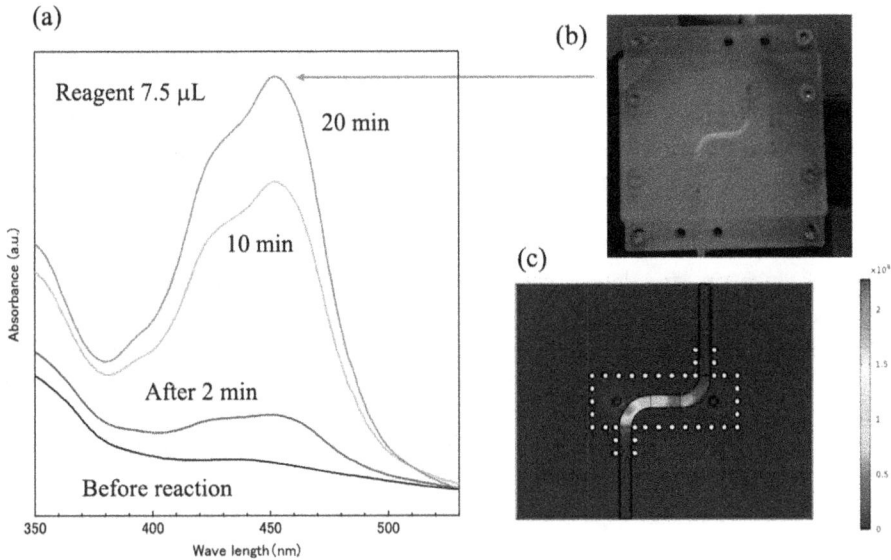

Figure 7.8. Time dependence of the fluorescence spectra of Ru complexes synthesized in a post-wall waveguide mounted microfluidic channel. The peak intensity of fluorescence emission increases with microwave irradiation time from before reaction to 20 min irradiation. The PDMS and ITO structures allow direct monitoring of chemical reactions during microwave irradiation. Fluorescence emission of reagents reacting along the flow path can be seen; temperature rise distribution simulated by COMSOL. The red glowing area shows the temperature increases due to the concentration of microwaves. It is found that the experimental results are well reproduced.

7.8 Summary

In this chapter, an example of a microchemical system implementing droplet control is presented, followed by an example of implementing a microwave irradiation mechanism. As mentioned repeatedly, microchemical systems are designed to perform desired chemical reactions in small quantities in a controlled manner by combining various elemental technologies. The combination of the droplet-controlled system and microwave irradiation mechanism introduced here enables the creation of multifunctional microchemical systems. For this purpose, a method was introduced whereby the microwave irradiation mechanism, which had previously been constructed using PTFE or metal for microwave shielding, could be converted to PDMS and produced in large quantities. The PDMS used as the structural material is a silicone resin commonly used in microfluidic devices because of its high transparency and easy fabrication from a mould. The post-wall waveguide was used as the microwave irradiation structure, which enables the connection of the channels from between the post-walls. In addition, ITO-sputtered glass plates are used on the top and bottom of the waveguide, so that the inside of the channel can be observed during microwave irradiation. In other words, it is easy to combine with the droplet transport system described above.

The evaluation of the chips fabricated by PMDS can be confirmed by electromagnetic field and temperature rise simulations, S-parameter measurements using

VNA and ethylene glycol temperature rise experiments. The measured frequency response of the fabricated chips showed that the return loss was low, although there were differences from the simulation. As a result of the temperature rise experiment of ethylene glycol using an optical fibre thermometer at a microwave output of approximately 4 W, a temperature rise to approximately 100 °C was recorded in 70 s, and the solvent could be heated in a short time using the fabricated chip. In addition, as this chip uses PDMS as the structural material, various flow path patterns can be fabricated using a mould and can be observed *in situ*. Furthermore, as PDMS has low autofluorescence, *in situ* detection is possible in combination with a spectroscopy system. In addition, as the microwaves above and below are shielded by ITO-sputtered glass plates, it is possible to fabricate a chip combining a microwave heating mechanism and units for mixing, separation and detection, which could be applied to sequential chemical reactions such as combinatorial chemistry.

Parts of this chapter have been reproduced from [15]. Copyright 2023 The Japan Society of Applied Physics. All rights reserved.

References

[1] Wade L G 2011 *Organic Chemistry* 8th edn (Englewood Cliffs, NJ: Prentice-Hall)
[2] Theberge A B, Mayot E, Harrak A E, Kleinschmidt F, Huck W T S and Griffiths A D 2012 Microfluidic platform for combinatorial synthesis in picolitre droplets *Lab Chip* **12** 1320–6
[3] Ugi I, Meyr R, Fetzer U and Steinbrückner C 1959 Versammlungsberichte *Angew. Chem.* **71** 373–86
[4] Ugi I and Steinbrückner C 1960 Über ein neues Kondensations-Prinzip *Angew. Chem.* **72** 267268
[5] Ugi I 1962 The α-addition of immonium ions and anions to isonitriles accompanied by secondary reactions *Angew. Chem. Int. Ed. Engl.* **1** 8–21
[6] Keng P Y *et al* 2012 Micro-chemical synthesis of molecular probes on an electronic microfluidic device *Proc. Natl Acad. Sci.* **109** 690–5
[7] Hessberger T, Braun L B, Henrich F, Müller,C. C, Gießelmann F, Serra C and Zentel R 2016 Co-flow microfluidic synthesis of liquid crystalline actuating Janus particles *J. Mater. Chem. C* **4** 8778
[8] Li L, Wu E, Jia K and Yang K 2021 Temperature field regulation of a droplet using an acoustothermal heater *Lab Chip* **21** 3184–94
[9] Takahashi M, Yamaguchi A, Utsumi Y, Takeo M, Amaya S, Sakamoto H and Saiki T 2022 Micro stirrer with heater mounted on SAW actuator for high-speed chemical reaction *J. Photopolym. Sci. Technol.* **35** 147
[10] Hellman A N, Rau K R, Yoon H H, Bae S, Palmer J F, Phillips K S, Allbritton N L and Venugopalan V 2007 Laser-induced mixing in microfluidic channels *Anal. Chem.* **79** 4484–92
[11] Srikanth S, Dudala S, Jayapiriya U S, Jayapiriya J, Mohan J M, Raut S, Dubey S K, Ishii I, Javed A and Goel S 2021 Droplet-based lab-on-chip platform integrated with laser ablated graphene heaters to synthesize gold nanoparticles for electrochemical sensing and fuel cell applications *Sci. Rep.* **11** 9750
[12] Baek S, Min J and Park J -H 2010 Wireless induction heating in a microfluidic devices for cell lysis *Lab Chip* **10** 909–17

[13] Comer E and Organ M G 2005 A microreactor for microwave-assisted capillary (continuous flow) organic synthesis *J. Am. Chem. Soc.* **127** 8160–7

[14] Utsumi Y, Yamaguchi A, Matsumura-Inoue T and Kishihara M 2017 On-chip synthesis of ruthenium complex by microwave-induced reaction in a microchannel coupled with post-wall waveguide *Sens. Actuators B* **242** 384–8

[15] Tanaka R, Nakano T, Fujitani K, Kishihara M, Yamaguchi A and Utsumi Y 2023 Development and evaluation of microwave microfluidic devices made of polydimethylsiloxane *Jpn. J. Appl. Phys.* **62** SG1027

[16] Fujitani K, Kishihara M, Nakano T, Tanaka R, Yamaguchi A and Utsumi Y 2021 Development of microfluidic device coupled with post-wall waveguide for microwave heating at 24.125 GHz *Sens. Mater.* **33** 4399–408

[17] Takeuchi M, Kishihara M, Fukuoka T, Yamaguchi A and Utsumi Y 2020 On chip synthesis of Au nanoparticles by microwave heating *Electron. Commun. Jpn.* **103** 49–55

[18] Cui W, Yesiloz G and Ren C L 2020 Microwave heating induced on-demand droplet microfluidic systems *Anal. Chem.* **93** 1266–70

[19] Wiesbrock F, Hoogenboom R and Schubert U S 2004 Microwave-assisted polymer synthesis: state-of-the-art and future perspectives *Macromol. Rapid Commun.* **25** 1739–64

[20] Kappe C O 2004 Controlled microwave heating in modern organic synthesis *Angew. Chem. Int. Ed.* **43** 6250–84

[21] Nüchter N, Ondruschka B, Bonrath W and Gum A 2004 Microwave assisted synthesis: a critical technology overview *Green Chem.* **6** 128–41

[22] Roy R, Agrawal D, Cheng J and Gedevanishvili S 1999 Full sintering of powdered-metal bodies in a microwave field *Nature* **399** 668–70

[23] Galema S A 1997 Microwave chemistry *Chem. Soc. Rev.* **26** 233–8

[24] Nakamura T, Tsukahara Y, Sakata T, Mori H, Kanbe Y, Bessho H and Wada Y 2007 Preparation of monodispersed Cu nanoparticles by microwave-assisted alcohol reduction *Bull. Chem. Soc. Jpn.* **80** 224–32

[25] Oshima K 1971 Mechanism and applications of microwave heating *Shikizai* **44** 27

[26] Osepchuk J M 1984 A histroy of microwave heating applications *IEEE Trans. Microw. Theory Techn.* **32** 1200

[27] https://comsol.com

[28] Hirokawa J and Ando M 1998 Single-layer feed waveguide consisting of posts for plane TEM wave excitation in parallel plates *IEEE Trans. Antennas Propag.* **46** 625–30

[29] Deslandes D and Wu K 2001 Integrated microstrip and rectangular waveguide in planar form *IEEE Microw. Wireless Compon. Lett.* **11** 68–70

[30] Toepke M W and Beebe D J 2006 PDMS absorption of small molecules and consequences in microfluidic applications *Lab Chip* **6** 1484–6

[31] Kurdesau F, Khripunov G, da Cunha A F, Kaelin M and Tiwari A N 2006 Comparative study of ITO layers deposited by DC and RF magnetron sputtering at room temperature *J. Non-Cryst. Solids* **352** 1466–70

[32] Terzini E, Thilakan P and Minarini C 2000 Properties of ITO thin films deposited by RF magnetron sputtering at elevated substrate temperature *Mater. Sci. Eng. B* **77** 110–4

IOP Publishing

Nano–Microsystems
Science and applications
Akinobu Yamaguchi

Chapter 8

Optoelectrofluidics, platforms based on combination of electrokinetics and optics

Combination of various technologies based on electronics and optics with nano/microfluidics has attracted much attention because of extraordinary advances in lab-on-a-chip systems. Here, optoelectrofluidic technology, which has been recently developed and advanced as a novel manipulation scheme, can provide programmable manipulation of fluids or particles in microfluidic channels. Recent progress on the optoelectrofluidic manipulation of nanoobjects, including nanospheres, nanowires, nanotubes, and biomolecules, etc is described. Optelectrofluidic platforms for some potential applications such as particle separation, structure forming, molecular detection, clinical diagnostics, and their future directions, are discussed.

8.1 Introduction

Due to the increase of needs for high performance manipulation of micro/nano-scale objects in Nano-Microsystems for a variety of applications, advances in micro/nanomanipulation technologies have been developed. The fundamental manipulations such as trapping, separation, concentration, transportation, and assembly have been achieved by numerous techniques based on various forces such as mechanical, optical, electrical, and magnetic forces. In micro-total analysis systems or point-of-care testing based on lab-on-a-chip, in particular, rapid and spectroscopic characterization of biological makers in tissue and fluid samples is also a major issue in not only medical diagnosis but also environmental analysis. Measurement and characterization of dynamic structural changes in biological systems are critical to the understanding of cell progress. Such efforts have attracted the attention of both academic researchers and engineers focussed on developing biosensors.

The optoelectrofluidic system is one of the candidates which can solve the problems and provide clues to respond to the demands [1–17]. Here, optoelectrofluidics refers to study of the motions or particles or molecules and their interactions with optically induced electric field and the surrounding fluid due to the optically induced electrohydrodynamics. The concept of the optoelectrofluidics includes electrothermal (ET) vortex, electrophoresis, dielectrophoresis, and electro-osmosis induced by combination of optical and electrical energy or by optical-electrical energy transfer. Thus, the optoelectrofluidic systems can also provide such fundamental manipulations by using optical, electronic, and fluidic techniques. Three typical types of optoelectrofluidic platforms are schematically illustrated in figure 8.1: (a) ET vortices due to the optical increase of local temperature of the liquid can facilitate mixing and reaction excitation and sensing for chemical reactions; (b) electrokinetic particle manipulation using optically induced virtual electrodes formed by an ultraviolet light pattern projected onto indium tin oxide (ITO); and (c) an image projected onto a photoconductive layer in an optoelectronic tweezers device. Thus, the optoelectrofluidic platforms/systems enable the manipulation, characterization, reaction control or measurements of analytes. As a result, the optoelectrofluidic platforms shed a light on the significant applications in various area such as biotechnology, chemical synthesis, analytical chemistries, and medical diagnosis.

Below, we emphasize a few salient points of optoelectrofluidics which will increasingly dramatically improve their functionalities. In particular, optical characterization is the strongest and most useful to achieve label-free spectroscopic detection based on surface-enhanced Raman scattering (SERS) spectroscopy and tip-enhanced Raman scattering (TERS) spectroscopy [12–35]. Optical imaging using fluorescence spectroscopy is also frequently used and widespread. Fluorescence from a good fluorophore is very efficient, and allows for routine single-molecule detection because the dye bound to the molecule of interest is used as a label. Therefore, fluorescence is currently a well-established technique. The most attractive aspect of SERS is its high specificity, proving a unique 'fingerprint' of the molecule. This feature makes it easier to distinguish the specific characteristic signature from any spurious background signals. It also provides a possibility of high-level multiplexing which displays simultaneous monitoring of many different probes or tags. Thus, a

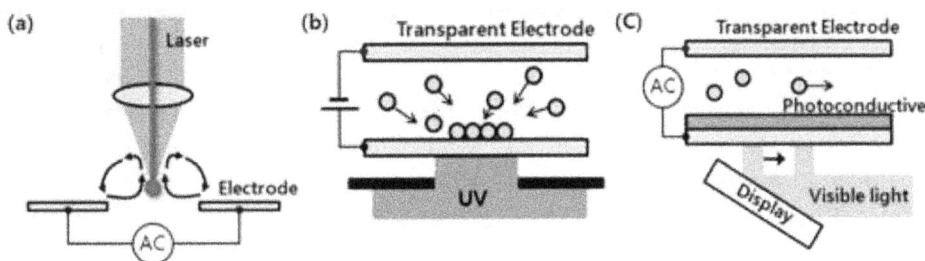

Figure 8.1. Three typical types of optoelectrofluidic platforms. (a) ET vortices due to the optical increase of local temperature of the liquid. Electrokinetic particle manipulation using optically induced virtual electrodes formed by (b) an ultraviolet light pattern projected onto ITO or (c) an image projected onto a photoconductive layer in an optoelectronic tweezers device. Reprinted from [2] with permission from the Royal Society of Chemistry.

system incorporating a SERS-based detection system will be one of the platforms based on combination of electrokinetics and optics, for chemical and biological applications.

In this chapter, we will describe the progress of optoelectrofluidic platforms. The fundamental principles and history of them will be outlined. Some cases and several applications will be then summarized. For example, ET vortices induced by a strong infrared (IR) laser projected into an electric field have be utilized to concentrate microparticles and molecules. In 2005, optoelectronic tweezers were reported [10]. The optoelectronic tweezers operate by using photoconductive materials to induce an electric field by the optical decrement of electrical resistance on a partially illuminated area. As shown below, there are a wide variety of research reports. There are many researches and developments on not only fundamental studies but also engineering applications, and interesting research fields have developed such as biosensors and nanostructure creation.

8.2 Overviews of principles

Optical manipulation techniques of nano/microparticles have attracted much attention since the appearance of optical tweezers in 1970. Optical manipulation has been one of the most frequently used methods because one can directly trap and transport individual particles on demand. The optical field of a tightly focussed laser beam enables such manipulations. However, it is well known that a lower bound on the size to which light can be focussed is limited by the diffraction limit. In the Rayleigh regime, the trapping force is proportional to the volume of the particle. The high-power laser source is required for trapping nanoparticles. It may not be practical for trapping or transporting nanoparticles made of latex, by comparing with the case that metallic nanoparticles of similar size are manipulated by laser. To deal with the limitation of conventional optical tweezers, several types of optical manipulation methods and techniques have been developed. Methods based on localized surface plasmon resonance (LSPR), silicon waveguides, and electrical methods using electrokinetic mechanisms such as electrophoresis, dielectrophoresis (DEP), and AC electro-osmosis (ACEO) have been demonstrated. In particular, the electrical methods have been widely used for the manipulation of nano/micro-particles. Green and Morgan demonstrated the DEP-based manipulation of nano-scale latex beads using microelectrode arrays in 1999 [6]. Gold nanoparticles-decorated polystyrene beads were also concentrated and assembled onto the electrodes on demand by several AC electrokinetic mechanisms such as DEP, ACEO, and ET flows. Optoelectrofluidics has been suggested to combine their own advantages of optical and electrical manipulation technologies. Thus, many researchers have applied some supporting substrates such as polymeric microbeads capturing the target molecules to make it easier or possible to accurately manipulate molecules and measure not only their physical but also their chemical properties combining with SERS and other techniques as described above. In similar ways, various manipulation techniques and methods such as optical tweezers, magnetic tweezers, and atomic force microscopy have been applied for those purposes.

The movement of charged objects in an electric field is originated from the Coulomb force, resulting in electrophoresis. The Coulomb force, F_{Coulomb}, is described by [36–41]:

$$F_{\text{Coulomb}} = qE, \tag{8.1}$$

where q is the net charge of the particle and E is the applied electric field. This relationship equation is basically valid for both monopoles and dipoles, and is universal. Next, we consider the behaviour of dielectric particles in the electric field E, as schematically illustrated in figure 8.2. In the electric field E, a dielectric particle behaves as an effective dipole moment p proportional to the electric field, that is $p \propto E$. In the presence of an electric field gradient, the force, F, on a dipole is given by

$$F = (p \cdot \nabla)E. \tag{8.2}$$

By the combination of these equations, the force acting on a spherical particle, that is, the dielectrophoresis force, is obtained as the following [36–41]:

$$F_{\text{DEP}} = 2\pi r^3 \varepsilon_{\text{m}} \operatorname{Re}\left[f_{\text{CM}}\right] \nabla |E|^2, \tag{8.3}$$

where r and ε_{m} are radius of a spherical particle and the medium dielectric constant, respectively. $\operatorname{Re}\left[f_{\text{CM}}\right]$ is the real part of the Clausius–Mossotti factor, which is related to the particle dielectric constant ε_{p} and medium dielectric constant ε_{m} by

$$f_{\text{CM}} = \frac{\varepsilon_{\text{p}}^* - \varepsilon_{\text{m}}^*}{\varepsilon_{\text{p}}^* + 2\varepsilon_{\text{m}}^*}. \tag{8.4}$$

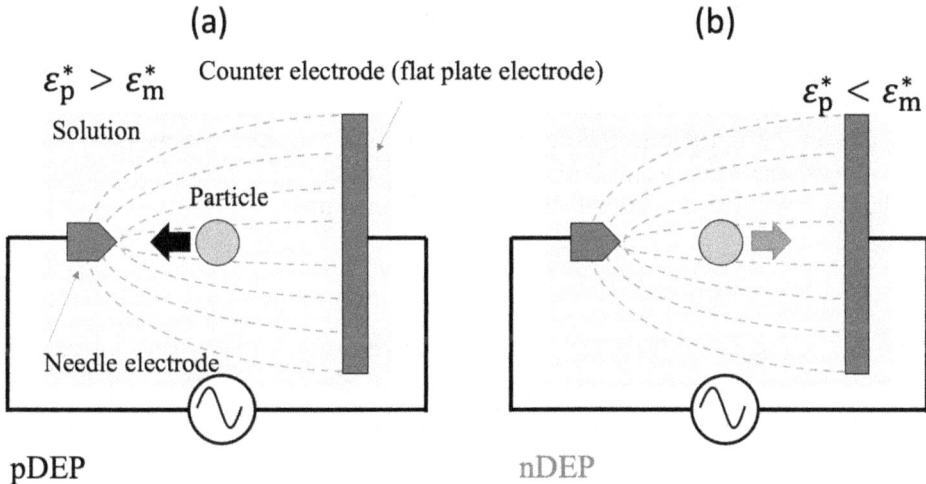

Figure 8.2. Schematic of alternative-current (AC) DEP: arrangement of flat and needle electrodes facing each other in a solvent, with microparticles and AC electric power applied between the electrodes. The microparticles have polarity and DEP occurs when they exist in a spatially non-uniform electric field. Here, the complex permittivity of the particles and the complex permittivity of medium are ε_{p}^* and ε_{m}^*, respectively. (a) Positive DEP (pDEP): particles move towards the needle electrode; (b) negative DEP (nDEP): particles move towards the flat electrode. Each dielectric constant has a frequency dependence, and the magnitude of each dielectric constant varies with frequency.

Here, the star (*) denotes that the dielectric constant, ε^*, is a complex quantity, and it can be related to the conductivity σ and the angular frequency ω. The standard formula is given by

$$\varepsilon_x{}^* = \varepsilon_x - i\frac{\sigma_x}{\omega}, \quad (x = \text{m, p}) \tag{8.5}$$

where the subscripts 'm' and 'p' correspond to the medium and particles, respectively. This equation (8.5) generalizes the fact that the dielectric constant has a frequency component. Dielectric constant is a response property of a material to electromagnetic waves, and its origin, as far as we are usually aware, is mainly due to the electron response in the material. Therefore, it has a material-specific frequency response component. The same can be said for magnetic permeability. Here, let us consider a single interface system, there is a single relaxation with a relaxation time that depends on the combined dielectric properties of the medium and the particle. This relaxation time is given by

$$\tau = \frac{\varepsilon_p + 2\varepsilon_m}{\sigma_p + 2\sigma_m}. \tag{8.6}$$

In a non-uniform electric field, as schematically illustrated in figure 8.2, a force is exerted on the dipole moment of the particle giving rise to movement DEP given by equation (8.1).

The DEP force is proportional to the volume of particle and the square of the electric field gradient. The DEP force essentially limits the rapid manipulation of nanoscale particles existing far from the edge of the virtual electrodes. Due to the limitation of DEP, the optically induced ACEO, which is a fluidic motion generated by the motion of ions within the electric double layer due to the tangential electric field, works very well for rapid concentration of micro/nanoscale particles and molecules in the optoelectrofluidic system. The fluids around the partially illuminated are in the optoelectrofluidic system flow along the surface of the photoconductive layer with a rectified slip velocity defined as [2, 3]

$$\langle v_{\text{slip}} \rangle_t = \frac{1}{2} \frac{\lambda_D}{\eta} \, \text{Re} \left[\sigma_q E_t^* \right], \tag{8.7}$$

where λ_D, η, and σ_q are the Debye length, fluid viscosity, and charges contained in the Debye layer, respectively. E_t^* is the tangential electric field. The ACEO flow is dominant at the relatively low-frequency conditions below about 10 kHz as described below in the actual system.

Here, an example of DEP reporting implementation is shown in figure 8.3. Green and Morgan demonstrated the DEP of submicrometer latex spheres [6]. Four captured video images of 557 nm spheres, suspended in 100 μM KCl, $K_m = 1.5$ mS m^{-1}, on 50 μm polynomial electrodes were taken at 3 s intervals. At an applied frequency of 500 kHz and a potential of 10 V peak to peak, the particles were observed to collect at the electrode edges under pDEP, as shown in figure 8.3(a). After switching the field to 5 MHz, the spheres experienced nDEP. As shown in the

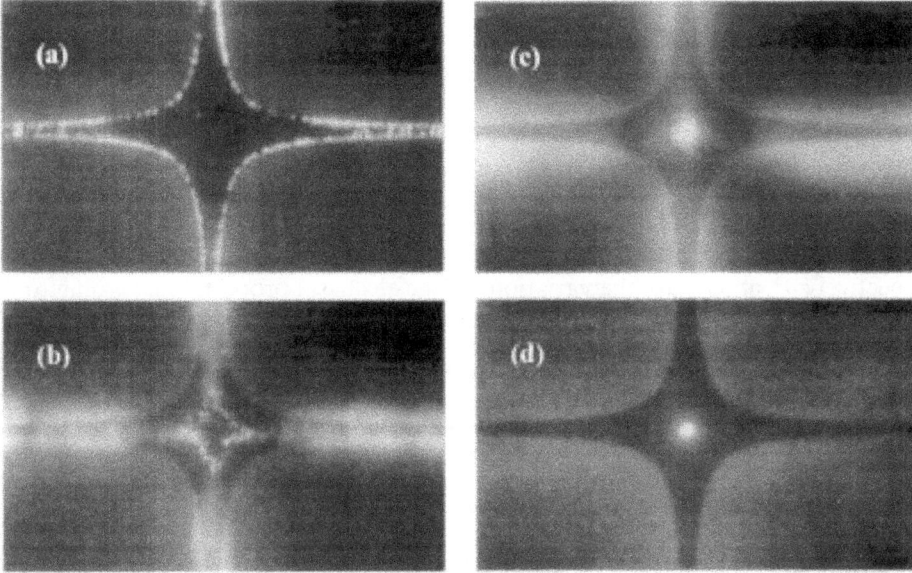

Figure 8.3. Captured video images of 557 nm diameter latex spheres collecting on 50 μm polynomial electrodes. At 500 kHz and 10 V peak to peak, the particles collect at the electrode edges as shown in (a). (b)–(d) Observation images taken at 3 s intervals after switching the frequency of the electric field applied to the electrodes to 5 MHz. It was found that the particles are repelled from the edges by nDEP forces. In (d), after 9 s, particles in the centre are trapped and held there indefinitely and the remaining particles have diffused back into the suspension. Reprinted from [6] with permission from the American Chemical Society.

sequential images of figures 8.3(b)–(d), the spheres close to the centre, in the area of the array where the electrodes were noticeably curved, were repelled from the electrode edge into the centre of the array. After 9 s, the spheres collected at the centre of the array, as shown in figure 8.3(d).

In an optoelectrofluidic device, the electrostatic interactions generated by induced dipole dielectric particles also influence the particle dynamics. The electrostatic force among the polarized particles can be governed by

$$F_{\text{DIP}} = \frac{12\pi r^6 \varepsilon_{\text{m}} \, \text{Re} \, [f_{\text{CM}}]^2}{d^4} [d_{ij}(E_i E_j) + (d_{ij}E_j)E_j + (d_{ij}E_j)E_i - 5d_{ij}(E_i d_{ij})(E_j d_{ij})], \quad (8.8)$$

where d_{ij} is the unit vector in the direction from the centre of the ith particle to the centre of the jth particle. In the optoelectrofluidic device, when two or more dielectric particles are out of level and close to each other at a certain vertical distance, they attract each other and form a chain in the direction of the electric field, while in the direction perpendicular to the field until they are kept apart from each other or meet another one which repels them in the opposite direction, when they are at the same level.

The effect of the temperature gradient generated by the light irradiation should also be considered. The ET effect was first demonstrated by Mizuno *et al* in 1995 [42, 43]. They induced ET vortex due to the layer-induced thermal gradient in the liquid.

Basically, the ET effect is induced by a temperature gradient created by a strong light source. The thermal gradient in the fluid results in a gradient in the fluid permittivity and conductivity. Therefore, a fluidic motion is induced by a body force due to an electric field, which is defined by:

$$\langle f_{ET} \rangle_t = \frac{1}{2} \text{Re} \left[\frac{\sigma_m \varepsilon_m}{\sigma_m + i\omega \, \varepsilon_m} (\kappa_\varepsilon - \kappa_\sigma)(\nabla T \cdot E)E^* - \frac{1}{2}\varepsilon_m \kappa_\varepsilon \, |E|^2 \, \nabla \, T \right], \quad (8.9)$$

where σ_m is the fluid conductivity, $\kappa_\varepsilon = (1/\varepsilon)(\partial \varepsilon / \partial T)$ and $\kappa_\sigma = (1/\sigma)(\partial \sigma / \partial T)$, respectively. κ_ε and κ_σ are the variations of the electrical properties according to the temperature T. E and E^* are the electric field and complex conjugate of E, respectively. The thermal gradients can be produced by Joule heating or by highly focused light. This ET effect to manipulate fluids, microparticles, and molecules was first demonstrated by Mizuno et al [42, 43]. Here, they have shown that not only dielectric constant but also permeability can be swept to particles with magnetic permeability, although there are examples of rapid measurement using magnetic force in Enzyme-Linked Immunosorbent Assay (ELISA).

Nevertheless, since electrical control using voltage is more direct than using magnetic force, and electrical control is easier to implement, the following section presents examples of research on voltage control. Below, for example, let us introduce some examples of dielectric constant-related implementations. By implementing a system like the one shown in figure 8.4 [7], it is also possible to seed cells at a desired location by DEP; Chiou et al have also successfully used the dielectric constant difference between living and dead cells to separate living and dead cells and fix only living cells at a given location [10]. The cells are then fixed in place. These methods provide very useful capabilities for studies that perform single-cell analysis, which is a significant important technique in comprehending many biological mechanisms as it looks at the spectrum of response of each individual cell under stimulation. Yasukawa et al demonstrated simple detection of surface antigens on living cells by applying distinct cell positions with nDEP. Using their fabricated devices, they succeeded in the rapid screening of the CD33 surface antigen expressed on human promyelocytic leukaemia cells (HL-60), as shown in figure 8.5 [8].

8.3 Optoelectrofluidic systems

Parts of this section have been reprinted from [15], copyright (2019), with permission from Elsevier.

8.3.1 Optoelectrofluidic manipulation using the optically induced AC electric field

For optoelectrofluidic manipulation, there are two typical approaches. One is direct change of liquid properties by light, the other is change of surface conductivity by light. In the case of the former, local temperature increase of a fluid due to the strong electromagnetic field by illumination provides change of electrical conductivity and permittivity of the fluid depending on its temperature. ET vortices were first demonstrated by Mizuno et al [42, 43] Using a strong infrared (IR) laser source,

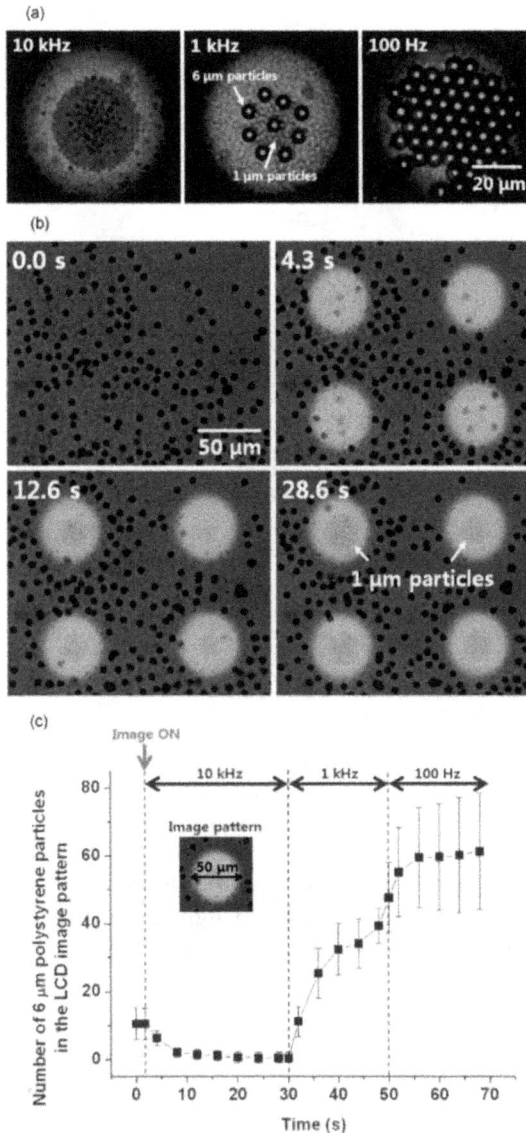

Figure 8.4. Demonstration of optoelectrofluidic concentration. (a) Microscopic image of the 1 and 6 μm-diameter particles concentrated into the illuminated area, corresponding to the bright red area, at the frequency of 10 kHz, 1 kHz, and 100 Hz. (b) Selective concentration of 1 μm particles from the mixture of the particles using the frequency dependence of AC electrokinetics. (c) The change of the number of 6 μm-diamter polystyrene particles with changing frequency from 10 kHz to 100 Hz. Reprinted from [7] with permission with the Royal Society of Chemistry.

they applied it to transporting DNA molecules in 1995. A light source is used to make only the partially illuminated area of the surface become more conductive than other areas and to form a non-uniform electric field in the liquid sample, resulting in several electrokinetic phenomena as shown in figure 8.1(b). They have

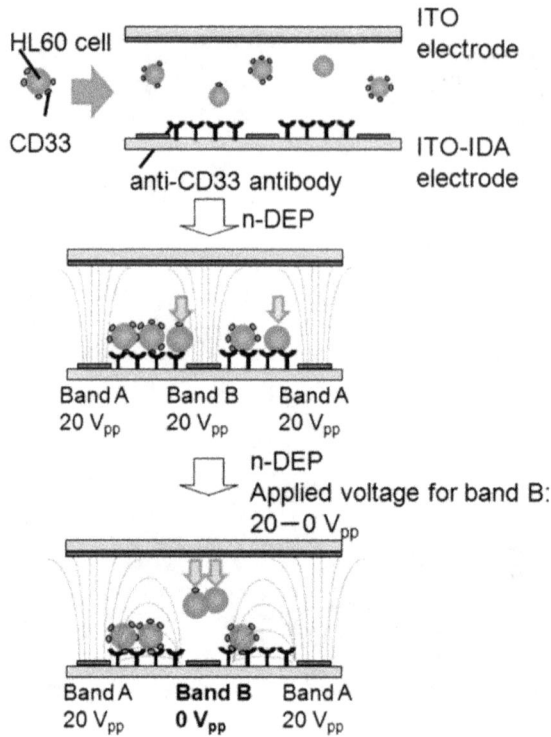

Figure 8.5. Schematic diagram of principle of how nDEP operation identifiees HL-60 cells with specific surface antigens. Reprinted from [8], copyright (2012) with permission from the American Chemical Society.

demonstrated the electrokinetic pattering of microbeads under a non-uniform electric field formed by an ultraviolet (UV) light pattern projected onto an ITO transparent electrode surface, as schematically shown in figure 8.1(c). Hwang and Park also have manipulated proteins and fluorescent dyes [2, 4]. They investigated the frequency-dependent concentration effect of molecules depending on several electrokinetic mechanisms. Their schematic of optoelectrofluidic concentration of microparticles using the lab-on-a-display setup is shown in figure 8.4(a). They have demonstrated the size-dependent microparticle separation as well as the local concentration and assembly of microparticles originated from the image-driven AC elecrtokinetics. They also have characterized the size- and frequency-dependent phenomena of the optoelectrofluidic particle concentration and self-assembly by the combination of several AC electrokinetic mechanisms and electrostatic interactions. Figure 8.4(b) shows the simulation results of the tangential electric distribution and fluidic motions generated by the optically induced ACEO in their device. Figure 8.4 (c) shows the typical experimental results of optoelectrofluidic concentration according to the particle size and the AC frequency condition. The concentration and separation of micro-/nanoparticles are strongly dependent on particle size and AC frequency. For example, the 1 μm beads fill the gap spacing between the 6 μm ones, which position at the equilibrium point of the hydrodynamic drag force by the

AC electro-osmotic flow and the electrostatic repulsive interactions among them, as shown in figures 8.4(a) and (b). In figure 8.4(c), the time evolution of particle concentrations is plotted while the AC frequency is switched from 10 kHz to 100 Hz. As a result, they demonstrated the two-dimensional particle assembly under the optically induced AC electric field. Their optoelectrofluidic manipulation allows us to rapidly concentrate and pattern micro-/nanoparticles which have a specific size by controlling the electric field. Regarding particle dispersion and pattering formation in mixed liquids with different particle sizes, it is necessary to consider not only ACEO but also the interaction between particles.

As Psaltis et al described [1], optofluidics refers to a class of optical systems that are synthesized with fluidics. Optoelectrofluidics is a device or system integrated by adding electronics to optofluidics. Fluidics have unique properties which cannot be induced in solid equivalent states. The following are other examples.

8.3.2 Optoelectronic tweezers (OETs)

Chiou et al demonstrated that photoconductive layer on a plate electrode made it possible to control an electric field only with a weak conventional light source and to spatially modulate the light pattern with a display device in a simple and easy way [10]. The OET platform served as a momentum to attract much attention to the optoelectrofluidic technologies, providing a solution for disposability and interconnection issues in the integrated systems which can manipulate multiple cells using a microelectrode array.

8.3.3 Combination with Fourier transform infrared (FTIR) spectroscopy

The application of FTIR spectroscopic imaging to microfluidics has been demonstrated. Chan et al demonstrated Attenuated Total Reflection (ATR)-FTIR spectroscopy imaging using an infrared focal plane array detector with microfluidic chip to elucidate the chemical reaction and optimize reactive processes within small-volume environments [11]. Their method is based on the combination of an inverted prism-shape ATR crystal with a poly(dimethylsiloxane)-based microfluidic mixing device. As shown in figure 8.6, the system was integrated. Using the system, they succeeded in direct measurement and imaging of the mixing of two liquids of different viscosities and the imaging and mixing of H_2O and D_2O with consecutive H/D isotope exchange.

The microfluidic system with the FTIR imaging can give the possibility of obtaining chemical and spatial information from dynamic systems in microfluidics, offering novel opportunities in this field. Chan et al fabricated the microfluidic channels by using the moulding technique of SU-8 maser and standard soft lithographic fabrication methods. The channels were enclosed by self-adherence of the structured polydimethylsiloxane (PDMS) layer with the ATR crystal surface, as schematically shown in figure 8.6. The channels were 1000 μm wide and 50 μm deep. The microfluidic systems connected to the precision syringe pumps (flow rates were set between 25 and 75 μl h^{-1}) were installed in a continuous scan spectrometer in conjunction with a large sample compartment chamber. Images were measured using a focal plane array detector containing 64×64 pixels with an 8 cm^{-1} spectral

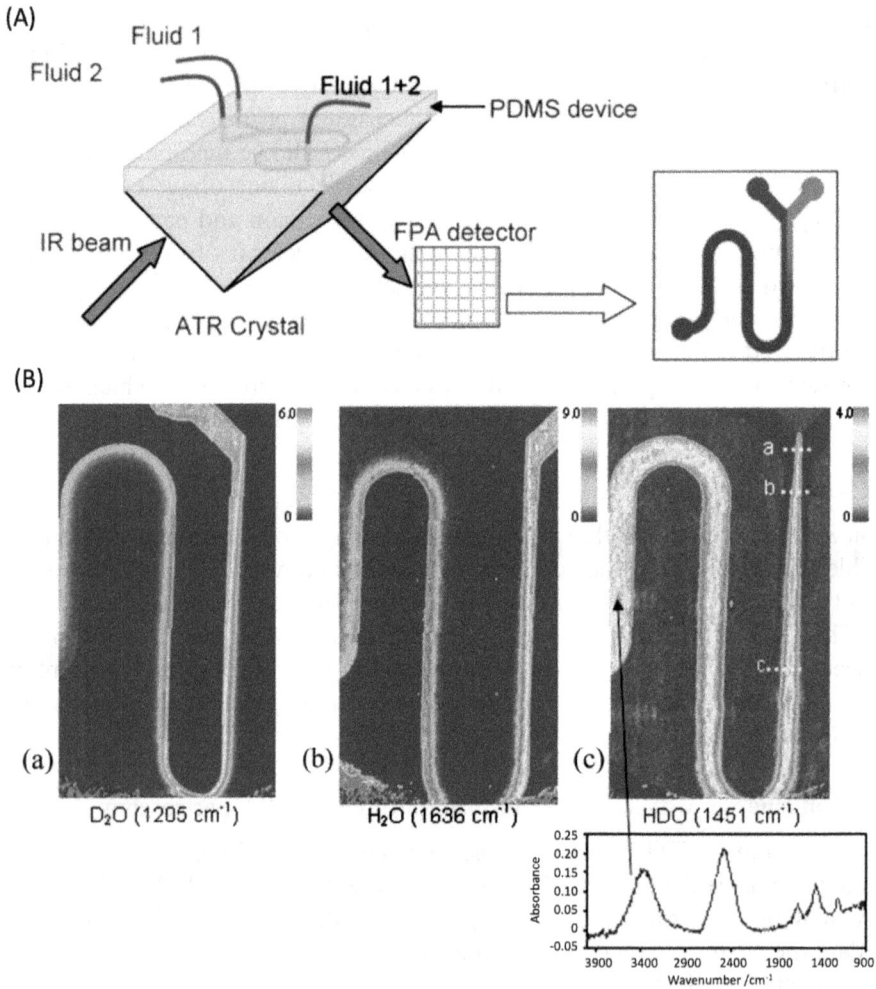

Figure 8.6. (A) Schematic diagram of an ATR-FTIR imaging system. A planar chip-based microfluidic device is integrated with the ATR-FTIR spectroscopic imaging system. (B) ATR-FTIR images of the flow pattern of (a) D_2O (1205 cm^{-1}), H_2O (1636 cm^{-1}) and HDO (1451 cm^{-1}) in the microfluidic channel, respectively. Reprinted from [11] with permission from the Royal Society of Chemistry.

resolution and a spectral range of 4000–900 cm^{-1}. Here, the details of the experiment will be omitted. Interested readers should refer to the literature [11]. Below, the results of their experiments are summarized. To demonstrate the feasibility of ATR-FTIR imaging of microfluidic flows, a simple Y-junction mixer was used, as shown in the inset of figure 8.6. Within microfluidic environments, laminar flow is dominant rather than turbulent flow, such that mixing occurs by diffusion only. They demonstrated ATR-FTIR imaging of two miscible fluids, water and PEG200. They revealed the mixing proceeded proportional to the concentrations of these substances. As a result, diffusion-based mixing of the two fluid streams was clearly confirmed.

8.3.4 Combination with surface-enhanced Raman scattering

Plasmonics has recently attracted much attention, encompassing all areas of research and engineering associated with the fundamental studies for physical/chemical mechanisms, fabrication, and applications of plasmon supporting structures [12–35]. Recent advances in nanotechnology and nanoscience have opened up new possibilities in the design and fabrication of metallic structures with features in the nanoscale. SERS is one of candidates that can achieve single molecular detection and multiplex fingerprint-type label-free detection [18, 19]. In particular, surface plasmon-polaritons (SPPs) at planar interfaces play a major role in all applications for molecular detection using electromagnetic field enhancements. SERS and surface-enhanced fluorescence (SEF) are also included in the effect associated with SPPs. In this section, the mechanism and evaluation are outlined.

SERS-based molecular analysis techniques demonstrate highly attractive characteristics in terms of sensitivity, cost, portability, controllability, and multiplexing [18, 19]. SERS has often been used to resolve various molecules down to a single molecular sensitivity. In medical diagnosis and environmental analysis, the use of SERS is a favourable analytical technique due to its rapid analysis with its ability to provide label-free detection due to spectroscopic characterization for strange analytes. As described in other chapters, with the development of microfluidic devices or systems for various applications, which include point-of-care testing, environmental and food analyses. Sensor technology based on SERS analysis has attracted attention significantly because of its easy operation coupled with label-free spectroscopic measurement of minute quantities of analyte. SERS techniques require a noble metal nanostructure which is capable of achieving the essential strong, localized surface plasmon resonance (LSPR).

One candidate for fulfilling the requirements of an automated label-free molecular detection system is a microfluidic chip with SERS-active particles embedded in the microfluidic channels. The active operation using the external inputs such as DEP, ACEO and OET etc can actively control aggregation and dispersion of the SERS-active functional particles in the microfluidic channels. In addition, the application of these effects also can provide an autonomous mixing function in the microfluidic channel. Microfluidic channels generally have a low Reynolds number and laminar flow, which means that a separate mixing mechanism is often required for encouraging chemical reactions. This mixing mechanism associated with particle agglomeration and dispersion by AC electric field excitation is an excellent method that can achieve both molecular detection based on the SERS effect and chemical reaction enhancement. The optoelectrofluidic systems integrated by electrically controllable SERS function are therefore superior for both analysis and reaction enhancement, being a system killing two birds with one stone.

From the above-mentioned features, the optoelectrofluidic systems have attracted the attention of academic researchers and engineers focussing on biosensor applications and nano- and micro-medicine development. An attractive aspect of the electrically controllable SERS platform is that the SERS spectrum can be tracked in real time by changing measurement conditions such as voltage, frequency and DC bias. This is very useful for chemical reaction analysis and rapid measurements. Several research groups have reported interesting studies on these systems, which are presented here. For example, Cherukulappurath *et al* fabricated pearl chain structures with nanometre-sized gaps and nanoparticles in wells based on dynamic dielectrophoretic aggregation of metal nanoparticles, as shown schematically in figure 8.7 [13]. Han *et al* also used dynamic ACEO and hydrodynamic flow to nanoparticle aggregation [3]. Hwang *et al* [2, 7] and Chrimes *et al* [12] performed *in situ* dynamic measurements of SERS signals using an optoelectrofluidic SERS platform coupled to an active microfluidic platform that integrated dielectrophoretic forces by controlling the spacing between silver nanoparticles (AgNPs) in the microchannel. A part of the experimental results conducted by Chrimes *et al* is shown in figure 8.8 [12] In these studies, nanoparticles were used as they wee. However, as will be shown later, optical or fluid flow assistance was required to generate the SERS-active site, as the DEP force was too small to cause nanoparticle aggregation. Barik *et al* experimentally demonstrated dielectrophoretic concentration of biological analytes on the surface of a gold nanohole array, which concurrently acted as a plasmonic sensor and mechanism for amplifying electric field gradients [14].

As discussed below, DEP is proportional to the cube of the radius of the moving particles and is controlled in the high-frequency range, making it a bit difficult to make SERS measurements while performing agglomeration and dispersion in a microchannel with simple and low-cost electrical control. Therefore, a microsystem using polystyrene (PS) particles with relatively large diameter and small density and provided in monodisperse form as shown below becomes useful.

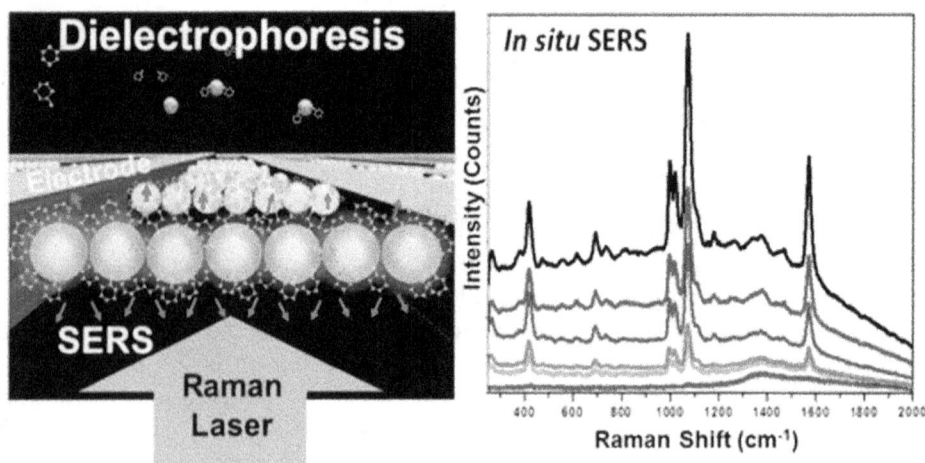

Figure 8.7. Conceptual diagram of a system that increases detection sensitivity by creating nanoparticle aggregates through DEP to increase the SERS-active hotspot. The nanoparticle aggregate structure is created between the electrodes so that molecules adsorb there and exhibit SERS; DEP-enabled assembly of metal nanoparticles in the form of pearl chains with nanometre-sized gaps enables the *in situ* SERS measurement of benzenethiol in less than 2 min without the requirement of long incubation times were achieved. Reprinted with permission from [13], copyright (2014) American Chemical Society.

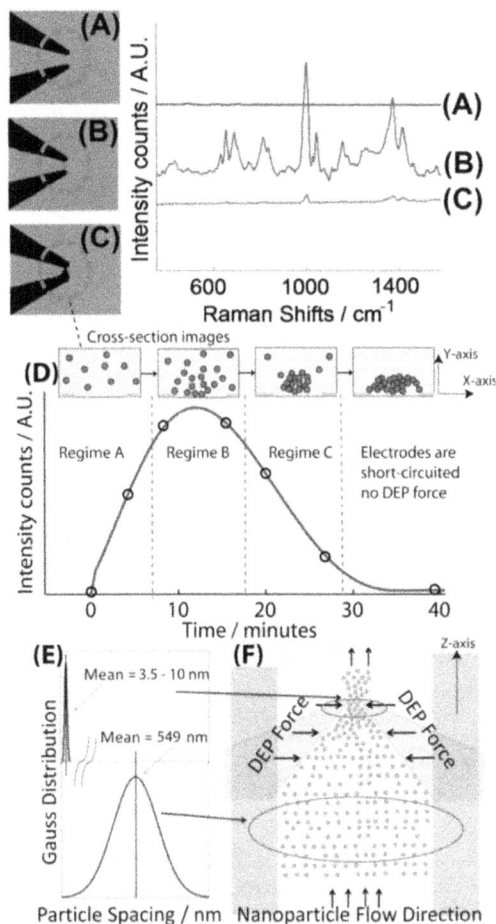

Figure 8.8. Dielectrophoretic induced nanostructures were fabricated to control the spacing of AgNPs flowing in a liquid channel. The detection sensitivity was increased by increasing the number of hotspots emitting SERS in a controlled manner, while avoiding irreversible aggregation of the AgNPs. The fabricated system successfully detected dipicolinic acid (2,6-pyridinedicarboxylic acid) (DPA), a biomarker for anthrax. Three main AgNPss configuration regimes, showing the resulting aggregation of particles and their corresponding SERS spectra for: (A) no DEP force, (B) combination of DEP and flow forces, controlled particle focusing, and (C) DEP force applied for a long time, large particle aggregation. (D) Intensity of the 1005 cm^{-1} peak over a period of 40 min; insets show cross-sectional images of the electrode tip regions for each of the three regimes. (E) The distribution is assumed to be Gaussian both far from and near the electrode tips. (F) Schematic (not to scale) of the nanoparticle' distributions far from and near to the tips. Reprinted with permission from [12], copyright (2012) American Chemical Society.

Here, Yamaguchi *et al* demonstrated our ability to perform dynamic aggregation of Au-nanoparticles (AuNPs)-decorated polystyrene (PS) beads in a micro-well, and highly sensitive detection of 4, 4'-bipyridine (4bpy) in a short time period [15–17]. Figure 8.9 shows a schematic of an optoelectrofluidic platform integrated with the AuNPs-decorated PS beads. PS particles are easily chemically modified, and their

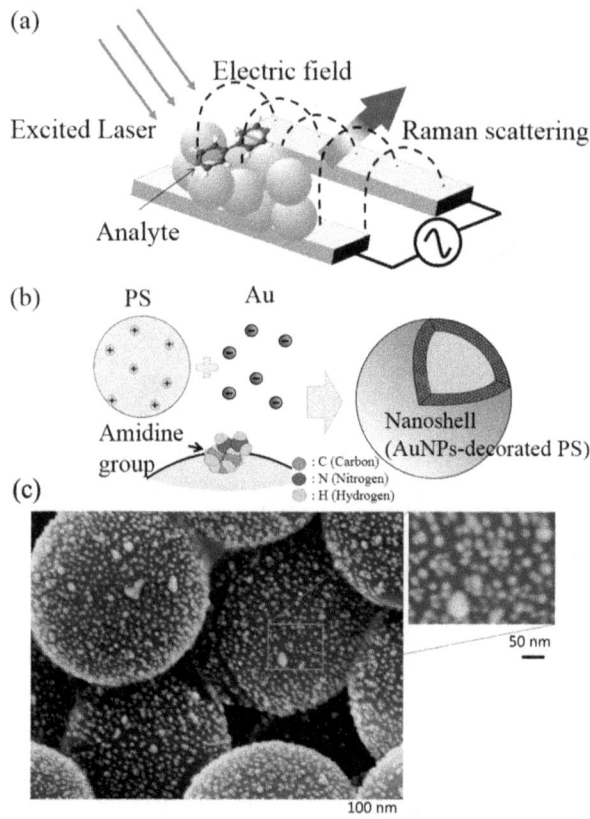

Figure 8.9. (a) Schematic of mechanism for increasing the SERS signal by actively creating an agglomerated structure of nanoshell particles of polystyrene particles decorated with AuNPs by applying an AC electric field to the electrodes. (b) Schematic diagram of the nanoshell structure. (c) SEM observation image of nanoshell.

surface potential and other properties can be controlled by functional devices. In this example, PS particles terminated with amide groups were used to create nanoshells by mixing PS particles with positively charged surfaces and negatively charged AuNPs, as schematically illustrated in figure 8.9(b). The coverage can be controlled by controlling the mixing ratio of AuNPs to PS particles. A SEM image of a nanoshell entirely coated with AuNPs is shown in figure 8.9(c). It can be seen that it is in good agreement with the schematic diagram.

Besides being easy to actively control the aggregation and dispersion of PS beads, another advantage of using PS beads is that they are easy to be chemically modified. One example, the introduction of a cross-linker, which alternated coupling of a peptide onto the PS beads using the carbodiimide method, improved in the dielectrophoresis aggregation of AuNPs-decorated PS beads, as shown in figure 8.10. The cross-linker enabled molecular recognition, via SERS spectra, and the capture of the observed analyte [16].

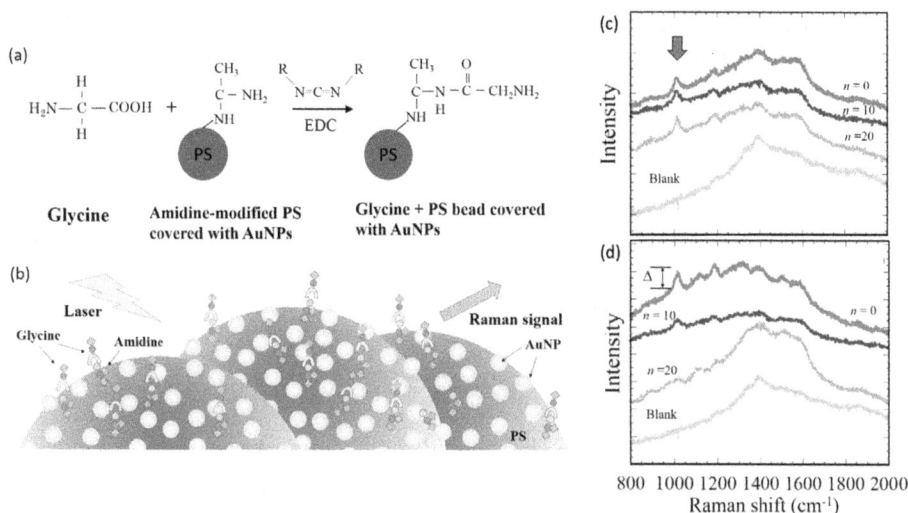

Figure 8.10. (a) Carbodiimide method for binding of glycine onto amidine-modified PS bead. (b) Schematic of the chemical reaction and process of PS beads binding glycine through the linker bridge. (c) Characteristic Raman spectra of glycine bound to AuNP-decorated PS beads after 0, 10, and 20 washings with flowing ultrapure water and of a blank (ultrapure water with the aggregation of the AuNP-decorated PS beads and no analyte). (d) Characteristic Raman spectra of glycine using the aggregation of AuNP-decorated PS beads without linker bridges on the Au electrode after washing with ultrapure water. Reproduced from [16], copyright (2016) with permission from Elesevier.

8.3.5 DLVO theory

When agglomeration operations are performed by DEP or electro-osmosis flow, it is necessary to consider the interaction between particles. Particles dispersed in a liquid are subject to repulsive interactions due to the electric double layer and van der Waals attraction due to intermolecular forces, which must be considered in the agglomeration and dispersion process. To obtain the understanding of mechanisms of these phenomena, we considered the hetero two spheres and evaluated the coagulation potential of AuNP–AuNP, AuNP–PS, PS–PS, AuNP–glass substrate and PS–glass substrate using Derjaguin–Landau–Verwey–Overbeek (DLVO) theory [36, 37]. They explained the dispersion and aggregation phenomena of hydrophobic colloidal particles as the potentials between their particles with the respect to stability of hydrophobic colloidal solutions based on the interaction between the electric double layers when two interfaces approach. The electrostatic part of the DLVO interaction is calculated by the mean-field approximation in the limit of low surface potentials, when the potential energy of the surface electroelements is much smaller than the thermal energy scale $k_{\mathrm{B}}T$, where k_{B} is the Boltzman constant and T is the absolute temperature. In a fluid of the relative dielectric constant ε_r containing monovalent ions of concentration n, the electrostatic potential takes the following form of a shield Coulomb potential, or Yukawa potential, in two spheres of radius a

with charges z (expressed in units of electric elementary quantities) that are separated by a certain distance d.

$$U(d) = \frac{Z^2}{\beta} \lambda_B \left(\frac{e^{\kappa a}}{1 + \kappa a} \right)^2 \frac{e^{-\kappa d}}{d}, \qquad (8.10)$$

where κ is the inverse of the Debye–Hückel screening length, given by $\kappa^2 = 4\pi\lambda_B n$, and λ_B is the Bjerrum length, being the separation at which the electrostatic interaction between two elementary charges is comparable in magnitude to the thermal energy scale $k_B T$. The Bjerrum length is given by [36, 37]

$$\lambda_B = \frac{e^2}{4\pi\varepsilon_0\varepsilon_r k_B T}. \qquad (8.11)$$

Here, e and ε_0 are the elementary charge and the vacuum permittivity, respectively. For water at room temperature, $T \approx 293$ K, $\varepsilon_r \approx 80$, so that $\lambda_B \approx 0.71$ nm.

To simplify the system, under the assumption that the two spheres have the same radius a and the distance between them is d, as schematically illustrated in the inset of figure 8.11, the interaction free energy between them is given by [36, 37]

$$U(d) = \frac{64\pi k_B T a n}{\kappa^2} \tanh^2\left(\frac{Ze\psi_0}{4k_B T} \right) e^{-\kappa d} - \frac{A_H a}{12d}, \qquad (8.12)$$

where ψ_0 and A_H are the potential on the surface of colloidal particles and Hamaker constant, respectively. Here, to simplify the estimation and phenomena, the Hamaker constant A_H, of the Au/PS microparticle is assumed to be 2.18×10^{-19} J, which is the same as that of Au because the outer shell is almost consisting of AuNPs. Later, we describe the Hamaker constant. Since the Au/PS microparticles are suspended in the distilled water, according to the reaction $2H_2O \rightarrow H_3O^+ + OH^-$, we temporarily assume $n = 8.3$ μM and $z = 1$, as the Debye length, $1/\kappa$ is $\sim$100 nm. $\psi_0 = -40$ mV is derived from the Au surface. Figure 8.11 shows the

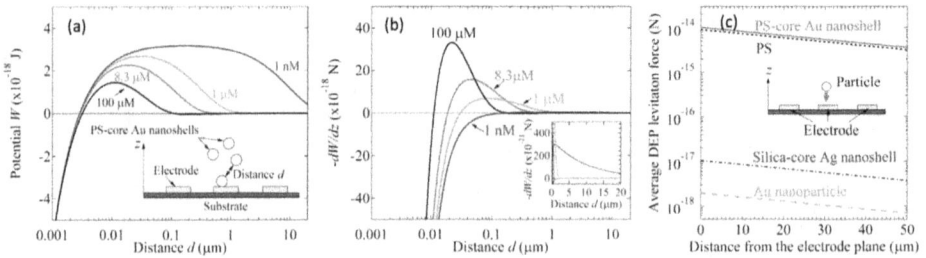

Figure 8.11. (a) Potentials evaluated by DLVO theory as a function of distance d between nanoshells (PS-core: radius $r = 500$ nm, Au nanoparticle: 20 nm) in water under the assumption of ion densities of 1 nM, 1 μM, 8.3 μM, 100 μM and 1 mM. (b) Forces derived from the gradient of potential, consisting of attractive van der Waals force and repulsive electrostatic double-layer force, versus distance d for two identical nanoshells. The inset shows the linear dependence of the forces on the distance d. (c) Estimation of height dependence of the average DEP forces on PS-core Au nanoshell, Silica-core Ag nanoshell and AuNPs. Reproduced from [17], copyright (2017) with permission from Elsevier.

DLVO interaction energies evaluated by equation (8.12) in the cases of $n = 1$ nM, $1\,\mu$M, $8.3\,\mu$M, $100\,\mu$M and 1 mM. As a result, the calculated results indicate that the Au/PS microparticles are not coagulated but dispersed, except $n = 1$ mM. The force, F, exerted on the particle is given by the gradient of interaction potential energy $U(d)$ given by equation (8.12); that is $F = -\nabla U$. When discussing the magnitude of the force, we can calculate F.

8.3.6 Dielectrophoresis and electro-osmotic flow

An example is shown in figure 8.12, where a SERS detection experiment for 4, 4′-bipyridine (4bpy) was carried out by introducing a nanoshell into an IDT electrode. Figure 8.12(a) shows an IDT electrode, fabricated using the device fabrication procedure described in chapter 3. The liquid solution including nanoshells and 4bpy as an analyte is confined by PDMS well and an AC electric field is applied during the Raman spectroscopy, as schematically illustrated in figure 8.12(b). When an AC electric field is applied, nanoshells gather at the centre of the electrode, as shown in figures 8.12(c) and (d). When the SERS measurement is then carried out, the SERS spectrum of 4bpy is obtained, as shown in figure 8.12(e). The SERS spectra were obtained for different concentrations of 4bpy and the concentration dependence of the SERS intensity at 1600 cm^{-1} is plotted in figure 8.12(f). Thus, the concentration-dependent change in SERS intensity indicates that calibration is possible. The detection time was less than 30 s, indicating that sensitive, rapid and label-free measurements can be achieved. Applications in environmental and medical analysis can be expected.

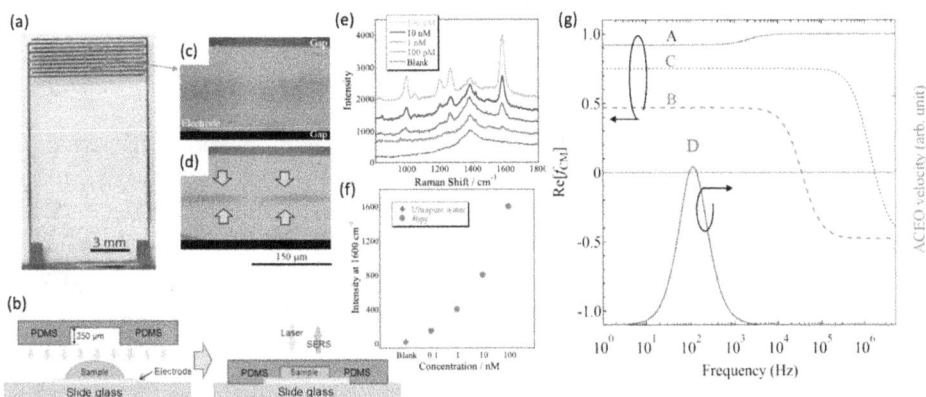

Figure 8.12. (a) Optical micrograph of fabricated interdigital electrodes. (b) Schematic of device preparation and SERS measurement. Optical micrographs of the sample well containing a solution of 1 mM 4,4′-bipyridine (4bpy) and AuNps-decorated PS beads (Au/PS nanoshells) (c) before and (d) during the application of electric field at 1 kHz and 10 Vpp (voltage of peak to peak). (e) SERS spectra of 4bpy within electric-field-induced aggregation of Au/PS nanoshells. (f) Variation in the intensity of the Raman peak at 1582 cm^{-1} of 4bpy as a function of 4bpy concentration. (g) Real part of Clausius–Mossotti factor and fluid flow velocity induced by AC electro-osmotic effect as a function of AC electric field frequency. The black dotted (A), purple broken (B), and pink dashed (C) lines are calculation lines. The red solid line (D) denotes the ACEO velocity for the nanoshell in the experiment of [15]. Reprinted from [15], copyright (2016) with permission from Elsevier.

Thus, the optoelectrofluidic system could be easily fabricated, and SERS measurement system could be achieved. Unlike other examples, it is possible to build SERS-active structures at a very fast rate. Is DEP the only force that can move these nanoshells? In general, the polarization rate and the AC electric field response of particles in liquid strongly depend on the particle size, the dielectric constant and the frequency of the electric field. The dielectric constant and conductivity of particles also depend on the surface treatment of the particles. The general experimental conditions associated with the Clausius–Mossotti factor are so complex that it is necessary to apply DEP forces to recognize the actual behaviour of the particles. The case described below is similar to that reported by Saucedo-Espinosa *et al* [41]. Therefore, a frequency as low as 1 kHz is expected to explain the dielectrophoretic-induced agglomeration mainly at the centre of the electrode. The frequency dependence of the particle displacement observations suggests that the crossover frequency is expected to be below about 5 kHz in the following system shown in figure 8.12. However, the electric field gradients in the vicinity of the electrode are also known to play a significant role in the flocculation behaviour, so the dependence of flocculation on the electric field gradient should be considered. In addition, the phenomenon of ACEO effect is also expected to contribute to the localization of particles near the centre of the electrodes due to long range electrohydrodynamic flows.

Here, in order to understand the phenomena, we should not reject all possibilities and consider the effects that are induced when an AC electric field is applied. That is, in this case, the DEP effect and the ACEO effect; we consider that either the DEP or the ACEO effect, or both, induce agglomeration by the application of an AC electric field. When the AC electric field E is applied in the suspending media, the DEP force exerted on the dielectric particle with a diameter of a is given by [38–41]

$$F_{\mathrm{DEP}} = 2\pi a^3 \varepsilon_{\mathrm{m}} \operatorname{Re}[f_{\mathrm{CM}}] \nabla |E|^2, \tag{8.13}$$

where ε_{m} and $\operatorname{Re}[f_{\mathrm{CM}}]$ are the permittivity of the medium and real part of the Clausius–Mossotti factor, respectively. This equation (8.13) is the same as equation (8.1), assumed to be $r = a$. In the case shown in figure 8.12, the AuNPs-decorated PS beads can be considered as a spherical particle with one concentric shell. When the AuNPs-decorated PS beads levitate at the height h from the interdigital transducer electrodes in the application of the AC electric field, the dielectrophoretic force given by equation (8.13) should be replaced with the corrected force [39, 40]:

$$F_{\mathrm{DEP\_Z}} = 2\pi A a_1^3 \epsilon_{\mathrm{m}} \operatorname{Re}[f_{\mathrm{CM}}] V^2 \exp(-4\pi h/w) p(f), \tag{8.14}$$

where V is the applied root mean square (RMS) voltage. The parameters w and A are the periodicity of the parallel electrode array and a constant inversely proportional to $w^3(=-176/w^3)$, respectively. For parallel electrodes of 150 μm widths and gaps, the parameters $w = 600$ μm and $A = -8.148 \times 10^{11}$ m^{-3} are obtained. The function $p(f)$ is introduced to correct for the electrode polarization effect as a function of an applied voltage frequency f. According to Huang *et al* [40], $p(f)$ is estimated that the value was changed from 0.3 to 1 depending on the frequency. Here, $p(f) = 0.8$ for simplifying the evaluation of dielectrophoretic force. Clausius–Mossotti factor which is defined in equation (8.4). ε_p^* and ε_m^* denote the complex permittivities of the particle and the media with $\varepsilon_p^* = \varepsilon_p - j\frac{\sigma_p}{\omega}$ and $\varepsilon_m^* = \varepsilon_m - j\frac{\sigma_m}{\omega}$, respectively. ε_p and ε_m are the permittivities of the particle and medium, respectively. σ_p and σ_m are the conductivities of the particle and medium, respectively. The complex permittivies are described as a function of the angular frequency, ω. The real and imaginary parts of the complex Clausius–Mossotti factor f_{CM} are, respectively, given by

$$\mathrm{Re}\,[f_{CM}] = \frac{\omega^2(\varepsilon_p - \varepsilon_m)(\varepsilon_p + 2\varepsilon_m) + (\sigma_p - \sigma_m)(\sigma_p + 2\sigma_m)}{\omega^2(\varepsilon_p + 2\varepsilon_m)^2 + (\sigma_p + 2\sigma_m)^2} \qquad (8.15a)$$

and

$$\mathrm{Im}[f_{CM}] = \frac{\omega(\sigma_m - \sigma_p)(\varepsilon_p + 2\varepsilon_m) - (\varepsilon_p - \varepsilon_m)(\sigma_p + 2\sigma_m)}{\omega^2(\varepsilon_p + 2\varepsilon_m)^2 + (\sigma_p + 2\sigma_m)^2}. \qquad (8.15b)$$

To confirm the reproducibility of the experimental result, the real part of the Clausius–Mossotti factor f_{CM} is calculated. If the DEP forces govern the displacement and agglomeration of particles, the sign of the Clausius–Mossotti factor f_{CM} should invert at the crossover frequency, as shown in figure 8.2. It is possible to agglomerate or disperse particles by DEP by giving a frequency higher or lower than the frequency of the crossover frequency. For the experimental example given here as an example, the Clausius–Mossotti factor f_{CM} calculated as a function of the AC voltage frequency is shown in figure 8.12(g). The black dotted (A), purple dashed (B) and pink dashed (C) lines were calculated by substituting the parameters in table 8.1.

For normal PS beads in distilled water, the crossover frequency is about 2 MHz [15], corresponding to condition C. PS beads not decorated with AuNPs in ultrapure water were controllable by the application of an AC electric field between 1 and 10 kHz, which was close to condition B. Comparing B and C, the crossover frequency in B is smaller, at about 35 kHz. The calculated crossover frequencies of B and C are larger than the experimentally measured values; the crossover frequency at which the DEP strength disappears from a positive system to a negative system and vice versa using condition A, which assumes PS beads decorated with AuNPs in a KCl solution, is not indicated. Therefore, the displacement of the particles by the AC electric field may not be induced by dielectrophoresis, but by another effect, or by a combination of both effects.

Table 8.1. Summary of parameter for calculation of Clausius–Mossotti factor.

	ε_p (F m^{-1})	ε_m (F m^{-1})	σ_p (S m^{-1})	σ_m (S m^{-1})
A	$-45 \times 10^6 \varepsilon_0$	$50\varepsilon_0$	5.00	0.14 (KCl 10 mM)
B	$2.5\varepsilon_0$	$78\varepsilon_0$	2.00×10^{-4}	5.50×10^{-5}
C	$2.5\varepsilon_0$	$78\varepsilon_0$	1.00×10^{-2}	1.00×10^{-3}

$\varepsilon_0 = 8.854\,187 \times 10^{-12}(\mathrm{m}^{-1})$

Here, as another contribution, let us consider the ACEO effect. ACEO is nonlinear electrokinetic phenomena of induced-charge-electro-osmotic flow around electrodes applying an AC electric field [2, 3, 15, 38–41]. The basic scaling velocity of time-averaged ACEO flow is given by

$$\langle u \rangle \propto \frac{\varepsilon V^2}{\eta(1 + \delta)L\left[\frac{\omega}{\omega_c} + \frac{\omega_c}{\omega}\right]^2},$$
(8.16)

where V, ε, and η are the applied voltage, the permittivity and viscosity of the liquid, respectively. The electrode spacing and ratio of the diffuse-layer compact-layer capacity are given by L and δ, respectively. Here, $\omega_c \propto \frac{D(1+\delta)}{\lambda L}$, where λ it the Debye screening length and D is a characteristic ionic diffusivity. In this study, we assumed $L = 150\ \mu$m (actual device design), $\delta = 1$ (assumed constant), $D = 6.02 \times 10^{24} (\text{m}^{-3})$, $\varepsilon_m = 50\varepsilon_0$, and $\lambda = 2.44$ nm considering the experimental condition shown in figure 8.12.

The results of the above calculations are shown in figure 8.12(g). The red coloured 'D' curve shows the calculation results when assuming ACEO effect contributes the particle displacement in the optoelectrofluidic system of figure 8.12. As a result, the peak frequency is about 120 Hz. Below 10 kHz, the fluid velocity is expected to increase with decreasing AC electric field frequency. The fluid flow is considered to contribute the particle displacement. The ACEO contribution plays an important role in moving the particles. In a relatively higher frequency region, the DEP dominates the system. In a lower frequency region, the ACEO-induced flow contributes the displacement of particles in the system. Therefore, the combination between the DEP and ACEO-induced flow enables us to control the aggregation of the AuNPs-decorated PS beads in the optofluidic device with micro-electrodes. When the AuNPs-decorated PS beads are aggregated using AC electric voltage application as shown in figure 8.12(c), they form aggregates due to DLVO interactions. At this time, SERS is induced as the specimen molecules in the liquid solution are adsorbed while mixing.

8.4 Optofluidic systems with SERS-active structure

Parts of this section have been reprinted from [27], with permission from the Royal Society of Chemistry.

8.4.1 Development of higher-order nano/micro-scale structure

As mentioned above, a system combining AC electric field application, laser excitation, and microfluidics would be very useful. However, there are many elements that need to be built in and made to work; since SERS activity is attractive because it can provide label-free molecular sensing, it may be possible to build and use a microfluidic system carrying only SERS activity. Therefore, a system in which SERS-active structures are pre-created using AC electric field application, DLVO interaction, etc in concert and incorporated into a microfluidic system is presented below.

$$U = \pi\varepsilon_r\varepsilon_0 R\left[2\psi_P\psi_S\left\{ \frac{1 + \exp(-\kappa h)}{1 - \exp(-\kappa h)} \right\} + \left(\psi_P^2 + \psi_S^2\right)\ln\left\{1 - \exp(-2\kappa h)\right\} \right] \tag{8.19}$$
$$- \frac{A_{132}R}{6h}[1 + (14\kappa h)]^{-1},$$

where ψ_P and ψ_S denote the zeta potentials of the particle and substrate, respectively. Here, zeta potential of the glass substrate is adopted to be -66 mV. Equation (8.19) can provide height dependence of potential energies for AgNp–glass substrate, AuNP–glass substrate and PS–glass substrate, and so on. The force between particle and substrate can be calculated by differential potential using the relation $F = -\partial U/\partial z$. Here, a particle travelling in a droplet experiences not only DVLO interactions but also other forces. In the vertical direction, DVLO interactions, hydrodynamically derived from radial flow and Marangoni flow, and sedimentation (gravitation) forces act to determine the particle height in the fluidic flow profile. Considering the depositional process in equilibrium, which is the sedimentation dominating state, the sedimentation of a spherical particle under the constant flow is described by [27, 37]

$$\frac{\partial\boldsymbol{u}}{\partial t} = \frac{\sigma - \rho}{\sigma + \rho/2}g, \tag{8.20}$$

where $\boldsymbol{u}$ is velocity vector of the particle, that is, $\partial\boldsymbol{u}/\partial t$ represents the acceleration. σ and ρ are mass densities of particle and media, respectively. g is the acceleration due to gravity, that is, 9.8 m s^{-2}. For example, the mass densities of Au, PS and water are 19.3 $\times 10^3$, 1050 and 997 kg m^{-3}, respectively. Substituting these densities into equation (8.20), the gravitation forces working AuNPs and PS are estimated to be about 5.86×10^{-18} and 3.98×10^{-17} N, respectively. By comparison of gravitation force and DVLO force, it is found that the gravitation forces working AuNPs and PS were larger than the DVLO repulsion forces between particles and substrate. The sedimentation and deposit of AuNPs and PS lead to inevitable consequences, overcoming the DVLO repulsive force. In the actual case, described below, PS particles repel each other and AuNPs fill the space between them, creating a stacked structure consisting of AuNPs and PS particles. This structure is hereafter referred to as Au3D; in the case of Ag, it is called Ag3D. Therefore, Au3D formation is caused by the colloidal deposit resulting from the competition among sedimentation, radial flow, Marangoni flow and DVLO interactions. Furthermore, when an AC electric field is applied in addition to the competition condition, Au3D structures can be created by targeting arbitrary areas, depending on the electric field application conditions. For example, the Au3D was formed by the condition when the mixed solution consisting of particle number ratio 1300:1 (Au:PS) was dropped in a 6 mm diameter PDMS well with an electrode with 100 µm width and AC voltage of 10 V applied at a frequency of 1 kHz, as shown in figure 8.14 [28]. The SEM images of figures 8.14(a)–(d) indicate a higher-order layer stack nanostructure (Au3D) on the middle of the electrode. In addition, the formation conditions are varied: electrode width, frequency, and particle number ratio, Au3D can be formed onto the desired area, for example, the middle and edge of electrode. An optical micrograph of the constructed Au3D structure and the photoluminescence (PL) spatial mapping result is shown in figure 8.14(e). As the edge of the Au3D structure has

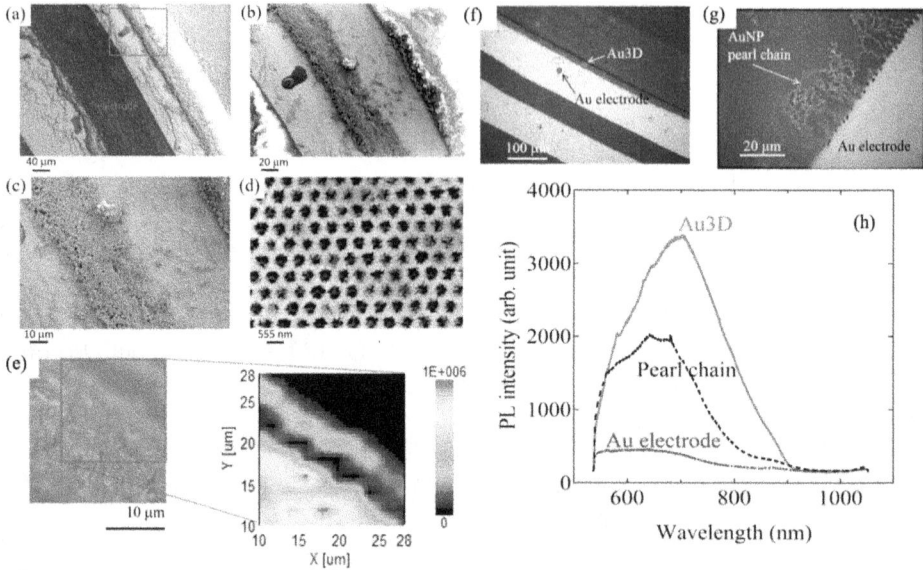

Figure 8.14. SEM images of the novel electrode with the formation of the higher-order layer stack nanostructure. The red square area shown in (a) is magnified in (b). SEM images with high magnification for (c) the electrode and (d) higher-order layer stack porous nanostructure. (e) Optical photograph of higher-order layer stack nanostructure. Mapping of the integrated PL value from 543 to 808 nm in wavelength with respect to the red square area shown in the optical photograph. Optical photographs of the electrode with (f) higher-order 3D Au porous nanostructure and (g) pearl chain structure consisting of AuNPs. (h) Comparison of PL spectra for Au3D, pearl chain and flat-plane Au electrode.

fewer Au stack layers and the number of layers increases towards the centre, the PL distribution also tends to reflect this structure, with greater strength in the centre. Other conditions formed the nanostructures shown in figures 8.14(f) and (g). Figure 8.14(a) shows the higher-order layer stack nanostructure is formed within the inner electrode, while the pearl chain structure is induced between electrodes, as shown in figure 8.14(f). It was found that a pearl chain structure is formed instead of Au3D, which is a stacked structure in figure 8.14(g). This shows the same result as in Cherukulappurath *et al* [13], corresponding to nanostructure formation in the absence of PS beads. Figure 8.14(h) shows the comparison of the PL spectra at the higher-order layer stack nanostructure (Au3D), pearl chain, and flat surface of Au electrode. As shown in figure 8.14(h), the PL intensity of the Au3D is the largest. This Au3D also enables the SERS detection of 1 nM 4bpy [28]. These SERS measurements were performed by a simple Raman spectrometer without any microscope (field is about 2 mm × 2 mm). Thus, the Au3D provides a clue to enhancing the performance of SERS detection of analyte.

8.4.2 Method and procedure to fabricate a coffee-ring-type Ag3D structure

Plasmon excitation is essential for SERS expression, and noble metals with high conductivity and chemical stability are often used. Silver is chemically unstable compared to gold, but interacts well with biogenic materials, so the creation of SERS-active structures using AgNPs and biosensor applications is an important research area when considering their use in biosensors and other applications. In addition, they are cheaper than gold, making them suitable for use in disposable testing chips. Therefore, the creation of silver nanostructures is also presented below. In the similar way, Ag3D can be prepared using convective self-assembly [26, 27], namely the coffee-ring effect. Here, we introduce the method and procedure to fabricate the coffee-ring-type Ag3D structure below.

Sulfate PS latex beads of diameter 600 nm (Invitrogen Inc., 8% w/v) were concentrated by centrifugation and washed to yield a 10%w/v suspension. Aqueous suspensions of colloidal AgNPs of diameter 40 nm were synthesized using a standard citrate reduction protocol, Lee–Meisel method. A dispersion of AgNPs was prepared at 1.00 mM and concentrated 100-fold by centrifuging at 6600 rpm for 60 min. The concentrations of PS and AgNPs were estimated as 8.4×10^{14} particles l^{-1} and 3.07×10^{16} particles l^{-1}, respectively. AgNPs (27 μl) were mixed with PS (1 μl) to prepare a mixed solution of ratio 1000:1 (AgNP:PS).

A silicone sheet with ϕ 6 mm through hole was put onto an Ag3D-coated ultraviolet-cleaned glass slide (Matsunami Glass Ind., Ltd), and a well (28 μl volume) was made. 5 μl of the mixed solution of AgNPs and PS was dropped into the well and dried at 75% relative humidity for about a day. A PS colloidal crystal was fabricated by convective self-assembly, where AgNPs were simultaneously accumulated in the PS gaps. The structure was then soaked in dichloromethane to remove the PS. Thus, Ag3D was successfully obtained. Figure 8.15(a) shows a highly magnified SEM image of Au3D just after prepation. Thus, the details are omitted, but basically, the Ag3D structure can be formed only by the competition among the DLVO interaction, sedimentation, and advection-accumulation effects described earlier without AC electric field application.

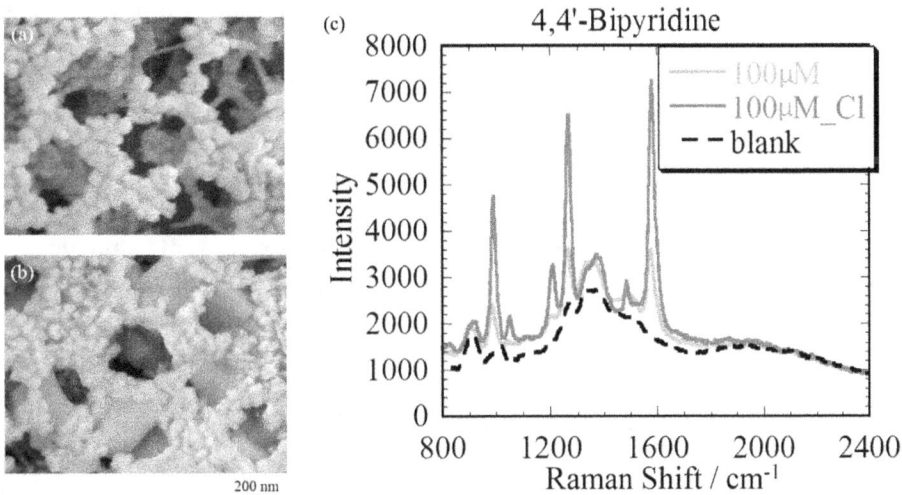

Figure 8.15. Highly magnified SEM images of Ag3D (a) before and (b) after adding the aqueous solution of sodium chloride, respectively. SERS spectra of 4bpy with and without the aqueous solution of conductivity than gold. (c) The measured SERS spectrum using sodium chloride. The concentration of 4bpy is 100 μM. Reproduced from [27] with permission from the Royal Society of Chemistry.

8.4.3 Chloride activation

To enhance the SERS spectra obtained using the Ag3D, chloride activation, which enhanced the Raman intensity of 4bpy by a factor of two to three, was performed as follows [26]. After 35 µl of 4bpy aqueous solution was dropped into a well on the Ag3D, 2 µl of NaCl (50 mM) was added to the Ag3D. As shown in figure 8.15(c), characteristic enhanced Raman peaks attributed to 100 µM 4bpy (located at 1000, 1250 and 1580 cm^{-1}) were clearly observed. This phenomenon is called 'chloride activation'. The blue and brown solid lines correspond to the SERS spectra of 100 µM 4bpy before and after chloride activation, respectively. The black dotted line represents the SERS spectrum of the blank (ultrapure water). As evident from figure 8.15(c), chloride activation enabled us to increase the intensity of the SERS spectra. Although the chloride activation in colloidal systems was a popular phenomenon, the mechanism had not been understood until the investigations by Prucek *et al* [44, 45]. In colloidal state, Prucek *et al* revealed that re-crystallization of AgNPs in a highly concentrated NaCl environment provided a considerable enhancement of the SERS signals of adenine by shift of LSPR due to the crystal growth. This Ag3D was fixed on the glass substrate under arid conditions, not in colloidal state. By comparing the SEM images of Ag3D before and after adding the NaCl solution, as shown in figures 8.15(a) and 8.15(b), respectively, it was found that Ag crystal growth occurred. This re-crystallization is attributable to the oxidative etching process due to the polyol reaction investigated by Wiley *et al* [46] The defects inherent in twinned nuclei of silver led to their selective etching and dissolution by chloride and oxygen from air, leaving the single crystals to grow. Therefore, chloride activation-induced rapid-crystallization process towards larger silver crystals provides an enhancement of SERS signals with excitation in the 785 nm wavelength region.

The SERS activity of silver is generally higher than that of gold in the colloidal system, because silver has higher electrical conductivity than gold. There is a possibility of obtaining a stronger SERS effect using Ag3D than Au3D if the conditions of chloride activation are optimized, i.e. the amount of the aqueous solution of NaCl added to the Ag3D versus the concentration of the aqueous solution of NaCl, and selection of suitable excitation laser wavelength.

8.4.4 Optofluidic system for molecular sensing integrated with Au3D structure

Implementing Au3D in a microfluidic device can provide a molecular sensor based on label-free SERS [29]. Here, a flow channel structure with steps is fabricated, as shown in figures 8.16(a) and (b), and Au3D is mounted in the reaction area in the microchannel. The microfluidic PDMS chip can be fabricated using the soft replica moulding technique, as described in chapter 3. A mask with a microfluidic pattern was generated using computer-aided design (CAD) software. A master mould was fabricated by spin-coating an SU-8 3005 photoresist (Nippon Kayaku) onto a silicon wafer to prepare a 7 µm thick layer. After development, the master mould was washed in isopropyl alcohol (IPA; Wako Pure Chemical Industries). In the same way, to prepare a 250 µm thick layer, an SU-8 100 photoresist (Nippon Kayaku) was

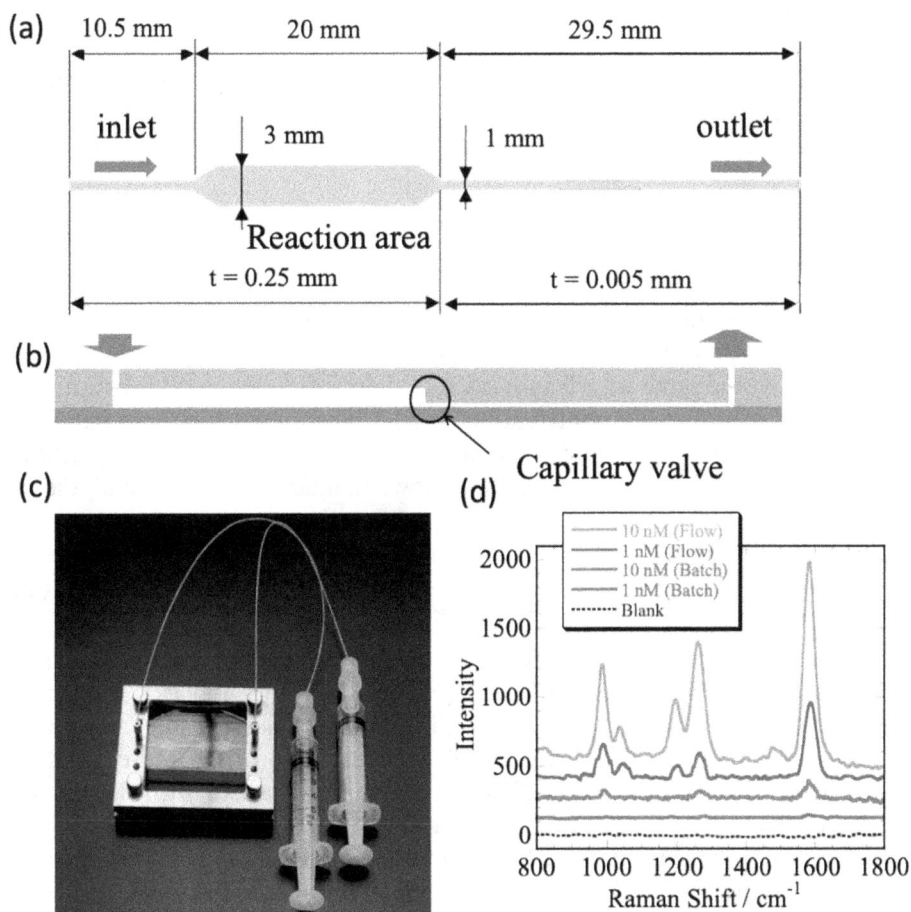

Figure 8.16. (a) Design diagram of the microfluidic structure. (b) Cross-sectional schematic diagram. (c) Optical photograph of the fabricated SERS sensor-mounted flow system microfluidic channel, with the Au3D structure formed and placed in the Reaction area shown in (a). Here, the Au3D structure, consisting porous aggregate of AuNPs, is shown in figure 8.15, for example. (d) SERS spectra measured during flow using this microfluidic system. Reprinted from [29], copyright (2013) The Japan Society of Applied Physics. All rights reserved.

spin-coated at 1000 rpm for 30 s onto the same wafer, pre-exposure baked at 65 C for 30 min and 95 °C for 90 min, and then a mask was placed over the coated wafer and they were exposed to UV light (700 mJ cm^{-2}). After development, the complete master was produced by washing with IPA. A degassed mixture of a Sylgard 184 elastomer (Dow Corning Toray) and its curing agent at a 10:1 ratio was moulded on the master. After baking at 60 °C for 2 h, the replica was removed from the mould. Two 1 mm holes for tubing connection were punched for fluidic access at the inlets and outlet. An optical photograph of the completed systems is shown in figure 8.16(c).

In situ and flow SERS measurements in a microfluidic device with Au3D were performed using 4bpy. Using this *in situ* optofluidic microsystem, SERS spectra were immediately observed in flow format by injecting 4bpy aqueous solution. As shown in figure 8.16(d), *in situ* SERS measurements could be made while the sample was flowing in the flow system, indicating that the measurements are very sensitive compared to the batch-type measurement device with no liquid flow. Let us consider this mechanism as follows. First, enhanced factor (EF) of the SERS can be defined as [18, 19]

$$EF = \frac{I_{SERS}/N_{SERS}}{I_{NR}/N_{NR}}, \tag{8.21}$$

where I_{SERS} is the intensity of the surface-enhanced Raman signal, N_{SERS} is the number of molecules contributing to I_{SERS}, I_{NR} is the intensity from the normal Raman signal at the same wavelength (as I_{SERS}) on a non-enhanced substrate, N_{NR} is the number of molecules contributing to I_{NR}. The values for these parameters were calculated as follows. Here, I_{SERS} and I_{NR} are decided by intensity height at 1600 cm^{-1} of 1 nM 4bpy and 10 mM 4bpy shown in figure 8.16(d). Second, the ratio of N_{SERS}/N_{NR} is 1 nM/10 mM, since

$$\frac{N_{SERS}}{N_{NR}} = \frac{N_A \times C_{SERS}}{N_A \times C_{NR}} \times \frac{V_{SERS}}{V_{NR}}, \tag{8.22}$$

and V_{SERS} is equal to V_{NR}, where N_A is the Avogadro constant, C is the concentration of 4bpy (C_{SERS}=1 nM, C_{NR}=10 mM), V_{SERS} and V_{NR} are the total volume in laser spot. As a consequence, EF in batch was estimated by equations (8.21) and (8.22) to be EF = 6.7 × 10^6 in figure 8.16(d). In the case of flow measurement, EF is estimated be 4.0 × 10^7. EF in flow measurement is almost six times larger than that in batch measurement [29].

This improvement was attributed mainly to the fact that Au3D had large surface for adsorption since the peak intensity was related to the number of molecules. The Au3D's surface area (S_{Au3D}) in laser spot was evaluated as follows. The number of holes, N_{hole}, due to removal of PS beads was calculated:

$$N_{hole} = \mu \frac{A_{Laser} \cdot t}{V_{PS}}, \tag{8.23}$$

where μ is filling factor (0.74) in hexagonal close-packed structure, A_{Laser} is the apparent area of laser spot ($\pi \times 2.5^2$ μm^2), t is Au3D's thickness (5 μm), V_{PS} is volume of single PS (4/3 × π × 0.3^3 μm^3). Therefore, N_{hole} is 642. Since surface area, $S_{PS} = 4\pi r^2$, which was produced by single PS is 1.13 μm^2, where r is PS radius (300 nm). Therefore, the effective surface area associated with the SERS activity can be estimated to $S_{Au3D} = S_{PS} \times N_{hole}$= 726 μm^2. This is 37-fold compared with Au2D's surface area which is approximately equal to laser spot.

Thus, this introduction of three-dimensional nanostructures, Au3D can greatly increase the number of hotspots that contribute to SERS activity. Therefore, Au3D can achieve very sensitive molecular sensing. In addition, such a porous structure can also be expected to have a molecular enrichment effect, making it suitable for applications such as environmental analysis.

8.5 Optoelectrofluidic printing

Here, we introduce a few recent research trends and provide a perspective on future systems. Gi *et al* demonstrated a novel optoelectrofluidic printing system that facilitates both the optoelectrofluidic patterning of microparticles and mammalian cells and harvesting of the patterned microparticles encapsulated with poly(ethylene glycol) dicarylate (PEGDA) hydrogel sheets [47]. Different types of microparticle were on demand manipulated within the hydrogel precursor solution by optically induced DEP. They demonstrated that HepG2 cells in the PEGDA precursor were patterned with circular and hexagonal shape under the manipulable conditions using the optically induced nDEP. Under a voltage of 20 V at 20 kHz, the HepG2 cells were repelled from the bright area projected through the liquid crystal display (LCD) forming a circular and a hexagonal shape, while they were randomly distributed regardless of the projected image in the absence of voltage application. Thus, on-demand arrangements of microparticle groups within a hydrogel sheet were achieved by combining basic functions of the optoelectrofluidic system. Their optoelectrofluidic printing system facilitates both the on-demand patterning and the harvesting of the patterned hydrogel sheets, resulting in solving the problem that practical applications of patterned microparticles have been limited due to the impossibility of harvesting.

8.6 Electrochemical reaction monitoring

Originally, SERS was attributed to the discovery of a peculiar Raman scattering effect at a silver electrode during an electrochemical (EC) reaction [18–20]. In recent years, electrochemistry and SERS have been combined, and research has been conducted on highest occupied molecular orbital (HOMO) and lowest unoccupied molecular orbital (LUMO) level manipulation of molecules and reaction dynamics of redox reactions by the SERS effect during electrochemical reactions, as shown in figure 8.17. In particular, there has been a recent upsurge in research on temporal and spatial imaging of molecular bonding and vibrational displacements during electrochemical (EC) reactions in combination with TERS [21–25, 48–54].

The most notable recent research trend is the redox reaction of CO_2 on copper and copper oxide surfaces, which is of great interest not only because it can immobilize CO_2, but also because it can synthesize ethylene and methanol from CO_2. (CO_2 reduction reaction, called CO2RR) [55–57]. A common method used is to follow the copper redox reaction by Raman spectroscopy while performing electrochemical reactions and cyclic voltammetry. de Ruiter *et al* performed time-resolved Raman spectroscopy measurements and revealed the redox reaction mechanism on the copper surface, as shown in figure 8.18 [55]. However, it is unclear how the structure and morphology of the catalyst surface and the compositional transitions of the electrolyte and the activity of the surface and the selectivity of the C_1, C_2 and C_3 compounds are determined and controlled. Therefore, it is important to combine electrochemical reaction control systems with time- and space-resolved FTIR and Raman spectroscopy measurements. Herzog *et al* reported that there is a reaction change by varying the pulse length of the electrochemical reaction [56]. In order to advance these studies, a highly sensitive system is needed to measure chemical changes and electronic state transitions at surfaces and

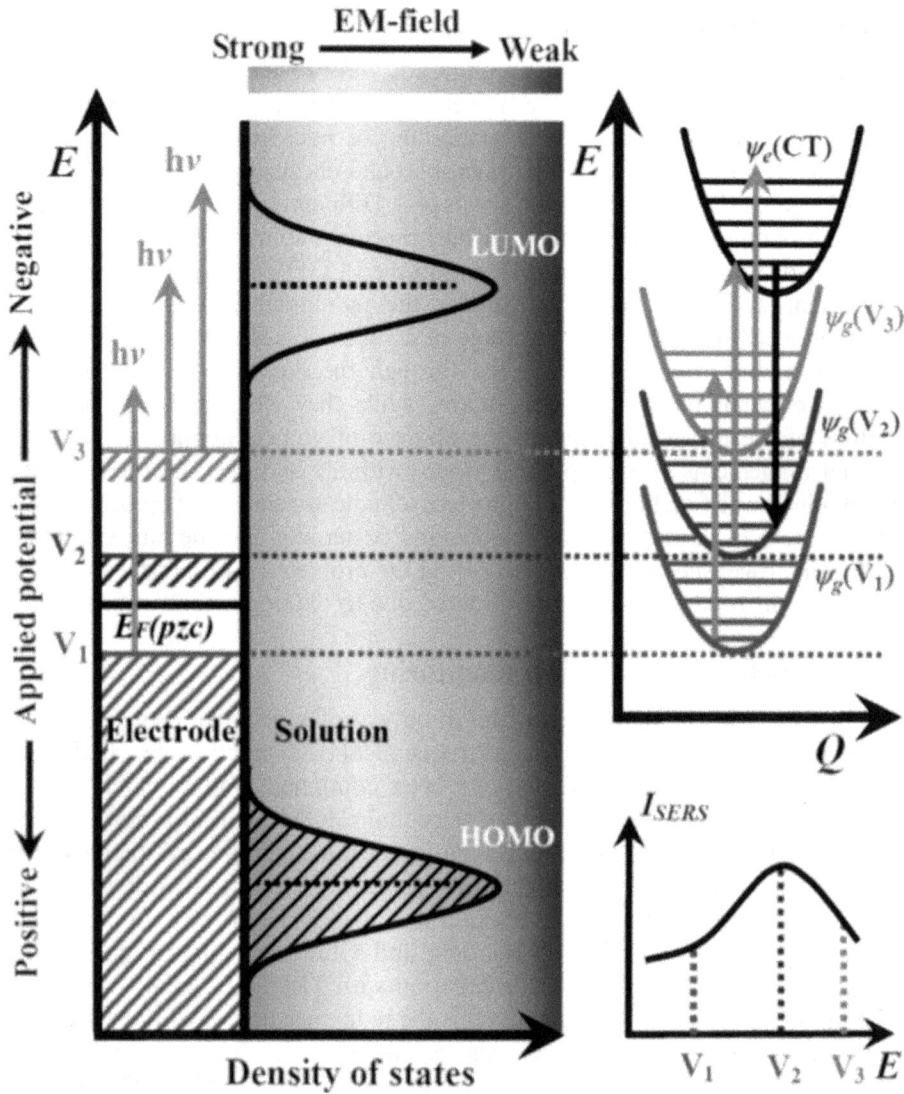

Figure 8.17. Schematic diagrams of the photon-driven charge transfer from a metal electrode to an adsorbed molecule in the EC-SERS system. Reprinted from [52] with permission from the Royal Society of Chemistry.

interfaces while quantitatively and systematically changing physical and chemical parameters by creating ideal experimental systems in a controlled system. Therefore, the EC-TERS and EC-SERS systems are attracting attention.

However, the equipment and know-how to realize both EC-TERS and EC-SERS are not easy to obtain. In response to this demand, Saegusa and Yamaguchi *et al* have recently developed a chip that can easily realize EC-SERS in a microchemical system, as shown in figure 8.19(a) [58]. In this system, gold nano coral (GNC) structure is

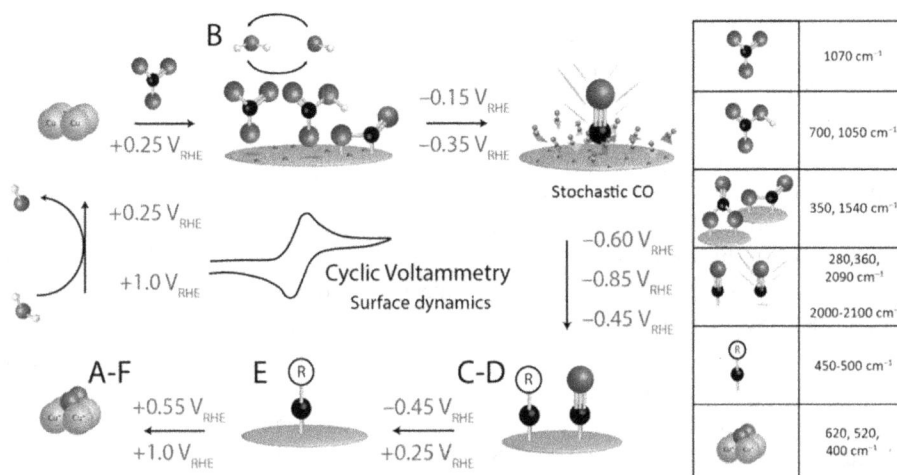

Figure 8.18. Schematic overview of surface dynamics during Cyclic voltammetry (CV) on electrodeposited Cu as observed with time-resolved SERS. Reprinted from [55], copyright (2022) with permission from the American Chemistry Society.

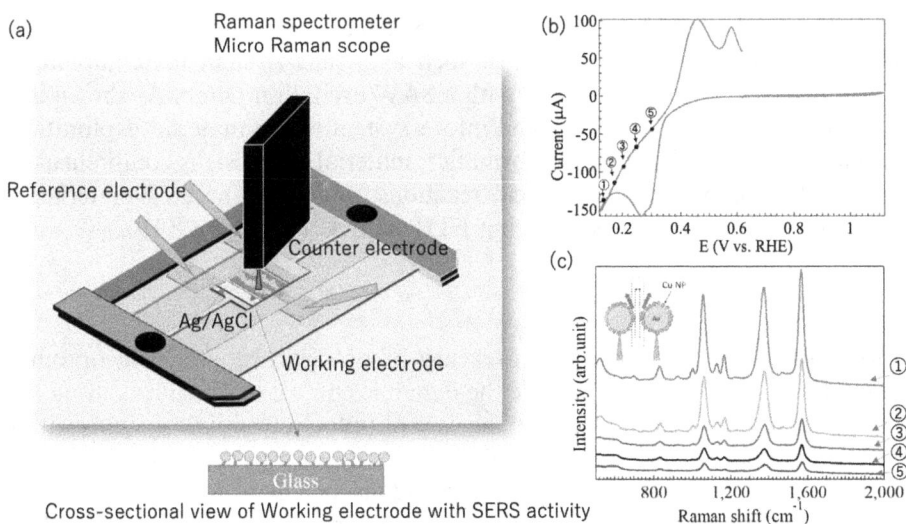

Figure 8.19. Conceptual diagram of an electrochemical reaction microchemical system with gold nanobean structures mounted on the working electrode. Cross-sectional schematic diagram of the working electrode is shown. The system can be easily integrated into a micro-Raman spectrometer or portable Raman spectrometer. SERS spectra at points 1–5 in CV as a result of cyclic voltammetry of a copper acetate solution after applying mercaptobenzoic acid as a SAM film on the working electrode. The Raman spectra of mercaptobenzoic acid and copper oxide are observed, which vary both in time and in the position of the CV. The inset is a schematic diagram showing the formation and growth of copper particles on the GNC structure electrode and the change in SERS intensity.

mounted on the working electrode, resulting in an electrochemical reaction chip with SERS activity. The nanostructure GNC can generate a strong electric field gradient, which improves the sensitivity and reaction efficiency of the electrochemical reaction. The chip can be easily combined with microfluidic channels and the electrodes can be freely designed and fabricated; although inferior to the Au3D structure in terms of SERS sensitivity, the GNC structure is very useful because of its large area and free design [30–33]. Saegusa *et al* have published on a flow-based SERS molecular sensor by incorporating this GNC structure into a microfluidic channel [30]. Here, the GNC structure was incorporated into an electrochemical microchemical system. Mercaptobenzoic acid was applied to the working electrode as a surface atomic monolayer (SAM) film and the cyclic voltammetry (CV) of a copper acetate solution was measured, resulting in the CV shown in figure 8.19(b). The SERS spectrum can be obtained at any point in the CV. The SERS spectra measured at the corresponding points in the CV are shown in figure 8.19(c). It can be seen how the SERS spectra change depending on the CV. In addition to mercaptobenzoic acid, SERS spectra derived from copper oxides can also be detected, successfully tracing the redox reaction of copper and copper oxides on the GNC electrode surface.

Thus, this EC-SERS chip is expected to provide very good clues for the search for reaction conditions for application, while carrying out basic research on CO2RR. As described above, the Nano-Microsystem can be used to simultaneously carry out basic and applied research. It is expected to be developed into an integrated system combining a wide variety of elements, such as a microsystem for combinatorial chemical reactions or a combination with a SAW excitation pump. As shown in the figure, the platform will be developed into a system where material exploration is performed by AI-based material informatics, material synthesis is comprehensively performed by combinatorial chemical reactions, and the generated materials are analyzed by an evaluation system using FTIR and SERS (figure 8.19).

8.7 Summary

In this chapter, we have given an overview of optoelectrofluidic and optofluidic systems. These platforms can provide the dynamic control of specimens using laser, electric field, or their combination. In the case of optoelectrofluidic systems with the AuNPs-decorated PS beads, ACEO fluid flow induced by the AC electric field application can manipulate the aggregation and dispersion of the beads on demand. The *in situ* SERS measurements of analyte can be simultenously performed during the control of the beads. These platforms demonstrate the feasibility of developing integrated medical nano/micromachines that are multifunctional and controllable. The reactions at interfaces play a significant role in electrochemical reactions. A gold nano coral structure fabricated on a working electrode enables *in situ* tracing the chemical reaction processes during precisely controlled electrochemical reactions by surface enhanced Raman scattering (SERS). The EC-SERS chip unveiled the reattachment of 4-mercaptobezoic acid by electrochemical manipulation, redox reactions and formation of copper nanoparticles in aqueous copper acetate solutions were tracked by *in situ* SERS measurements while controlling electrochemical reactions. Further design, integration, and improvement of these systems and combination with other systems can become a novel platform where spontaneous material exploration, synthesis, and analytical evaluation are repeated.

References

[1] Psaltis D, Quake S R and Yang C 2006 Developing optofluidic technology through the fusion of microfluidics and optics *Nature* **442** 381

[2] Hwang H and Park J -K 2011 Optoelectrofluidic platforms for chemistry and biology *Lab Chip* **11** 33–47

[3] Han D and Park J-K 2016 Optoelectrofluidic enhanced immunoreaction based on optically-induced dynamic AC electroosmosis *Lab Chip* **16** 1189

[4] Hwang H and Park J-K 2013 Optoelectrofluidic platforms for chemistry and biology *Lab Chip* **11** 33

[5] Camacho-Alanis F and Ros A 2015 Protein dielectrophoresis and the link to dielectric properties *Bioanalysis* **7** 353–71

[6] Green N G and Morgan H 1999 Dielectrophoresis of submicrometer latex spheres. 1. Experimental results *J. Phys. Chem.* B **103** 41

[7] Hwang H and Park J -K 2009 Rapid and selective concentration of microparticles in optoelectrofluidic platform *Lab Chip* **9** 1999–206

[8] Yasukawa T, Hatanaka H and Mizutani F 2012 Simple detection of surface antigens on living cells by applying distinct cell positioning with negative dielectrophoresis *Anal. Chem.* **84** 8830

[9] Mark D, Haeberle S, Roth G, von Stetten F and Zengerle R 2010 Microfluidic lab-on-a-chip platforms: requirements, characteristics and applications *Chem. Soc. Rev.* **39** 1153–82

[10] Chiou P Y, Ohta A T and Wu M C 2005 Massively parallel manipulation of single cells and microparticles using optical images *Nature* **436** 370

[11] Chan K L A, Gulati S, Edel J B, de Mello A J and Kazarian S G 2009 Chemical imaging of microfluidic flows using STR-FTIR spectroscopy *Lab Chip* **9** 2909–13

[12] Chrimes A F, Khoshmanesh K, Stoddart P R, Kayani A A, Mitchell A, Daima H, Bansal V and Kalantar-zadeh K 2012 Active control of silver nanoparticles spacing using dielectrophoresis for surface-enhanced Raman scattering *Anal. Chem.* **84** 4029–35

[13] Cherukulappurath S, Lee S H, Campos A, Haynes C L and Oh S-H 2014 Rapid and sensitive *in situ* SERS detection using dielectrphoresis *Chem. Mater.* **26** 2445–52

[14] Barik A, Otto L M, Yoo D, Jose J, Johnson T W and Hyun Oh S 2014 Dielectrophoresis-enhanced plasmonic sensing with gold nanohole arrays *Nano Lett.* **14** 2006–12

[15] Yamaguchi A, Utsumi Y and Fukuoka T 2019 Aggregation and dispersion of Au-nanoparticle-decorated polystyrene beads with SERS-activity using AC electric field and Brownian movement *Appl. Surf. Sci.* **465** 405–12

[16] Yamaguchi A, Fukuoka T, Kuroda K, Hara R and Utsumi Y 2016 Dielectrophoresis-enabled surface enhanced Raman scattering glycine modified on Au-nanoparticles-decorated polystyrene beads in micro-optofluidic devices *Colloids Surf., A* **507** 118–23

[17] Yamaguchi A, Takahashi R, Fukuoka T, Hara R and Utsumi Y 2016 Dielectrophoresis-enabled surface-enhanced Raman scattering on gold-decorated polystyrene microparticle in micro-optofluidic devices for high-sensitive detection *Sens. Actuator B: Chem.* **230** 94–100

[18] Le Ru E C and Etchegoin P G 2009 *Principles of Surface-Enhanced Raman Spectroscopy* (Amsterdam: Elsevier)

[19] Procházka M, Kneipp J, Zhao B and Ozaki Y 2024 *Surface- and Tip-Enhanced Raman Scattering Spectroscopy -Bridging Theory and Applications* (Singapore: Springer)

[20] Langer J *et al* 2020 Present and future of surface-enhanced raman scattering *ACS Nano* **14** 28–117

[21] Liu P, Chulhai D V and Jensen L 2017 Single-molecule imaging using atomistic near-field tip-enhanced Raman spectroscopy *ACS Nano* **11** 5094

[22] Jones T 2023 Fabrication of nanostructured electrodes for electrochemical surface-enhanced Raman spectroscopy (E-SERS): a review *Mater. Sci. Technol.* **39** 2287

[23] Zeng Z-C, Huang S-C, Wu D-Y, Meng L-Y, Li M-H, Huang T-X, Zhong J-H, Wang X, Yang Z-L and Ren B 2015 Electrochemical tip-enhanced Raman spectroscopy *J. Am. Chem. Soc.* **137** 11928

[24] Yokota Y, Hayazawa N, Yang B, Kazuma E, Celine F, Catalan I and Kim Y 2019 Systematic assessment of benzenethiol self-assembled monolayers on au(111) as a standard sample for electrochemical tip-enhanced Raman spectroscopy *J. Phys. Chem.* C **123** 2953

[25] Kang G, Yang M, Mattei M S, Schatz G C and Van Duyne R P 2019 *In situ* nanoscale redox mapping using tip-enhanced Raman spectroscopy *Nano Lett.* **19** 2106

[26] Yamaguchi A, Fukuoka T and Utsumi Y 2018 Study on fabrication of molecular sensing system using higher-order nanostructure for environmental analysis and food safety *Electron Comm. Jpn.* **101** 38–44

[27] Hara R, Takahashi R, Fukuoka T, Utsumi Y and Yamaguchi A 2014 Surface-enhanced Raman spectroscopy using a coffee-ring-type three-dimensional silver nanostructure *RSC Adv.* **5** 1378–84

[28] Yamaguchi A, Fukuoka T, Hara R, Kuroda K, Takahashi R and Utsumi Y 2015 On-chip integration of novel Au electrode with a higher order three-dimensional layer stack nanostructure for surface-enhanced Raman spectroscopy *RSC Adv.* **5** 73194–201

[29] Takahashi R, Fukuoka T, Utsumi Y and Yamaguchi A 2013 Optofluidic devices with surface enhanced raman scattering active three-dimensional gold nanostructure *Jpn. J. Appl. Phys.* **52** 06GK12

[30] Saegusa S, Tanaka T, Naya M, Fukuoka T, Amano S, Utsumi Y and Yamaguchi A 2023 Integration of gold-nanofève-based-SERS active nanostructure and microfluidic devices *IEEJ Trans. Sens. Micromach.* **143** 120–5

[31] Yamazoe S, Naya M, Shiota M, Morikawa T, Kubo A, Tani T, Hishiki T, Horiuchi T, Suematsu M and Kajimura M 2014 Large-area surface-enhanced raman spectroscopy imaging of brain ischemia by gold nanoparticles grown on random nanoarrays of transparent boehmite *ACS Nano* **8** 5622

[32] Shiota M *et al* 2018 Gold-nanofève surface-enhanced Raman spectroscopy visualizes hypotaurine as a robust anti-oxidant consumed in cancer survival *Nat. Commun.* **9** 1561

[33] Tanaka T, Saegusa S, Naya M, Fukuoka T, Aamano S, Utsumi Y and Yamaguchi A 2022 Evaluation of surface-enhanced Raman scattering substrate consisting of gold nanoparticles grown on nanoarrays of boehmite fabricated using magnetron sputtering process *J. Photopolym. Sci. Technol.* **35** 249

[34] Zhang R *et al* 2013 Chemical mapping of a single molecule by plasmon-enhanced Raman scattering *Nature* **498** 82

[35] Urbieta M, Barbry M, Zhang Y, Koval P, Sánchez-Portal D, Zabala N and Aizpurua J 2018 Atomic-scale lightning rod effect in plasmonic picocavities: a classical view to a quantum effect *ACS Nano* **12** 585

[36] Israelachvili J N 2011 *Intermolecular and Surface Forces* 3rd edn (London: Academic)
[37] Landau L D and Lifshitz E M 1963 translated from the Russian by J B Sykes and J S Bell Electrodynamics of continuous media *Course of Theoretical Physics* **8** (Oxford: Pergamon Press Ltd)
[38] Jones T B 1995 *Electromechanics of Particles* (Cambridge: Cambridge University Press) Jones T B 2003 Basic theory of dielectrophoresis and electrorotation *IEEE Eng. Med. Bio. Mag.* **22** 33–42
[39] Wang X-B, Vykoukal J, Becker F F and Gascoyne P R 1998 Separation of polystyrene microbeads using dielectrophoretic/gravitational field-flow-fractionation *Biophys. J.* **74** 2689–701
[40] Huang Y, Wang X B, Becker F F and Gascoyne P R 1997 Introducing dielectrophoresis as a new force field for field-flow fractionation *Biophys. J.* **73** 1118–29
[41] Saucedo-Espinosa M A, Rauch M M, LaLonde A and Lapizco-Encinas B H 2016 Polarization behavior of polystyrene particles under direct current and low-frequency ($<$ 1 kHz) electric fields in dielectrophoretic systems *Electrophoresis* **37** 635–44
[42] Mizuno A, Nishioka M, Ohno Y and Dascalescu L D 1995 Liquid microvortex generated around a laser focal point in an intense high-frequency electric field *IEEE Trans. Ind. Appl.* **31** 464–8
[43] Mizuno A, Nishioka M, Tanizoe T and Katsura S 1995 Handling of a single DNA molecule using electric field and laser beam *IEEE Trans. Ind. Appl.* **31** 1452–7
[44] Prucek R, Panáček A, Fargašová A, Ranc V, Mašek V, Kvítek L and Zbořil R 2011 Recrystallization of silver nanoparticles in a highly concentrated NaCl environment–a new substrate for surface enhanced IR-visible Raman spectroscopy *Cryst. Eng. Comm.* **13** 2242
[45] Prucek R, Panáček A, Soukupová J, Novotny R and Kvítek L 2011 Reproducible synthesis of silver colloidal particles tailored for application in near-infrared surface-enhanced Raman spectroscopy *J. Mater. Chem.* **21** 6416
[46] Wiley B, Herricks T, Sun Y and Xia Y 2004 Polyol synthesis of silver nanoparticles: use of chloride and oxygen to promote the formation of single-crystal, truncated cubes and tetrahedrons *Nano Lett.* **4** 1733
[47] Gi H J, Han D and Park J -K 2017 Optoelectrofluidic printing system for fabricating hydrogel sheets with on-demand patterned cells and microparticles *Biofabrication* **9** 015011
[48] Willets K A 2019 Probing nanoscale interfaces with electrochemical surface-enhanced Raman scattering *Curr. Opin. Electrochem.* **13** 18
[49] Fu B, Van Dyck C, Zaleski S, Van Duyne R P and Ratner M A 2016 Single molecule electrochemistry: impact of surface site heterogeneity *J. Phys. Chem.* C **120** 27241
[50] Oyamada N, Minamimoto H and Murakoshi K 2019 *In situ* observation of unique bianalyte molecular behaviors at the gap of a single metal nanodimer structure via electrochemical surface-enhanced Raman scattering measurements *J. Phys. Chem.* C **123** 2570
[51] Zhou L *et al* 2022 Electrically tunable SERS based on plasmonic gold nanorod-graphene/ion-gel hybrid structure with a low voltage *Carbon* **187** 425
[52] Wu D-Y, Li J-F, Ren B and Tian Z-Q 2008 Electrochemical surface Raman spectroscopy of nanostructures *Chem. Soc. Rev.* **37** 1025
[53] Banholzer M J, Millstone J E, Qin L and Mirkin C A 2008 Rationally designed nanostructures for surface-enhanced Raman spectroscopy *Chem. Soc. Rev.* **37** 885
[54] Willets K A and Van Duyne R P 2007 Localized surface plasmon resonance spectroscopy and sensing *Annu. Rev. Phys. Chem.* **58** 267
[55] de Ruiter J, An H, Wu L, Gijsberg Z, Yang S, Hartman T, Weckhuysen B M and van der Stam W 2022 Probing the dynamics of low-overpotential CO_2-to-CO activation on copper electrodes with time-resolved Raman spectroscopy *J. Am. Chem. Soc.* **144** 15047

[56] Herzog A, Luna M L, Jeon H S, Rettenmaier C, Grosse P, Bergmann A and Cuenya B R 2024 Operando Raman spectroscopy uncovers hydroxide and CO species enhance ethanol selectivity during pulsed CO_2 electroreduction *Nat. Commun.* **15** 3986

[57] Timoshenko J *et al* 2022 Steering the structure and selectivity of CO_2 electroreduction catalysts by potential pulses *Nat. Catal.* **5** 259

[58] Saegusa S, Naya M, Fukuoka T, Tabata M, Sumitomo K and Yamaguchi A 2025 Microchemical system for simultaneous measurement of surface-enhanced Raman scattering and electrochemical reactions *Sci. Rep.* **15** 18574

IOP Publishing

Nano–Microsystems
Science and applications
Akinobu Yamaguchi

Chapter 9

Future systems

Before this chapter, examples of research implementations of various microchemical systems have been described. Progress has been made in elemental technologies for deployment in sensor applications and combinatorial chemistry, and development can be expected towards systems that can carry out more complex operations in unprecedented combinations. In addition, with the development of AI, machine learning and other advances, linkages with information science and complementary relationships with robotics can be expected. Here, as one outlet, in this chapter we consider Nano–Microsystems, which are expected to be researched and developed in the future.

9.1 Recent research trends

Nanomicrosystems have been developed in research and development as Lab-on-a-chip and micro-total analysis systems (μTAS). Research has developed from fundamental studies on fluid behaviour and interfacial phenomena at the nanoscale and microscale to cancer detection and cell analysis. As chemical analysis is possible with small amounts of specimen samples, it is also being developed for environmental analysis and medical devices.

For example, the selection of circulating tumour cells (CTCs) directly from blood as a real-time liquid biopsy has received increasing attention over the past decade. CTC analysis has the potential to be of great use in both research and clinical applications. Microfluidic systems provide a very suitable experimental environment for CTC analysis, as microfabrication can create microfluidic channels and implement a variety of functions; the ability to perform CTC analysis can lead to a better understanding of metastatic cascades, tumour evolution, patient heterogeneity and drug resistance [1–5]. Until now, the rarity and heterogeneity of CTCs have been technical challenges to their widespread use in clinical research, but microfluidics-based separation techniques have emerged as a promising tool to address these limitations. Recently, a fully automated and integrated microfluidic system for efficient CTSs detection was reported by Wang *et al* [4]. Lab-on-a-Disk (LOD) is

doi:10.1088/978-0-7503-3111-1ch9

also being introduced to the market as point-of-care-testing (POCT) equipment, not only for CTCs, and its use is gradually expanding [6, 7].

As described above, microfluidic systems are inextricably linked to basic and applied research. This is because the fundamental question is how to carry out systematic and quantitative experiments in a reproducible and precise experimental environment on phenomena of scientific interest to be clarified. On the flip side, once the experimental conditions are set up, the product can be produced at any time in a factory. Not only fluids, but also solid-state devices are considered below as Nano–Microsystems. The following will consider future research developments, although it will be very much a personal view on Nano–Microsystems.

9.2 How will Nano–Microsystems develop?

Microfluidic systems have made chemical laboratories in the palm of one's hand. By combining elemental technologies as unit chemical operations for the processing to be carried out in advance, such as semiconductor IC chips, it has become possible to carry out a given analysis or synthesis. At the same time, new tools have been provided related to the area of science of fluid behaviour at the nanoscale and microscale and nanoscience at the interface. Although they cannot all be introduced here, not only fluidic devices but also MEMS, spintronics, graphene- or diamond-based devices, quantum devices and semiconductor hetero-interface devices etc, can be conceptually extended and recaptured as Nano–Microsystems. From the above perspective, the future development of Nano–Microsystems can be considered to lead in the following three directions.

(A) **Platform for science**

Systems that provide an ideal an experimental environment for conducting science.

(B) **Analytical systems**

A role as an analytical tool (micro-total analysis systems: μTAS) and its evolved systems.

Rapid diagnostics, POCT and environmental analysis systems.

(C) **Combinatorial chemistry and material informatics**

Automated, controlled reagent small-volume, multi-product synthesis-type systems related to drug discovery and material informatics.

In the following, each of them is outlined in a little more detail.

(A) **Platform for science (use as an ideal experimental environment development platform for science)**

The application to various objectives such as CO2RR research [8, 9] and organ-on-a-chip [10] studies of human tissue function and molecular biology can provide a very good experimental environment, controllability and reproducibility, as described at the end of chapter 8. Once the system has been constructed, a platform can be built up in which the conditions of the reagents to be fed are quantitatively varied and combined with analytical instruments such as XPS, SEM, SERS, TERS, XRD, FTIR and synchrotron radiation analysis according to the physical properties to be investigated, as shown in figure 9.1. If the measurement programme is automated by computer control and the measurement results are also analysed automatically, it is possible to carry out measurement and analysis while automatically and exhaustively changing the conditions. An automatic system can also be built in which the analysis results and experimental conditions can be feature-extracted by AI.

(B) **Analytical systems (role as an analytical tool and its development)**

Pre-treatment is in fact very important for the construction of analytical tools specializing in enzyme-linked immunosorbent assay (ELISA), exosome analysis, virus detection or specific molecule detection. Blood contains a wide variety of substances, including red blood cells, white blood cells, platelets, hormones, various ions, exosomes and in some cases bacteria and viruses. In order to analyse only the objectives we are interested in, we need to combine

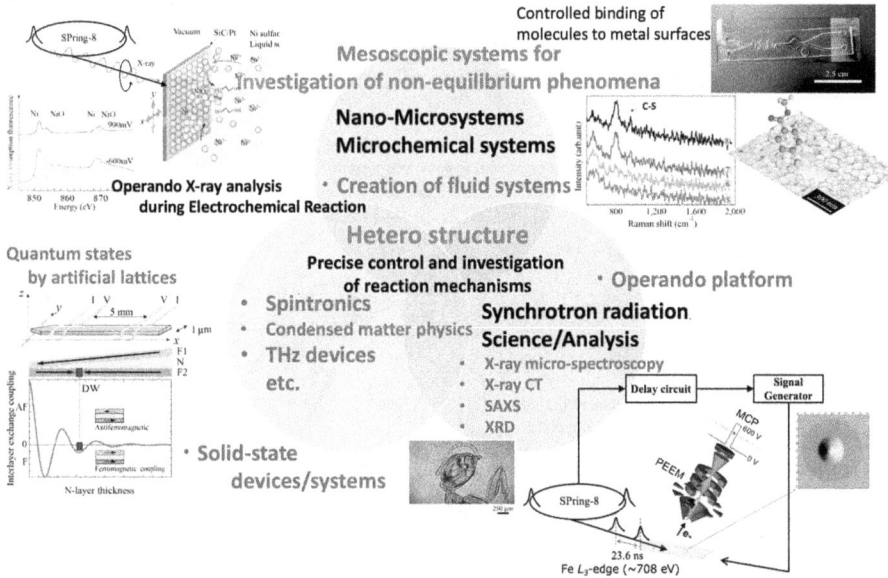

Figure 9.1. Schematic of platform for science. Systematic and quantitative scientific experiments are carried out by creating controlled systems and developing an ideal experimental environment. Form a platform by introducing Nano–Microsystems to various measuring instruments and other equipment. Here, systems focusing on hetero-interfaces are introduced. (Top left) operando measurement of synchrotron radiation absorption spectra during an electrochemical reaction at a solid–liquid interface Pt electrode. *In situ* analysis of electronic states during electrochemical reactions is possible [11]. (Bottom left) asymmetric fabrication of quantum interference effects by a metallic artificial lattice to spatially modulate ferromagnetic interlayer interactions mediated through nonmagnetic layer and control magnetization reversal [12]. (Top right) SERS detection of molecular adsorption states at the solid–liquid interface by electrochemical reactions *in situ* while reagents flow through a gold nano coral (GNC) structure fabricated in a microfluidic chip [22]. (Bottom right) direct observation of magnetic vortex swirling behaviour in magnetic discs by synchronizing synchrotron x-ray magnetic circular dichroism photoemission microscopy (XMCD-PEEM) and a pulsed magnetic field [13]. X-ray CT imaging of an ant by synchrotron radiation-based *x*-ray CT measurement system [14]. Biological and chemical reactions can also be viewed directly and live *in situ*.

unit chemistry operations or combine them with other pre-treatment processes, and the main prospect for practical use, as far as POCT instruments are concerned, for example, using LOD, centrifugation can be incorporated as an automated protocol ELISA is also possible in combination with microparticles, as shown in figure 9.2 for example. If the analytical performance and cost required by the market can be met, the system can be put into practical use and implemented in society. Once the pre-processing is completed, the system can be completed by incorporating an analysis mechanism. Whether this analysis mechanism requires Raman, XRD or just absorbance depends on the analysis required, but if the system is an ELISA in a medical POCT device, only absorbance is required, so a platform can be provided that can be used with a smartphone. Detection and analysis using smartphones is already possible, although it is still in the research and development phase, and there are reported cases where it has actually been integrated into Nano–Microsystems. With smartphone detection and information communication, the immediate connection to the digital space can be developed into big data analysis and epidemiological studies by collecting large amounts of patient data (figure 9.2).

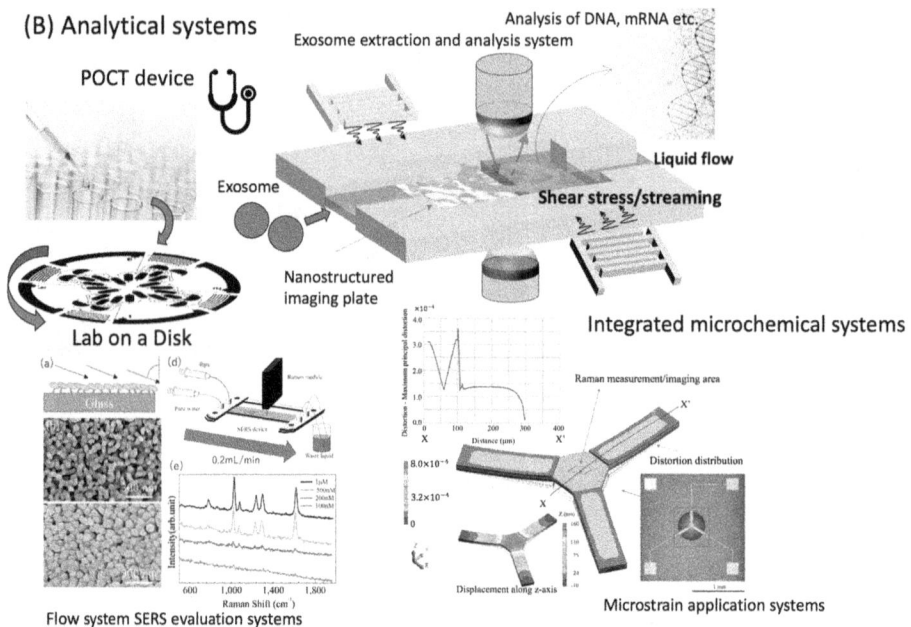

Figure 9.2. Schematic of analysis platform: Typical analysis system overview. For example, by LOD, example of rapid multi-sample, multi-parameter diagnostics in a hospital or clinic for the POCT. Schematic diagram of exosome content extraction and molecular analysis by a microchemical system integrating a microfluidic system implementing a GNC structure and a surface acoustic wave (SAW) application and mixing system [15, 16]. Micro-Electro-Mechanical-Systems (MEMS) device that applies strain from three directions to a thin-film sample, for example. The MEMS systems that can be incorporated and implemented in synchrotron radiation analysis systems such as micro-Raman spectroscopy and XMCD-PEEM to analyse electronic and chemical states during the application of strain [17].

(C) **Combinatorial chemistry and material informatics**

At the research stage of drug discovery and the exploration of functional materials, it is essential to automate the synthesis of reagents in small quantities and in large quantities, as well as the analysis and evaluation of these reagents. To form this platform, the foundations of (A) platform for science shown in figure 9.1 and (C) synthesis platform shown in figure 9.3 need to be built on and, where necessary, mechanisms for analyzing instruments or need to be combined with (B) analytical platform shown in figure 9.2. If the platform in which synthesis, analysis and evaluation are integrated is automatically controlled by computer, and if the calculated data and experimental data are harmonized by comparing them with the pre-calculated information of the material informatics, a more accurate database can be constructed, enabling the discovery and synthesis of new functional materials, etc. This will enable the discovery and synthesis of new functional materials, etc. If the created drug is administered to iPS cells or tissue created from iPS cells and its properties are evaluated in a Nano–Microsystem, the creation of

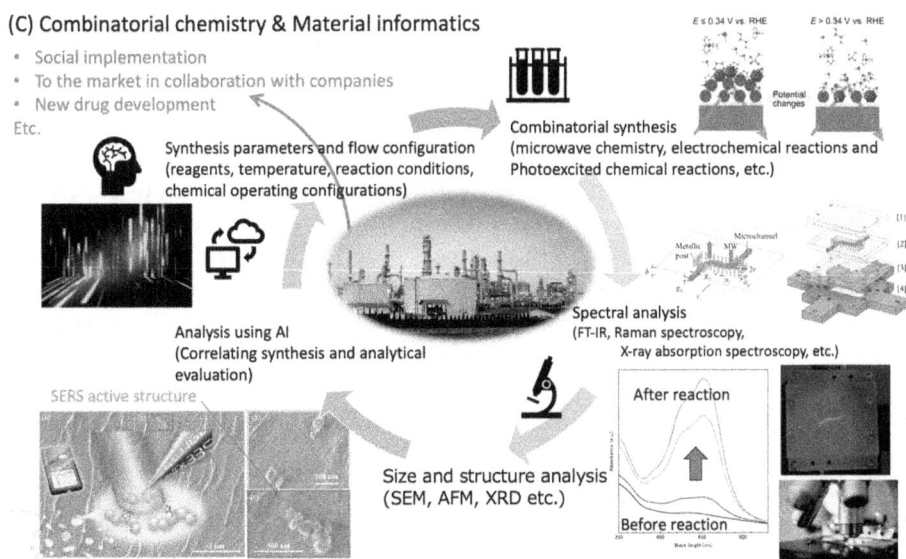

Figure 9.3. Conceptual diagram of the platform on combinatorial chemistry and materials informatics: combinatorial chemical synthesis is performed by microwave chemistry, electrochemical reaction, photo-excited chemical reactions, etc; spectroscopic analysis is carried out [11]; size and morphology, and crystal structure etc are evaluated; AI-based analysis is performed for correlating synthesis and analytical evaluation; restructuring the database, synthetic conditions and physical properties of synthesized materials is related one by one. Rebuilding a database of synthetic conditions and synthetic materials, conducting machine learning, integrating with material informatics, automatically predicting materials while building a combinatorial synthesis system: conducting synthesis, analysis and evaluation. Ultimately, the platform update the material database, and create new materials in an independent manner, resulting in developing into a self-sustaining material exploration system, etc. It is expected to be possible to synthesize nanoparticles, analyse them and to used them as functional particles [18–21].

new drugs may be possible in a short time. In addition, if blood and urine tests from patients to whom newly synthesized drugs are administered are conducted via POCT equipment, biological data and the effects of the drugs can be confirmed.

As described above, the three developments have a common basis and develop in a complementary manner. Thus, the field of Nano–Microsystems is a research area in which it is easy to enter for researchers specializing in basic science, for engineers close to product development, and for researchers and engineers in the field of informatics, where they can realize their ideas and produce results. It can also provide very accessible science and applications for users of microfluidic systems and POCT devices. SERS was introduced as a molecular sensor, but SERS is really a quantum effect occurring at room temperature, and if you know that sensors using quantum effects operate invisibly to analyse the environment and detect viruses, you will see this world in a completely different way.

Thus, Nano–Microsystems are reconsidered and three important qualities are presented. Microfabrication has made it possible for solids as well as fluids to be interacted with in a controlled manner, making it easier to design and construct systems according to the physical, chemical or biological phenomena that we wish to reveal. Computerized control has also become easier, and in data analysis, in conjunction with recent developments in AI, feature extraction has also become possible. Thus, it is now possible to do things that could not be done even if they had been thought of before, and moreover, it is now possible to carry out experiments under automatic control. If more research and development is carried out, it may become possible to carry out material searches and CTCs completely automatically.

Nano–microsystems, which are constructed as simplified models for the purpose of clarifying various physical, chemical and biological phenomena occurring in space and on Earth, may require a slight leap of imagination, but they can also be viewed as simplified models of the world we live in, as shown in figure 9.4. In other words, Nano–Microsystems constructed to clarify scientific problems are used to deepen scientific understanding by placing phenomena in ideal experimental systems. Through this research and development, new discoveries can be made, and assumptions can be confirmed, enabling progress toward engineering applications. Conversely, by using devices capable of precise analysis of unknown phenomena, such as environmental analysis or cell analysis, scientific understanding can be deepened. Furthermore, it is possible to construct Nano–Microsystems that perform combinatorial synthesis to apply the gained insights to the development of functional materials or new pharmaceuticals. As described above, Nano–Microsystems serve as a platform for conducting both basic science and applied engineering research. By utilizing such a platform, it is expected that not only basic research but also insights related to commercial development will be accumulated, contributing to advancements across various fields. Nano–Microsystems play a role in driving the cycle where new scientific and technological discoveries and developments act as rain, nurturing the seeds of new ideas and fostering their growth through the fertile soil and abundant rainfall they provide.

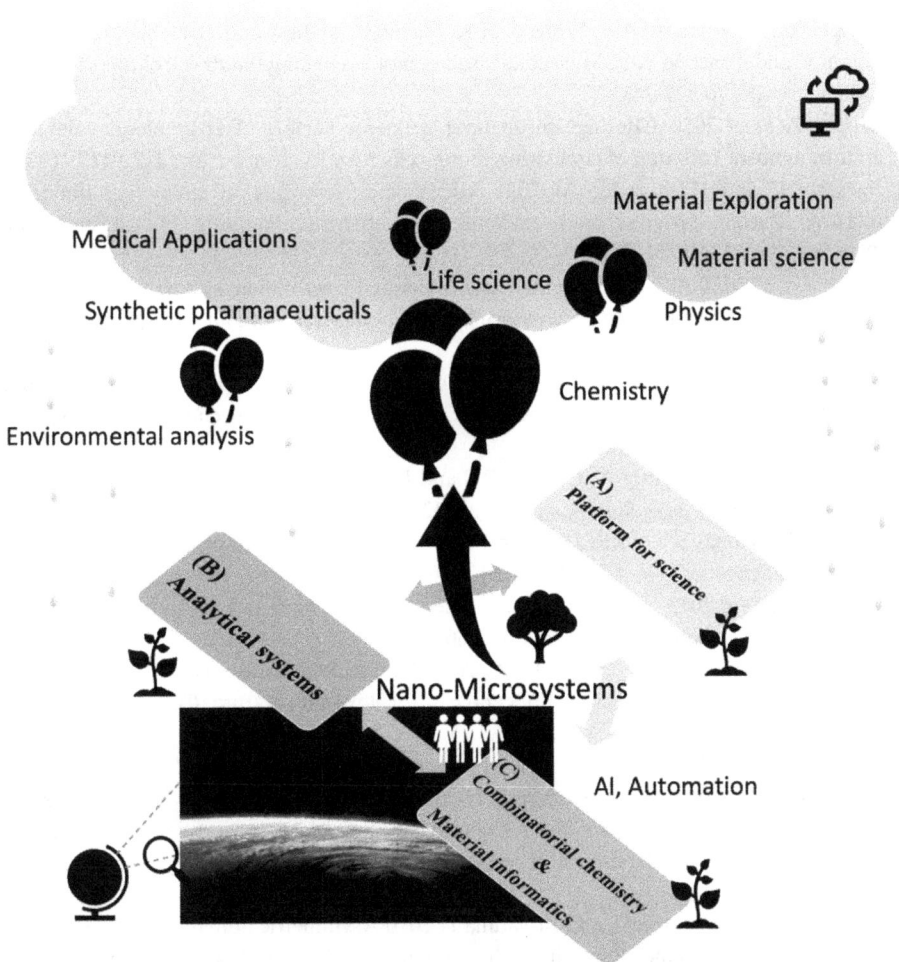

Figure 9.4. The Universe, the Earth and society can be considered as systems. Nano–Microsystems can provide an ideal experimental environment for precise, quantitative and systematic testing of specific phenomena. Specialized functions for the analysis of experiments can also be implemented. Chemical synthesis can also be carried out and combined with materials informatics to create systems. The system can be deployed in basic as well as applied research, contributing to the development of basic science and engineering technology. This figure shows a conceptualized diagram where developments related to Nano–Microsystems, like the development of water circulation systems, forest growth mechanisms and civilized societies, will advance the development of fundamentals and applications, which will become a blessing rain and further enrich the soil supporting the science and technology base for further development.

The Nano–Microsystem is a system that gives form to the readers' thoughts and may be a system they use without knowing it. Readers are encouraged to experience and enjoy their thoughts in the form of Nano–Microsystems.

References

[1] Au S H, Edd J, Stoddard A E, Wong K H K, Fachin F, Maheswaran S, Haber D A, Stott S L, Kapur R and Toner M 2017 Microfludic isolation of circulating tumor cell clusters by size and asymmetry *Sci. Rep.* **7** 2433

[2] Mishra A *et al* 2020 Ultrahigh-throughput magnetic sorting of large blood volumes for epitope-agnostic isolation of circulating tumor cells *Proc. Natl Acad. Sci. USA* **117** 16839–47

[3] Farshchi F and Hasanzadeh M 2021 Microfluidic biosensing of circulating tumor cells (CTCs): Recent progress and challenges in efficient diagnosis of cancer *Biomed. Pharmacother.* **134** 111153

[4] Wang J *et al* 2021 A fully automated and integrated microfluidic system for efficient CTC detection and its application in hepatocellar carcinoma screening and prognosis *ACS Appl. Mater. Interfaces* **13** 30174–86

[5] Hassanzadeh-Barforoushi A, Tukova A, Nadalini A, Inglis D W, Tsao S C-H and Wang Y 2024 Microfluidic-SERS technologies for CTC: a perspective on clinical translation *ACS Appl. Mater. Interfaces* **16** 22761–75

[6] Dincer C, Bruch R, Kling A, Dittrich P S and Urban G A 2017 Multiplexed point-of-care testing—xPOCT *Trends Biotechnol.* **35** 728–42

[7] Sunkara V, Kumar S, del Río J S, Kim I and Cho Y -K 2021 Lab-on-a-disc for pint-of-care infection diagnostics *Acc. Chem. Res.* **54** 3643–55

[8] Appel A M *et al* 2013 Frontiers, opportunities, and challenges in biochemical and chemical catalysis of CO_2 fixation *Chem. Rev.* **113** 6621–58

[9] Masel R, Liu Z, Yang H, Kaczur J J, Carrillo D, Ren S, Salvatore D and Berlinguette C P 2021 An industrial perspective on catalysis for low-temperature CO_2 electrolysis *Nat. Nanotechnol.* **16** 118–28

[10] Leung C M *et al* 2022 A guide to the organ-on-chip *Nat. Rev. Method Primers* **2** 33

[11] Yamaguchi A, Akamatsu N, Saegusa S, Nakamura R, Utsumi Y, Kato M, Yagi I, Ishihara T and Oura M 2022 *In situ* fluorescence yield soft X-ray absorption spectroscopy of electrochemical nickel deposition processes with and without ethylene glycol *RSC Adv.* **12** 10425–30

[12] Yamaguchi A, Kishimoto T and Miyajima H 2010 Asymmetric domain wall propagation in a giant magnetoresistance-type wire with oscillating interlayer exchange coupling *Appl. Phys. Exp.* **3** 093004

[13] Yamaguchi A, Hata H, Goto M, Kodama M, Kasatani Y, Sekiguchi K, Nozaki Y, Ohkochi T, Kotsugi M and Kinoshita T 2016 Real-space observation of magnetic vortex core gyration in a magnetic disc both with and without a pair tag *Japan. J. Appl. Phys.* **55** 023002

[14] Vella J R, Hao Q, Donnelly V M and Graves D B 2023 Dynamics of plasma atomic layer etching: Molecular dynamics simulations and optical emission spectroscopy *J. Vac. Sci. Technol.* A **41** 062602

[15] Saegusa S, Tanaka T, Naya M, Fukuoka T, Amano S, Utsumi Y and Yamaguchi A 2023 Integration of gold-nanofève-based-SERS active nanostructure and microfluidic devices *IEEJ Trans. Sens. Micromachin.* **143** 120–5

[16] Yamaguchi A, Takahashi M, Saegusa S, Utsumi Y and Saiki T 2024 Removable and replaceable micro-mixing system with surface acoustic wave actuators *Japan. J. Appl. Phys.* **63** 030902

[17] Yamaguchi A, Ohkochi T, Oura M, Yokomatsu T and Kanda K 2024 Consideration of experiment to introduce MEMS devices into spectroscopic systems for bending and three-point tension tests *Sens. Mater.* **36** 3465

[18] Fukuoka T, Mori Y, Yasunaga T, Namura K, Suzuki M and Yamaguchi A 2022 Physically unclonable functions taggant for universal steganographic prints *Sci. Rep.* **12** 985

[19] Yasunaga T, Fukuoka T, Yamaguchi A, Ogawa N and Yamamoto H 2022 Physical stability of stealth nanobeacon using surface-enhanced Raman scattering for anti-counterfeiting and monitoring medication adherence: Deposition on various coating tablets *Int. J. Pharmaceut.* **624** 121980

[20] Fukuoka T, Yasunaga T, Namura K, Suzuki M and Yamaguchi A 2023 Plasmonic nanotags for on-dose authentication of medical tablets *Adv. Mater. Interfaces* **10** 2300157

[21] Yamaguchi A, Yasunaga T, Namura K, Suzuki M and Fukuoka T 2025 Print evaluation of inks with stealth nanobeacons *RSC Adv.* **15** 4173–86

[22] Saegusa S, Naya M, Fukuoka T, Tabata M, Sumimoto K and Yamaguchi A 2025 Microchemical system for simultaneous measurement of surface-enhanced Raman scattering and electrochemical reactions *Sci. Rep.* **15** 18574

www.ingramcontent.com/pod-product-compliance
Lightning Source LLC
Chambersburg PA
CBHW080529220326
41599CB00032B/6253